INVESTIGATING BIOLOGY

An Introduction to the Tools and Techniques of a Biologist

VOLUME I

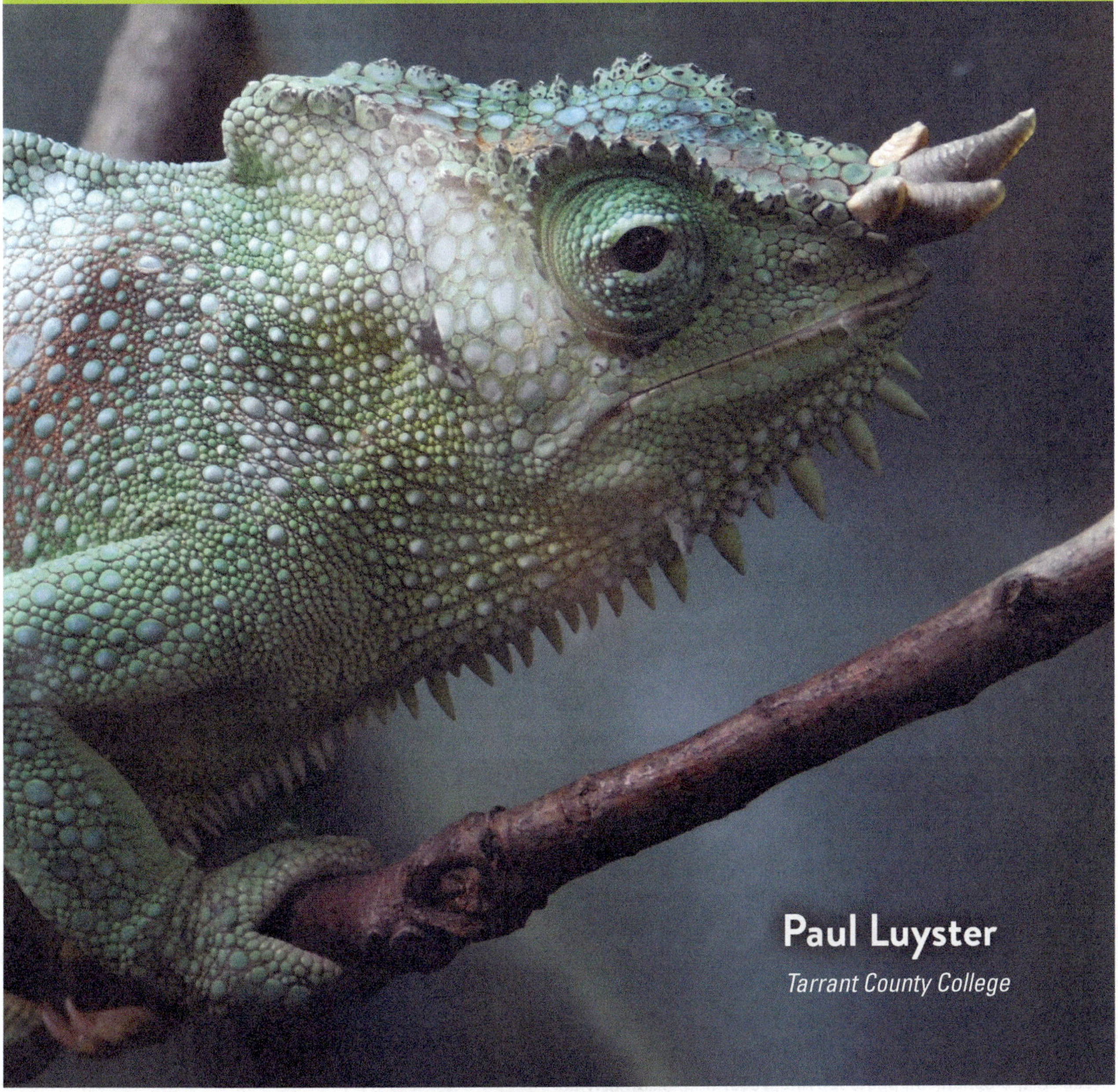

Paul Luyster

Tarrant County College

Kendall Hunt
publishing company

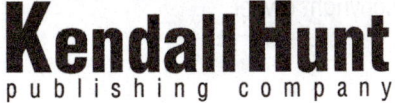

Cover image: © Paul Luyster.
All tables and images by Paul Luyster unless otherwise noted.

Kendall Hunt
publishing company

www.kendallhunt.com
Send all inquiries to:
4050 Westmark Drive
Dubuque, IA 52004-1840

Copyright © 2021 by Kendall Hunt Publishing Company
Preliminary Edition: Copyright © 2020 by Kendall Hunt Publishing Company

ISBN 978-1-7924-5347-2

All rights reserved. No part of this publication may be reproduced,
stored in a retrieval system, or transmitted, in any form or by any means,
electronic, mechanical, photocopying, recording, or otherwise,
without the prior written permission of the copyright owner.

Published in the United States of America

Contents

	Introduction	vii
	Acknowledgments	vii
	About the Author	viii
Lab 1	**An Introduction to the Biology Laboratory**	**1**
	Lab Skill 1.1 Interpreting Lab Safety Symbols	6
	Lab Skill 1.2 Keeping a Lab Notebook	8
	Lab Skill 1.3 Measuring Liquids with Different Lab Equipment	11
	Lab Skill 1.4 Using a Micropipette	13
	Pre-Lab Quiz 1	15
	Problem to Solve 1.1 Can I Use Scientific Reasoning to Solve a Puzzle?	17
	Problem to Solve 1.2 Can I Accurately Measure Liquids?	19
	Post-Lab Quiz 1	23
Lab 2	**Using Chemistry in the Biology Lab**	**25**
	Lab Skill 2.1 Understanding the Names of Chemical Compounds	36
	Lab Skill 2.2 Creating Solutions of Specific Concentration	37
	Lab Skill 2.3 Creating Dilutions of a Specific Volume	38
	Lab Skill 2.4 Measure Solute Concentration with a Hydrometer	39
	Lab Skill 2.5 Calculating pH from Molarity or Molarity from pH	40
	Pre-Lab Quiz 2	43
	Problem to Solve 2.1 Can I Make 500 ml of a 0.154 M Saline Solution?	47
	Problem to Solve 2.2 Can I Make Dilute Acid and Base Solutions?	50
	Post-Lab Quiz 2	55
Lab 3	**Biochemistry: Chemical Compounds of Living Organisms**	**57**
	Lab Skill 3.1 Using Indicators to Determine the Presence of Specific Compounds	66
	Lab Skill 3.2 Managing Heat in the Laboratory	68
	Pre-Lab Quiz 3	69
	Problem to Solve 3.1 Can I Observe Enzymes Digesting Macromolecules?	71
	Post-Lab Quiz 3	79
Lab 4	**Molecular Biology I: Understanding and Isolating DNA**	**81**
	Lab Skill 4.1 Using a Benchtop Centrifuge	91
	Pre-Lab Quiz 4	93
	Problem to Solve 4.1 Can I Recognize the Phases of Mitosis?	95

	Problem to Solve 4.2 Can I Synthesize a Peptide?	100
	Problem to Solve 4.3 Can I Collect and Isolate My DNA?	104
	Post-Lab Quiz 4	109
Lab 5	**Microscopy: Observing Life at the Cellular Level**	**111**
	Lab Skill 5.1 Properly Using a Light Microscope	120
	Lab Skill 5.2 Calculating Magnification and Estimating Specimen Size	121
	Lab Skill 5.3 Using the Oil Immersion Lens	122
	Lab Skill 5.4 Making a Wet Mount Slide	123
	Pre-Lab Quiz 5	125
	Problem to Solve 5.1 Can I Visually Identify Types of Cells?	127
	Problem to Solve 5.2 Can I Estimate the Size of One Human Cell?	134
	Post-Lab Quiz 5	137
Lab 6	**Cell Biology I: Cellular Membranes and Structures**	**139**
	Lab Skill 6.1 Using Dialysis Tubing	151
	Lab Skill 6.2 Using (More) Chemical Indicators	151
	Pre-Lab Quiz 6	153
	Problem to Solve 6.1 Can I Predict When Diffusion or Osmosis Will Occur?	155
	Problem to Solve 6.2 Can I Demonstrate Differences in Osmotic Pressure?	160
	Problem to Solve 6.3 Can I Recognize the Effects of Hypotonic, Hypertonic, and Isotonic Solutions on Cells?	165
	Problem to Solve 6.4 Can I Design a Cell?	171
	Post-Lab Quiz 6	175
Lab 7	**Molecular Biology II: Manipulating DNA**	**177**
	Lab Skill 7.1 Conducting DNA Gel Electrophoresis	184
	Lab Skill 7.2 Interpreting Electrophoresis Data	186
	Pre-Lab Quiz 7	187
	Problem to Solve 7.1 Can I Predict Restriction Enzyme Recognition Sites?	189
	Problem to Solve 7.2 Can I Verify Restriction Enzyme Recognition Sites by DNA Electrophoresis?	194
	Problem to Solve 7.3 Can I Interpret DNA Fingerprint Data?	199
	Post-Lab Quiz 7	201
Lab 8	**Cell Biology II: Photosynthesis and Cellular Respiration**	**203**
	Lab Skill 8.1 Separating Compounds Using Chromatography	213
	Lab Skill 8.2 Using a Spectrophotometer	214
	Pre-Lab Quiz 8	217
	Problem to Solve 8.1 Can I Observe Cellular Respiration and Photosynthesis in a Plant?	219
	Problem to Solve 8.2 Can I Observe Differences in Human CO_2 Production?	224
	Problem to Solve 8.3 Can I Determine the Pigments in a Leaf?	227

	Problem to Solve 8.4 Can I Create an Absorbance Spectrum for Different Plant Pigment Combinations?	233
	Problem to Solve 8.5 Can I Explain the Chemistry of Photosynthesis and Cellular Respiration?	237
	Post-Lab Quiz 8	241
Lab 9	**Cell Biology III: Fermenting Sugar to Produce Alcohol**	**243**
	Lab Skill 9.1 Preparing for an Experiment	248
	Lab Skill 9.2 Conducting an Experiment	249
	Lab Skill 9.3 Graphically Representing Data	250
	Lab Skill 9.4 Analyzing Data with Statistics	252
	Pre-Lab Quiz 9	253
	Problem to Solve 9.1 Can I Conduct an Experiment and Write a Scientific Report?	255
	Post-Lab Quiz 9	265
Lab 10	**Genetics I: Meiosis and Promiscuous Alien Sex**	**267**
	Lab Skill 10.1 Creating and Using Biological Models	277
	Pre-Lab Quiz 10	279
	Problem to Solve 10.1 Can I Use a Model to Demonstrate the Phases of Mitosis and Meiosis?	281
	Problem to Solve 10.2 Can I Model Nondisjunction and Interpret Nondisjunction Karyotypes?	287
	Problem to Solve 10.3 Can I Simulate Promiscuous Alien Sex?	289
	Post-Lab Quiz 10	295
Lab 11	**Genetics II: Understanding Inheritance**	**297**
	Lab Skill 11.1 Creating a Pedigree	312
	Lab Skill 11.2 Using a Chi Square Test to Compare Actual Results with Expected Results	315
	Pre-Lab Quiz 11	317
	Problem to Solve 11.1 Can I Solve Basic Genetics Problems?	319
	Problem to Solve 11.2 Can I Describe Genetic Characteristics of Humans?	323
	Problem to Solve 11.3 Can I Determine if Observed Phenotypic Ratios Match Predictions?	328
	Problem to Solve 11.4 Can I Win the Bloody Game?	334
	Post-Lab Quiz 11	339
Lab 12	**Microbiology: Viruses, Bacteria, and Protists**	**341**
	Lab Skill 12.1 Preventing Microbiological Contamination	355
	Lab Skill 12.2 Culturing Microbes in a Petri Dish	357
	Lab Skill 12.3 Preparing a Gram Stain	360
	Pre-Lab Quiz 12	363
	Problem to Solve 12.1 Can I Visually Recognize Types of Microbes?	365

Problem to Solve 12.2 Can I Observe the Ubiquity and Diversity of the Microbes Around Me? 371

Problem to Solve 12.3 Can I Identify the Types of Bacteria Living in My Mouth? 376

Problem to Solve 12.4 Can I Keep my Classmates from Infecting Me? 380

Post-Lab Quiz 12 385

Appendix A Metric and SI Units 387

Appendix B Scientific Notation and Significant Figures 389

Appendix C Atoms, Isotopes, and Ions 391

Appendix D The Periodic Table of the Elements and How to Use It 394

Appendix E How to Write a Scientific Paper 399

Appendix F An Application of Stoichiometry and Dimensional Analysis 403

Appendix G Values of the Chi Square Distribution 405

Introduction

This manual has been designed to expose introductory biology students to at least some of the major branches of biology, and perhaps pique their interest in an area they had not previously explored. The author readily acknowledges that experts in any one subdiscipline of biology may find the coverage of their particular field cursory and inadequate. This is necessarily so, as biology is just too broad a subject to deeply explore all of its subdisciplines in a two-semester introductory college course.

Likewise, the author does not claim to be an expert in all of the subdisciplines discussed in this manual. He hopes that these lab activities provide the student with an accurate representation of the field as well as some of the laboratory techniques employed in it. He encourages subject experts to share additional ways they might introduce students to their field.

The author hopes that colleagues and students who use this manual will consider it a dynamic teaching and learning tool that, like all scientific knowledge, is open to correction and revision. Scientific knowledge evolves, and published works need to be updated as new discoveries are made and theories revised. The author welcomes suggestions for ways to keep the information in this manual current, fun, and engaging. He hopes that this manual will provide students a rewarding experience as they begin to learn the tools and techniques of the contemporary biologist.

Acknowledgments

The author would like to thank Shane Marshall and Emily Rodriguez for their assistance in reviewing and editing this manual. Mr. Marshall's scientific knowledge, research training, keen eye for detail, and experience working with students proved to be very helpful as he provided feedback for these laboratory activities. Ms. Rodriguez also provided very helpful feedback, and she caught mistakes the author and Mr. Marshall missed. Her witty and humorous comments about the lab exercises helped keep the text grounded and practical, and her experience working with students helped the author keep the students' perspective in mind. The author wishes to sincerely thank both of them for generously donating so much of their time to help with this project, and for their passion for biology and biology education.

The author would also like to thank his wife who graciously and uncomplainingly put up with the author's long work hours and writing-induced grumpiness.

Thank you, Dawn. Every day you do indeed light up my world.

About the Author

Paul Luyster earned his bachelor's degree from Oral Roberts University, where he studied biology, chemistry, and philosophy. He completed a summer program at the University of Montana Biological Field Station, which is where his passion for biology really developed. He pursued a doctorate in systems ecology at the University of Georgia's Institute of Ecology and also earned a masters in science education from Piedmont College. He completed a graduate certificate in medical human anatomy and physiology from the University of Florida's College of Medicine, and he is finishing a PhD in biology at the University of North Texas. He has over 30 years of teaching experience. He has been employed by Tarrant County College in Fort Worth, Texas, since 2004. There, he has taught general biology for science majors, general biology for non-science majors, anatomy and physiology, microbiology, undergraduate biology research, and environmental biology field studies. He has received a number of teaching awards, including the NISOD Excellence Award and the Chancellor's Quarterly Employee Excellence Award, and he was a finalist for the national Bellwether Award for Teaching. He is an active member of the Human Anatomy and Physiology Society and is a National Academies Education Fellow. He has contributed to a number of print and digital publications, including Mastering A&P. He also sponsors a student organization for pre-nursing students and serves as the lead instructor for the Majors Biology courses at Tarrant County College, Fort Worth, TX. Most importantly, Mr. Luyster enjoys getting to know his students and seeing them succeed as they pursue their academic and professional goals.

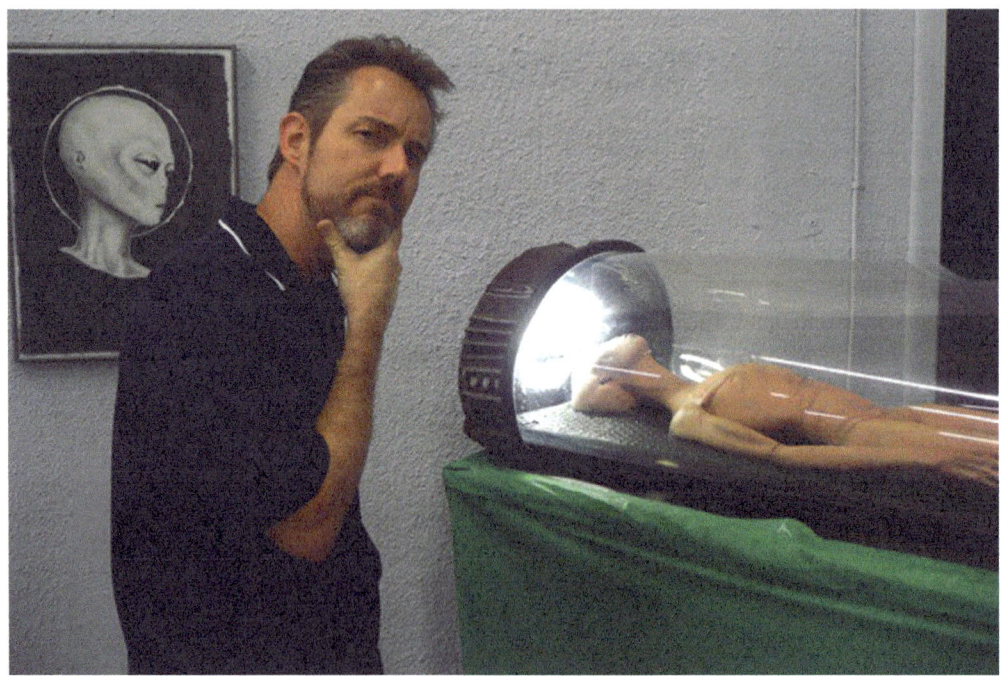

Laboratory Exercise 1

An Introduction to the Biology Laboratory

Concepts to Understand

After completing this lab the student should be able to:

1. Explain the difference between laboratory sciences and other disciplines.
2. Describe the personal protective equipment (PPE) and the safety practices that should be used in the biology laboratory.
3. Explain the steps of the scientific method.
4. Differentiate between peer-reviewed scientific literature and popular scientific articles.

Skills to Learn

After completing this lab, the student should be able to:

1. Keep a laboratory notebook.
2. Understand and use Metric and SI units.
3. Use laboratory equipment including mass scales, disposable pipettes, micropipettes, graduated cylinders, and beakers, to measure small quantities of biological samples.
4. Make and test a hypothesis.
5. Use simple statistical calculations to analyze quantitative data.

Matej Kastelic/Shutterstock.com

SEE ALSO

Appendix A: Metric and SI Units

Appendix B: Scientific Notation

Campbell Biology - Chapter 1, Section 1.3

JOVE (video) - An Introduction to the Micropipettor

Others:

Important Terms to Know

Accuracy	Metric units	Qualitative
Carcinogen	MSDS Sheet	Quantitative
Control	Peer-reviewed	SI units
Dependent variable	PPE	Standard deviation
Empiricism	Precision	Teratogen
Hypothesis	Prediction	Toxin
Independent variable	Protocol	Variance

Pre-lab Preparation and Background Information

Is "cupping" a beneficial medical treatment for muscle soreness and headaches?

Imagine a family member discovers a website claiming that "cupping" provides health benefits. (See Figure 1.1) They ask you if this is a good solution for their occasional aches and pains. Should they believe the claims? Is this a reliable source of accurate information? If we are in doubt, where can we find accurate information? How can we know which information to trust?

Today, all kinds of medical and scientific information are easily available on television, webpages, blogs, and print media. But are the claims always true? Some websites contradict each other. Which one is giving you accurate information? How can you distinguish good information from hype, opinion, or marketing? Since you are a science major, friends and family members may ask *you* for advice about a scientific or medical question. If a relative with chronic back and knee pain asked you if cupping was a good solution for them, what would you tell them?

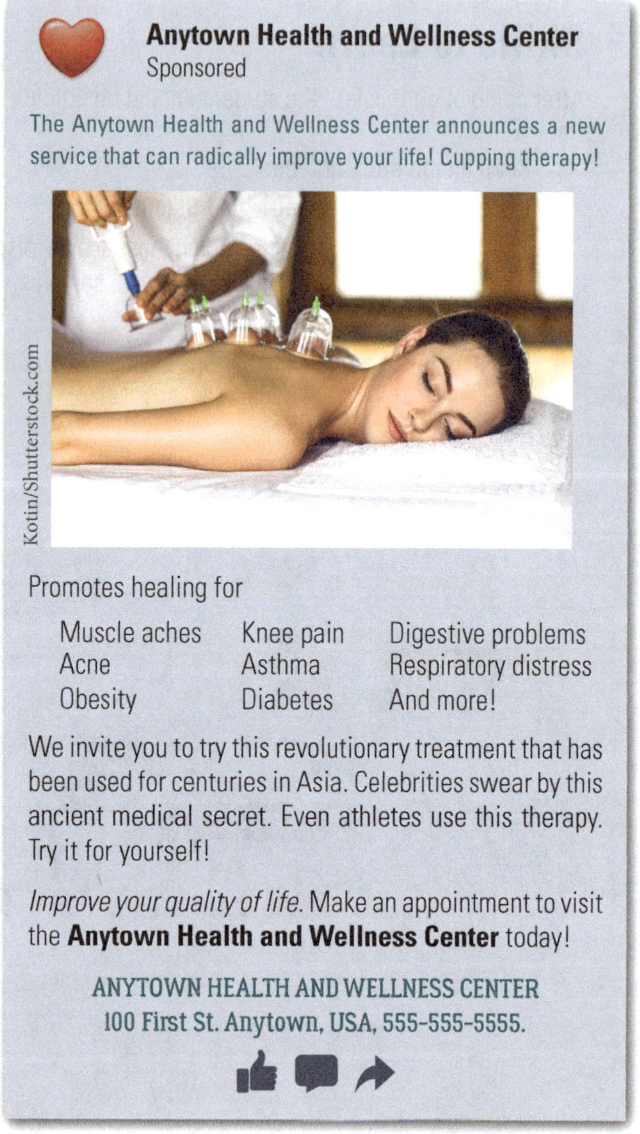

A Closer Look

What specifically does this webpage claim? Does it make sense? Is there any biological rationale that would suggest "cupping" the skin with suction would improve knee pain, digestive health, or diabetes? Who made this website, and why? Why did they use terms like "therapy" and "medical" in their posting? Does their claim that celebrities and athletes use this treatment make it legitimate? Does the claim that it is an "ancient medical secret" make it legitimate?

Finding Good Scientific Information

We live in what is rightly called the information age, where almost anything we want to know is easily discovered with a quick internet search. However, with easy access

Anytown Health and Wellness Center
Sponsored

The Anytown Health and Wellness Center announces a new service that can radically improve your life! Cupping therapy!

Promotes healing for

Muscle aches	Knee pain	Digestive problems
Acne	Asthma	Respiratory distress
Obesity	Diabetes	And more!

We invite you to try this revolutionary treatment that has been used for centuries in Asia. Celebrities swear by this ancient medical secret. Even athletes use this therapy. Try it for yourself!

Improve your quality of life. Make an appointment to visit the **Anytown Health and Wellness Center** today!

ANYTOWN HEALTH AND WELLNESS CENTER
100 First St. Anytown, USA, 555-555-5555.

Figure 1.1 When people search for health-related information, they often find websites like this. Is this a good source of information?

to massive amounts of information comes easy access to massive amounts of *misinformation*. There are many websites that provide inaccurate or even maliciously false information.

Many people assume most information on the internet is reliable, accurate, and trustworthy. But this is not always true. In fact, the information posted on freely accessible websites is usually there because the authors have a purpose for posting it. They may want to change your political views, sway your opinion on a social issue, or simply get you to buy something. When reading any article in electronic or print media, ask yourself the following questions:

- Why did the author post this information? Do they have an obvious or hidden agenda?
- Does the information present an objective perspective, or is it biased?
- What are the author's credentials? Are they professionally trained in this area?
- Is the author trying to get me to buy something? Is the information a marketing ploy?
- Do other experts in this area support this author's perspective?
- Is the article self-published by the author? Has the information been reviewed or fact-checked by an outside source?
- If data is provided, is it from an unbiased research study based on a controlled, scientific experiment? Has the data been subject to statistical analysis?
- How old is the information? Has more recent information been published that supports or refutes this author?

The best scientific or medical information can be found in well-respected, **peer-reviewed** journals. Experts in that field must evaluate all articles before they can be published in a peer-reviewed journal. The peer-review process is rigorous, and articles that have been peer reviewed are far more likely to be well supported by scientific evidence and have the backing of other experts. These journals are often only available through a subscription service.

Google Scholar: https://scholar.google.com is a search engine that allows users to access journal article abstracts (summaries) and some journal articles for free. Since many journals are not free, Google Scholar may only provide a starting point for finding good articles.

Library databases: Colleges and universities have access to electronic databases of thousands of articles published in peer-reviewed journals. Since the college or university pays for this access, you can find far more articles here than in Google Scholar.

Government websites: Government websites are subject to careful scrutiny, and articles posted on them are carefully checked for accuracy. Government websites like the Centers for Disease Control (CDC.gov) or National Institutes of Health (NIH.gov) are reliable websites for unbiased and well-researched information.

The Scientific Method—The foundation of laboratory science

The scientific method is a set of well-established steps scientists use to test ideas and develop theories (Figure 1.2).

Experimental science is **empirical**. This means claims must be based on reliable evidence that can be verified by others. Biologists rely on data gathered from experiments that have been conducted according to the scientific method. Biologists observe nature, pose questions about their observations, make "educated guesses" about what they suspect may explain this phenomenon, and conduct experiments to see if the data they collect supports that hypothesis.

In order to design an experiment, biologists determine which variables they will study. Experiments should have one **independent variable**, or variable that the researcher manipulates, and one or more **dependent variables**, variables that change in response to the manipulated variable. For example, if a team of biologists is studying the effects of different kinds of fertilizers on a specific type of crop, the independent variable might be the type of fertilizer used and the dependent variable might be the growth rate of the plant. However, all experiments must have a *control*, or something with which to compare the experimental results. In this case, the control would be a set of plants not treated with fertilizer. Without such a control, it would be impossible to know if the plant's growth rate is a result of the fertilizer or simply the normal growth rate of the plant.

Biological experimentation requires collecting **quantitative** and/or **qualitative** data. Quantitative data are numerical measurements using a set standard, such as standard units (**SI units**) of length, mass, or volume. Qualitative data are descriptive and may not always have a numerical value, such as when observing animal behavior. Quantitative data are

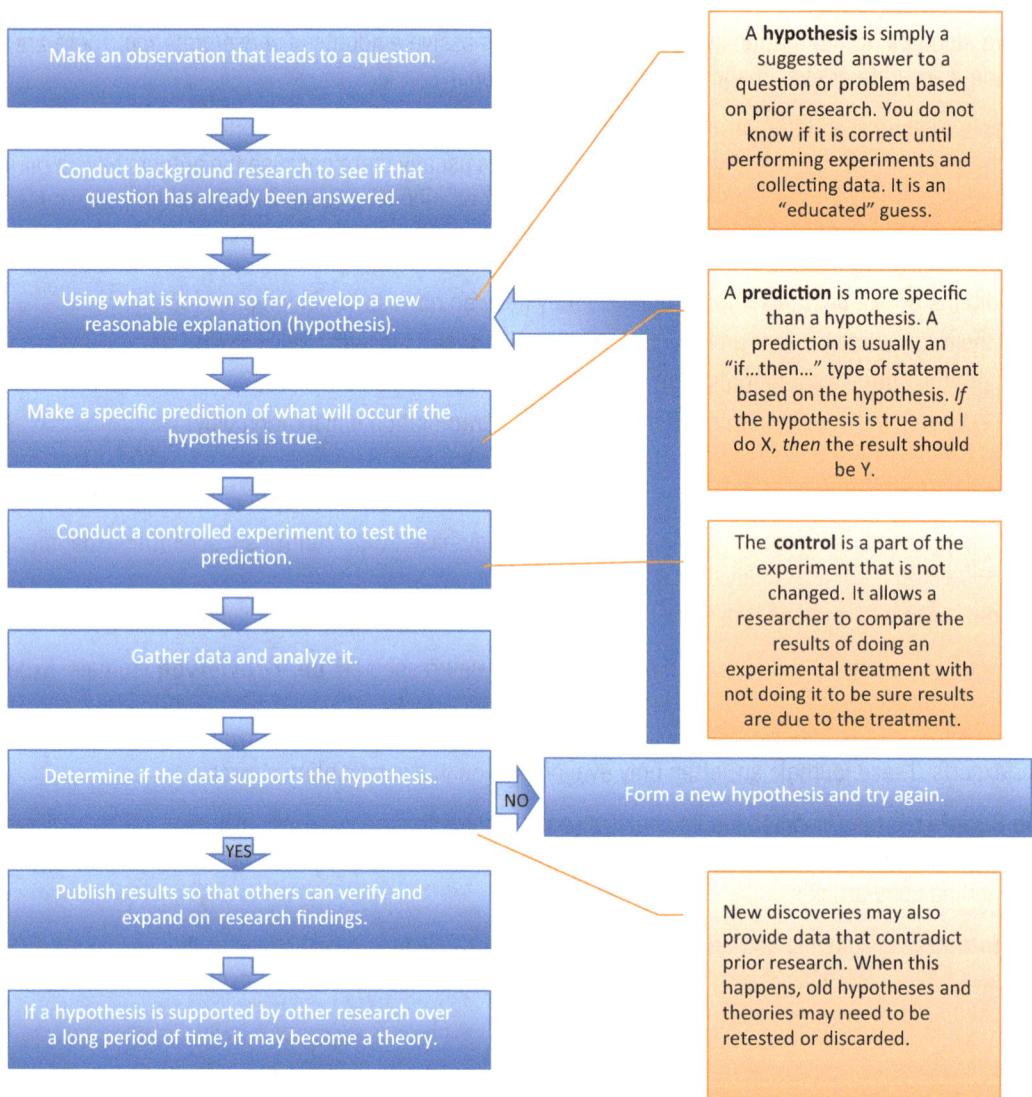

Figure 1.2 The Scientific Method is based on testing hypotheses, conducting repeatable experiments, and collecting an abundance of credible data.

often recorded in tables, presented in graphs, and subjected to statistical analysis. Statistical analysis helps determine if the data show "mathematically significant" relationships between variables.

In this lab, you will measure **variance** and **standard deviation** of data you collect. These statistical values represent how far apart the data points are when repeating a measurement. This also can be thought of as a measure of **precision**. But being precise is not the same as being **accurate**. Accuracy refers to how close to the correct value a measurement may be. For example, if 10 people all guessed your age to be 23, their guesses are very *precise* since they all guessed the same age. But if you are actually 34, these results are far from *accurate* (Figure 1.3).

Measuring accurately and precisely in the laboratory

In this course you will be using basic lab equipment to measure many things, especially small amounts of different kinds of chemicals. You will be using **Standard International (SI)** units and the **metric system**. If you are not familiar with these, see **Appendix A: Metric and SI Units**

Figure 1.3 In the first target, the arrow shots are fairly *precise* but not *accurate*. In the second target the shots are less precise but more accurate. In the third target they are both precise and accurate. Scientific measurements seek both precision and accuracy.

You will also be using scientific notation to record very large or very small numbers. If you are not familiar with scientific notation, see **Appendix B: Scientific Notation and Significant Figures.**

In biology, mass is commonly measured in kilograms (kg), grams (g), and milligrams (mg). Volume is commonly measured in liters (l), milliliters (ml), and microliters (μl). You should be familiar with these terms and abbreviations.

- The prefix *kilo-* means one thousand
  One kilogram (1 kg) = 1000 grams or 1×10^3 g
- The prefix *milli-* means one one-thousandth.
  One milliliter (1 ml) is 1/1000 of one liter or 1×10^{-3} l
- The prefix *micro-* means one one-millionth.
  One microliter (1 μl) is 1/1000000 of a liter or 1×10^{-6} l
- Switching things around
  1 liter = 1000 ml or 1×10^3 ml
  1 liter = 1000000 μl or 1×10^6 μl
  1 ml = 1000 μl or 1×10^3 μl

Water has a useful mass–volume relationship under standard pressure and temperature conditions.

One liter of water has a mass of one kilogram.

Therefore, <u>one milliliter</u> of water is equal to <u>one gram</u> of water.

Likewise, one microliter (μl) of water is equal to 0.001 grams (g) or 1 milligram (mg) of water.

Lab Skill 1.1 Interpreting Lab Safety Symbols

Biologists often run experiments in a laboratory where they can use special equipment and better control experimental variables. Some data are collected "in the field." Regardless of where the work is done, biology experiments should always be done safely and carefully. Here, we will cover principles of lab safety that should be used in any biology lab.

Lab Policies. The following are policies for this lab course. Other labs may vary.

- No food or drink is allowed in biological laboratories due to the potential presence of toxic chemicals. Eating and drinking in the lab is strictly prohibited.
- Tabletops should be wiped with disinfectant spray before and after use.
- Report any injuries, broken glassware, and spills to your lab instructor immediately.
- **Safety Data Sheets (SDS)** describing all chemicals used in the lab and their risks must be readily available in the lab.
- Follow instructions and precautions for handling **carcinogens** (chemicals that can cause cancer), **teratogens** (chemicals that can cause birth defects), and **toxins** (poisons).
- Biohazardous materials should always be disposed of in proper biohazard containers.
- Students may not enter or use the lab without instructor supervision.
- Lab materials may not be removed from the lab without instructor permission.
- Students must be able to interpret symbols for laboratory hazards.

Safety Symbols. Study the icons in Figures 1.4 and 1.5. Many compounds used in biology laboratories have risks. It is critical that you understand how to read the warning labels used in science laboratories.

Figure 1.4 Common symbols for toxins, corrosive substances, and biohazards from the Occupational Safety and Health Administration (OSHA) Hazard Communication Standard.

Figure 1.5 The National Fire Protection Association (NFPA) and National Fire Rating(NFR) labeling system uses color-coded diamonds and numbers 0-4 to identify health risk (blue), flammability (red), instability (yellow) and other specific hazards (white).

Laboratory safety equipment. Any person using the lab must know their location and use.

- **Fire extinguisher**, but ensure it is the right kind of extinguisher for the type of fire.
- **Eyewash** for rinsing chemicals or contaminants from eyes.
- **Fire blanket** for throwing over a burning object or person to extinguish the flame.
- **Glass box** for disposable slides and broken glass.
- **Sharps box** for disposing of scalpel blades or other sharp items.

Personal Protection Equipment (PPE). Anyone working in the lab must wear proper PPE.

- Safety glasses must be worn when using any liquid chemicals.
- Gloves must be worn when handling chemicals or specimens.
- Closed-toe shoes must be worn at all times to protect from sharps and chemical spills.
- Long hair should be tied back.
- Dangling jewelry (necklaces and bracelets and large earrings) should be removed.
- Pants or dresses that cover the legs are strongly recommended.
- Lab coats or aprons are recommended but not required.

Lab Skill 1.2 Keeping a Lab Notebook

One of the main tools of a biologist is a laboratory notebook. A lab notebook serves as a legal document for scientists working on grant-funded projects or in biological industries. If you plan to publish your research, your laboratory notebook serves as the official document to verify your procedures and results. It is also a record of your intellectual property and is the basis for establishing ownership of scientific discoveries and patent rights. Although it is unlikely that you will make a new scientific discovery in this course, it is helpful to develop these good record-keeping habits early.

You do not need to purchase a formal lab notebook for this course. This manual includes pages that will serve as your lab notebook. You will be expected to complete more of the notebook entries as you progress through the course. However, in more advanced biology courses, you may be required to purchase a lab notebook and complete all notebook entries yourself.

A lab notebook can be formatted a number of ways, but the following guidelines will help you keep a well-organized and professional looking lab notebook.

- Write everything in pen so that no one can alter your work. Write everything by hand so that the handwriting can be identified as yours.
- Write your name and date at the top of each new entry. This informs a reader that this is your work and when you did it. You should start a new entry each day you do a lab activity, even if an ongoing project takes multiple days.
- Write a clear and descriptive title that indicates what you are trying to accomplish that day. A title like "Lab 4" tells you nothing about what was done that day. A title such as "Isolation of nuclear DNA from human cheek cells" is much more informative.
- Give the purpose of the work you intend to do in lab that day. If your work is part of a long-term project, indicate which phase of the project is being described in this entry.
- Use headings to separate sections of your notebook entry. Each notebook entry should include a *Protocol* section, an *Observations and Data* section, and a *Conclusion* section. You may include more sections if needed, such as *Background Information, Related Research,* or *Statistical Analysis*.
- The **Protocol** section lists the steps that will help you accomplish your goal. You will often write out and use protocols developed by others until you learn enough lab techniques to create your own. When you write a previously prepared protocol in your lab notebook, you may need to adjust it for your experiment or for the resources and equipment you have available.
- Write out your protocol *before* you begin an experiment. The more specific and detailed it is, the better. It should be written clearly enough that when you start working in the lab, you can simply follow the steps you have written. Anyone else should be able to replicate your lab work by following your protocols. Keep your notebooks as a reference for how to do the techniques. You may need these protocols again.
- If you have to change anything in your protocol during the experiment, make a line through the original steps and write changes or comments in the margin. Do not completely erase or scratch out your original steps. You might need to refer back to your original plan and see exactly which changes you made.
- The *Observations and Data* section should include all of the information you gather while doing the lab. Make sketches of specimens, models, and microscope slide images. Use different colors to easily identify parts. Label drawings clearly. Make drawings large enough to be useful. See Figure 1.6.
- Record *quantitative* data with all appropriate units. Create graphs or tables of data to easily visualize trends. Show any calculations and clearly indicate how you obtained results.
- Include *qualitative* data as needed. However, when possible, use a standard or scale. For example, if you record a description such as "the mixture turned blue," it may be unclear what that means. How blue? You could improve this observation by identifying the mixture's color on a color chart. Stating "the mixture resembled #2 on the following color chart" is much more specific (Figure 1.7).
- The *Conclusion* section should summarize what your data implies and how it relates to your original goal or hypothesis. It should explain what you have learned from the experiment, and what the data implies. It is important to clearly state what your results mean in relation to your original hypothesis so readers can easily interpret your work. If you encountered problems, describe them here. Also include new questions raised or ideas for future research.

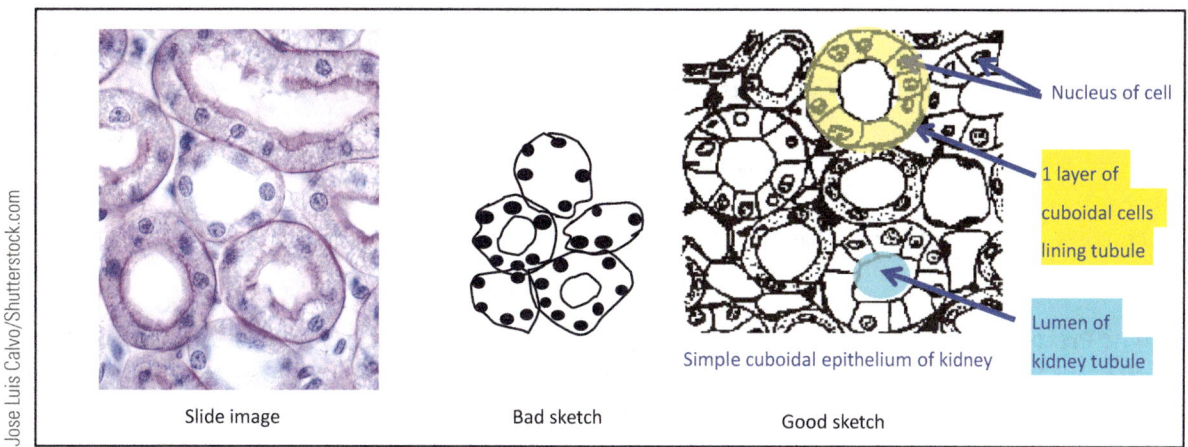

Figure 1.6 Comparison of good and bad notebook sketches.

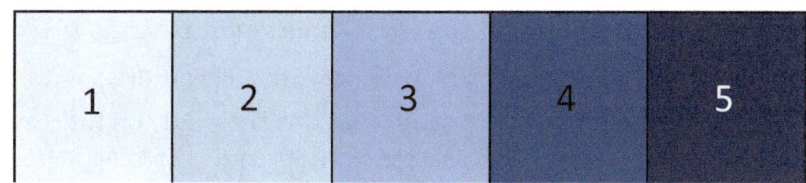

Figure 1.7 Color charts like this can be used to improve descriptive data about color.

In industry or grant-funded projects, the researcher and supervisor sign notebook pages to verify that the information is accurate. When publishing research or seeking patents, it is important to accurately document the name of the researcher that did the work, what they did, and exactly what data was collected. Read Figure 1.8 for a description of a typical lab notebook entry.

| Today's date | Your name |

Title: Provide a descriptive title of what you are doing in lab that day. You might need to refer to this notebook entry later, so a good title will make it easier to find specific protocols or data.

Purpose: Clearly state what you are trying to accomplish in the lab on this particular day. If you are testing a hypothesis, include it here. If you are learning how to do a lab technique, state that as well. Use headings as shown below.

The **Protocol** section should be written <u>before</u> working in the lab. The **Observation and Data** section should be written <u>during</u> the lab. The **Conclusion** section should be written <u>immediately after</u> the lab. It is likely that one day's entry will require multiple notebook pages. Don't crowd your work. Make it easy to read so you can refer back to it later.

Protocol

As clearly as possible, write out your plan <u>before</u> you start working in the lab. Leave enough room in the margins and between steps to make corrections and additions <u>as</u> you work.

- List specific pieces of lab equipment you will need and how many.
- If you will be using any chemicals or solutions, list them all. Include quantities and concentrations and appropriate units. Also note if any require special safety precautions.
- List each step of the procedure you plan to do. It also helps to note <u>why</u> you are doing them. You or any reader should be able to accurately repeat your procedures. You might also create a flowchart (workflow diagram) to summarize main steps.
- Leave some room between steps and in the margins.
- If you change any steps in the protocol, do not scratch out or erase the original step. Rather, draw a line through what you did not do and write changes in the margin. That way you have a record of all modifications.

Observations and Data

Carefully record all measurements and observations <u>as</u> you work. It is easy to forget them if you wait until later to write them down.

- Sketch models or specimens you might need to recall later including those you observe through a microscope. Clearly label everything you draw. Use colors to make labeling easier.
- Use graphs and tables to organize your data. Show all calculations and give appropriate units with all numerical data.

Conclusions

Summarize your results for the day.

- What do your data suggest? Do your data support your hypothesis? What do your results mean?
- If you had trouble with a procedure, what could be done to improve it?
- If you are working on a long-term project, what is the next step?

Sometimes research programs require signatures to verify the accuracy of the notebook entry.

| Researcher signature | Date | Supervisor/Instructor signature | Date |

Figure 1.8 How to use a laboratory notebook

Laboratory Exercise 1   An Introduction to the Biology Laboratory   11

Lab Skill 1.3 Measuring Liquids with Different Lab Equipment

You may have used common lab equipment like the ones illustrated in Figure 1.9. However, not all students know how and when to use them appropriately.

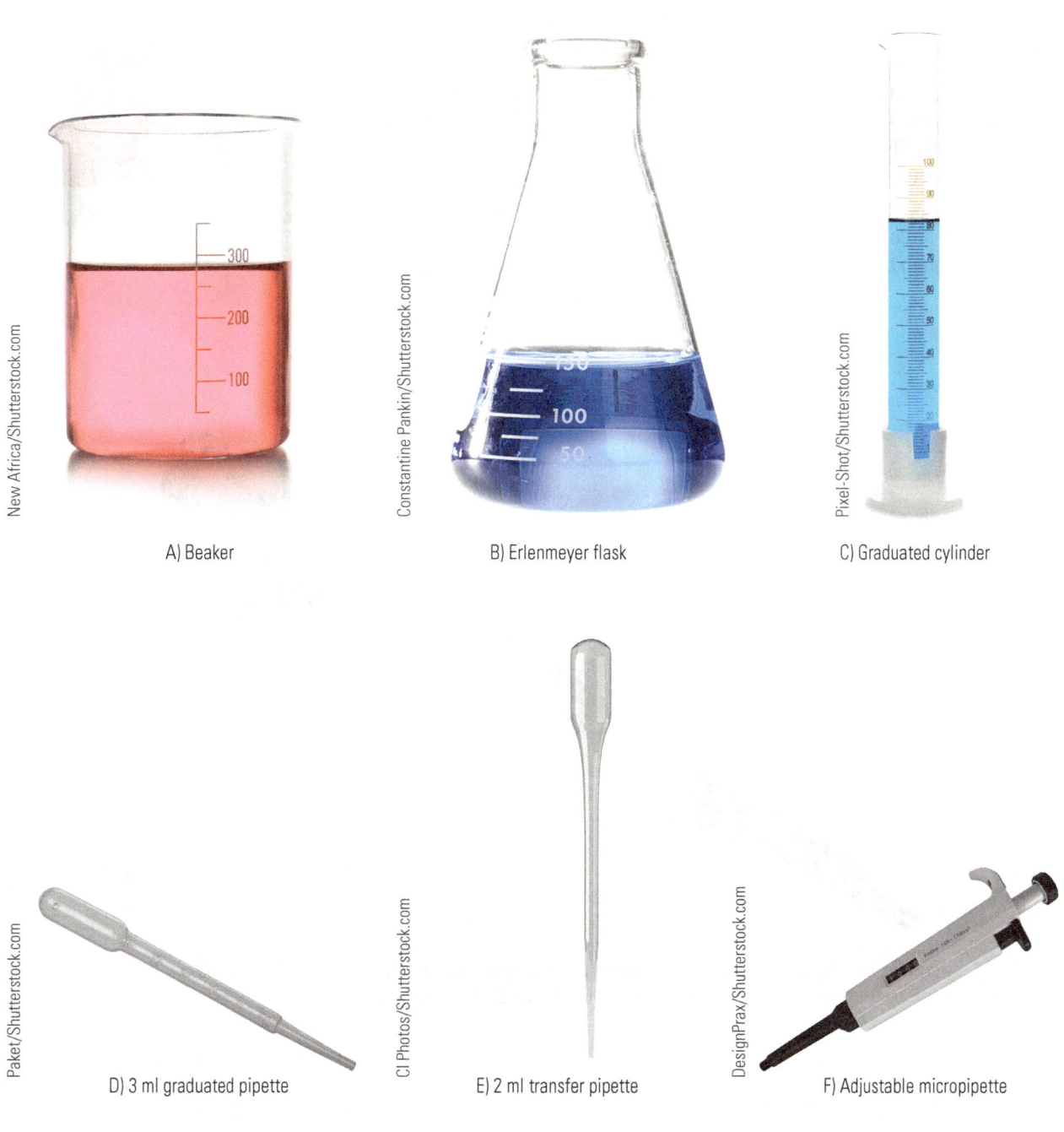

A) Beaker

B) Erlenmeyer flask

C) Graduated cylinder

D) 3 ml graduated pipette

E) 2 ml transfer pipette

F) Adjustable micropipette

Figure 1.9 Common lab measuring equipment.

When measuring large quantities of liquids, containers such as beakers and flasks are marked less precisely. It is impossible to measure to single milliliters with a container marked with 50 or 100 ml increments. However, when measuring small amounts, the precision needs to be much greater. If you need exactly 0.1 ml of a solution, you must use an instrument that is marked at that level of precision or greater. But as you use instruments with greater precision, the volume of liquid they can measure is reduced. For example, a micropipette (Figure 1.9F) can measure very precisely, but can usually only measure quantities less than 1 ml.

When using glassware like a graduated cylinder, it is important to measure the bottom of the meniscus, the curved surface of the liquid in the tube. For example, in Figure 1.10, the accurate volume indicated here is 7.6 ml. Not all of the graduations are numbered, so it is important to look at the markings carefully.

Disposable pipettes are commonly used in biology labs when accuracy is not critical. They are inexpensive and do not require cleaning as they can be thrown away after use. The 3 ml disposable pipette (Figure 1.9D) does have volume markings, but they are difficult to read, and are often difficult to use. Most 1 ml or 2ml disposable pipettes (Figure 1.9E) lack volume markings or are only marked at 1 ml. They are often called transfer pipettes because they are mainly used to move substances from one container to another, not to accurately measure quantities.

Figure 1.10 Reading a graduated cylinder.

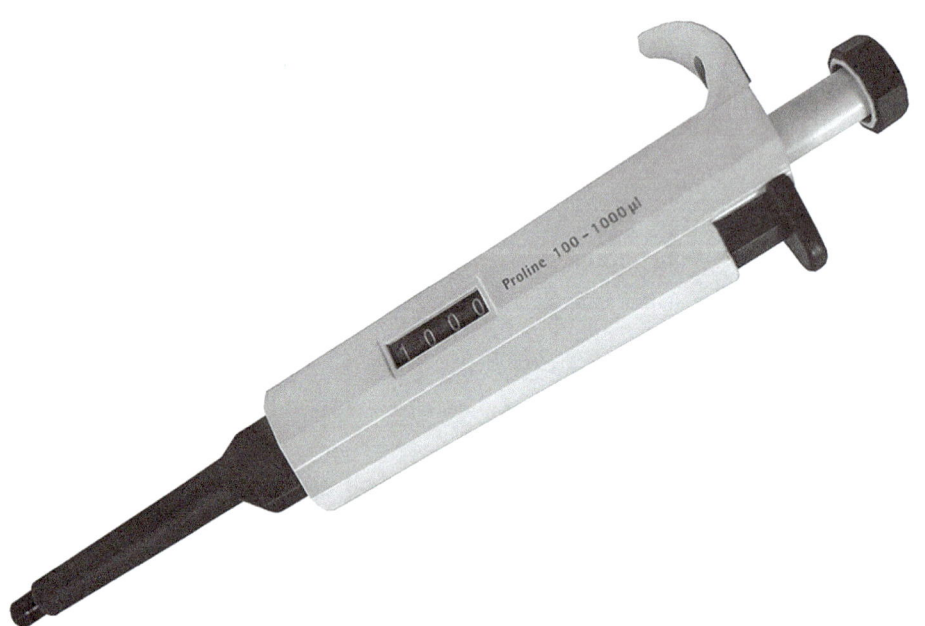

Figure 1.11 Adjustable micropipette. This micropipette may be used for volumes between 100 and 1000 µl. It is currently set for 1000 µl which is equal to 1 ml. The disposable tip is not shown.

Lab Skill 1.4 Using a Micropipette

The adjustable micropipette is highly precise and can accurately measure very small quantities. However, it requires some training to use. The steps are described briefly here.

1. Select a micropipette that is appropriate for the volume desired. The pipette usually has the maximum volume it can accept noted on the body (Figure 1.11).
2. Ensure the adjustment knob is unlocked. Turn the plunger to the volume desired. Ensure you do not go beyond the acceptable range for that pipette, as it will damage the pipette.
3. Select the proper tip for the pipette. Place the pipette into the top of the tip and tap gently to ensure a good seal. Keep the tip in the box until it is firmly on the micropipette. Do not touch the tip with your fingers.
4. Always hold the pipette with the tip pointed down. This ensures that no samples will get into the body of the micropipette. Your thumb should rest on the top of the plunger, and the pipette should "hook" over your hand.
5. Press the plunger down to the first stop.
6. Carefully place the tip into the solution you wish to collect. <u>Slowly</u> relax the pressure on the plunger and allow the liquid to be drawn into the tip. Watch carefully to ensure you do not draw in air. Wait a few seconds to ensure all liquid has been drawn into the tip.
7. Keeping the tip pointed down, withdraw the tip from the solution, and inspect the tip to ensure it does not contain bubbles or air at the end of the tip. If it does, start over.
8. Place the tip into the receiving container. Do not immerse the tip. Using your thumb, <u>slowly</u> press the plunger to the first stop. Then to ensure all liquid is expelled, press the plunger to the second stop. Wait a few seconds to ensure all liquid has been expelled.
9. Keep the plunger depressed and slowly raise the micropipette. Ensure all liquid is out of the tip. Then release the plunger slowly.
10. Use the ejector button to eject the tip into a waste or biohazard container as needed.

To get the best results when using a micropipette:

- Carefully select the pipette size. Do not use a micropipette for collecting volumes of liquids outside the stated range of the micropipette.
- Always verify the volume setting before extracting your sample.
- Adjust the micropipette volume with care. The gears that adjust the volume are fragile and can be damaged by turning the plunger too fast or too hard. Do not force it.
- Use the proper tip for the micropipette and sample size.
- If possible, set your elbow on the table while transferring the sample to improve stability.
- Only put the tip on the micropipette when ready to use it. Eject the tip immediately when done with a transfer.
- Never lay the micropipette down with a tip on it. It can contaminate the tip. If a sample is in the tip, it can flow into the micropipette body.
- Carefully inspect the tip during the transfer to ensure you did not draw in air bubbles or leave any sample in the tip.
- Keep the micropipette on the stand when not in use. This helps ensure the tip always points down.
- Take your time. It is better to be accurate and precise than fast.

Laboratory Exercise 1 An Introduction to the Biology Laboratory

Problem to Solve 1.1 Can I Use Scientific Reasoning to Solve a Puzzle?

For this activity, you will need to work with one lab partner. You will also need a notecard. One of you will set the code for an imaginary combination lock, and will be the *code maker*; the other will try to figure out that code, and be the *code breaker*. You will repeat this activity twice and then switch roles until each person has attempted to discover two of their partner's lock codes.

The code maker must choose four symbols for their code. A symbol may only be used once per code. The symbols you may choose from are shown below:

An acceptable code might be:

The code maker should sit across from the code breaker. **The code maker must write their code in the code breaker's lab manual on the next page. But it must be kept secret until the code is broken or the code breaker has failed.** The code maker should record the code in the top four boxes under FIRST CODE and immediately cover it with the notecard. (See Figure 1.12.)

The code breaker must attempt to guess the code by filling in symbols in the following boxes. Note that symbol *order* matters, but it is acceptable if the symbol is written upside-down.

After each guess, the code maker must indicate the number of *correct symbols in the right location*, and the number of *correct symbols in the wrong location* in the boxes provided. The code maker should not tell the code breaker which symbols are correct!

Figure 1.12 Code maker's view.

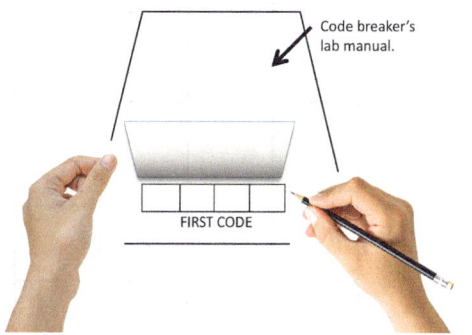

The code breaker should use those clues to try to figure out the secret code. The code breaker and code maker should repeat this process until the code has been broken or until all ten attempts have been made. The code maker should then create a new secret code and the code breaker should try again. Then the code maker and code breaker should switch roles and lab manuals and repeat the challenge until both partners have been code maker and code breaker twice.

18 Laboratory Exercise 1 An Introduction to the Biology Laboratory

FIRST CODE | **SECOND CODE**

The code maker fills in the orange boxes. The code breaker enters guesses in the green boxes.

Possible code symbols:

Problem to Solve 1.2 Can I Accurately Measure Liquids?

You have been tasked with measuring out four specific volumes of water: 15 ml, 2 ml, 0.5 ml, 500 µl, and 50 µl. You have several pieces of equipment at your disposal.

Which type of lab equipment will provide the best *accuracy and precision* for measuring different volumes of water? Remember that we can calculate the standard deviation of a set of measurements. The standard deviation indicates how closely grouped, or *precise*, they are.

Recall that under normal conditions 1 ml of water (volume) is equal to 1 g (mass). We will determine the *accuracy* of different measuring tools by comparing volumes of water obtained with those tools to the mass of the water as measured by a mass scale. You may choose from a variety of beakers, graduated cylinders, disposable transfer pipettes, and adjustable micropipettes of different sizes.

Hypothesis and Prediction

Before starting this activity, you will need to make a hypothesis. A hypothesis should always be made *before* doing an experiment. Recall that a hypothesis is an educated guess, or a statement of what you *think* is probably the right explanation for something. You may not know which instrument has the greatest accuracy or precision, but you should be able to make a good guess.

A prediction, on the other hand, is usually an "if-then" statement based on the hypothesis. *If* the hypothesis is true, *then* when I do this, I should get that as a result." In this case, a prediction might be that "If I measure X ml of water with a _____ (piece of lab equipment) and place it on a scale, the mass of that sample of water will be exactly _____ g."

The data from the experiment will either support the hypothesis, or it will not. The hypothesis should not be made after you already know the results! The point of the experiment is to find out if your "educated guess" was a good one. And if it wasn't, you have still learned something.

Create a hypothesis for which lab instrument will be most precise and/or most accurate and record it in the Purpose box on the lab notebook page. Then, based on your hypothesis, make a prediction for which instrument will have the smallest standard deviation.

Again, you will work with one lab partner to complete this activity. As one of you performs the protocol, the other will record data. Then you will switch roles.

The protocol for this activity has already been written for you on the following lab notebook page. You will need to follow the steps as directed and fill in the data table.

Date 8/23/21

Laboratory Exercise 1 — An Introduction to the Biology Laboratory

Title: Determining precision and accuracy when measuring liquids

Purpose: Learn to use lab equipment and compare their accuracy and precision. Learn to measure standard deviation. **Hypothesis:** The graduated cylinder will be the most accurate + precise. **Predictions:** If we use the graduated cylinder to measure, it will have the smallest standard deviation.

Materials needed

Beakers (multiple sizes) disposable transfer pipettes paper towels
graduated cylinders mass balance
adjustable micropipettes weigh boat

Protocol

1. Determine which lab partner will be "recorder" and which will be the "measurer."

2. Ensure the mass balance is on. Place it in front of the recorder so that the measurer cannot see the readings. Place a weigh boat on it and press "tare" or "zero."

3. The measurer must use the various pieces of equipment to pour the appropriate amount of water into the weigh boat as accurately as possible. The measurer may choose which pieces of lab equipment to use for each volume.

4. As the measurer attempts to place the correct amount of water in the weigh boat, the recorder must record the type of lab equipment used and the mass of the water in grams in the data table in *their* lab manual without letting the measurer see the results.

5. After each attempt, the weigh boat must be dried, placed back on the scale, and the scale zeroed.

6. The measurer must repeat each trial using the same volume and lab equipment three times.

7. Do this for all five volumes. A different piece of lab equipment may be used for measuring different volumes.

8. Then switch roles so that the recorder becomes the measurer and the measurer becomes the recorder and repeat the procedure.

Date: 8/23/21

Continued from previous page.

Data and Observations

Remember that this is the recorded mass of your lab partner's trials.

Sample	Volume desired	instrument used	predicted mass (g)	mass of sample (g) trial 1	trial 2	trial 3
1	15 ml	50 ml graduated cylinder	15 g	14.64 g	14.56 g	14.67 g
2	2 ml	10 ml graduated cylinder	2 g	1.94 g	1.79 g	1.84 g
3	0.5 ml	pipette	0.5 g	0.61 g	0.49 g	0.58 g
4	500 µl	pipette	0.5 g	0.54 g	0.55 g	0.5 g
5	50 µl	5-50 µl micropipette	0.05 g = 0.001g	0.07 g	0.05 g	0.05 g

Fill in the predicted mass for each trial. Recall that one ml of water has a mass of one gram.

Analysis

Find the Average Error for each sample run. Ignore negative signs. All % error values should be positive.

Sample	% error #1 (trial 1 − predicted mass) × 100 / predicted mass	% error #2 (trial 2 − predicted mass) × 100 / predicted mass	% error #3 (trial 3 − predicted mass) × 100 / predicted mass	Average Error (% error 1 + % error 2 + % error 3) / 3
1	2.4 %	2.93 %	2.2 %	2.51 %
2	3 %	10.5 %	8 %	7.17 %
3	22 %	2 %	16 %	13.33 %
4	8 %	10 %	0 %	6 %
5	40 %	0 %	0 %	13.33 %

Laboratory Exercise 1 An Introduction to the Biology Laboratory

Continued from previous page.

Analysis continued
Find the Standard Deviation

Sample	difference 1 (% error 1 − average error)²	difference 2 (% error 2 − average error)²	difference 3 (% error 3 − average error)²	Variance sum of differences ÷ 3	Standard Deviation square root of variance
1	0.0121	0.1764	0.0961	0.0948667	0.308
2	17.3889	11.0889	0.6889	9.722233	3.118
3	75.1689	128.3689	7.1289	70.222233	8.380
4	4	16	0	6.667	2.582
5	711.2889	0	0	237.0963	15.398

Conclusions

Did you and your lab partner use the same equipment to measure the liquids? Why or why not?

Yes, we did use the same equipment.

Based on your results, which instrument had the greatest accuracy (smallest % error)?

The 50ml graduated cylinder had the greatest accuracy based on our results

Which instrument had the greatest precision (smallest standard deviation)?

The 50 ml graduated cylinder had the greatest precision based on our results.

Did you and your partner get similar results for accuracy and precision? If not, why do you think they varied?

We got similar but different results. varied due to how much each poured

Was your hypothesis supported by the data?

Yes, my hypothesis was supported by the data.

If you need to accurately measure these quantities again in the future, will you use the same lab equipment you used today, or will you change your strategy? Explain your answer.

I believe I would use the same equipment but just be more precise with my pouring.

Name: Bethany   Date: _____

POST-LAB QUIZ 1: What have you learned?

1. In the code-breaking activity, you were informally using the scientific method to try to break the code. What served as the "hypothesis" in your code-breaking attempt?

2. In the code breaking activity, what data did you collect?

3. If you did not guess the code correctly on the first try, was your second guess more informed than your first? Why?

4. Were you able to crack the second code in fewer tries than the first? Explain why or why not.

5. How might this code-breaking activity compare in some ways to a pathologist trying to find a cure for a disease, or an ecologist trying to address an environmental problem?

Laboratory Exercise 1 An Introduction to the Biology Laboratory

6. What was your *independent* variable when measuring water samples? What was your *dependent* variable?

   The independent variable was the amount of water, the dependent

7. List at least three "rules" you must follow in order to properly use an adjustable micropipette.

   1. Push the button down before putting into solution
   2. always point down
   3. do not touch the tip with your hands

8. What does the standard deviation of a set of data tell you?

9. Which piece of equipment had the *lowest* accuracy? Which had the *lowest* precision? Should these instruments still be used in a lab? Why or why not?

   According to our data, the micropippete and the graduated pippete had the lowest accuracy and precision. I do believe they should be used still. We may have just used them at the incorrect time.

10. What is the mass in grams of 50 μl of water? What is the volume in μl of 5 g of water?

    The mass in grams of 50 μl of water is 0.5g & of 5g it is 0.005 μl.

Laboratory Exercise 2

Using Chemistry in the Biology Lab

Concepts to Understand

After completing this lab, the student should be able to:

1. Describe the atom and apply the Kinetic Theory of Matter to states of matter.
2. Explain ionic, covalent, and hydrogen bonding.
3. Use the periodic table of elements and the octet rule to predict chemical bond formation.
4. Identify synthesis, decomposition, and exchange reactions.
5. Explain the naming conventions commonly used in chemistry.
6. Identify solutions, solutes, and solvents, including acids and bases.

> This lab may be assigned as a take-home laboratory exercise. Students who are unable to complete the in-class activities should still be able to *predict the theoretical results* in the experimental section.

Skills to Learn

After completing this lab, the student should be able to:

1. Name ionic and covalent compounds.
2. Create a solution of a specific concentration.
3. Create a dilute solution of a desired volume.
4. Measure pH accurately with pH indicator, pH test strips, and a pH probe.
5. Measure solute concentration/density with a hydrometer.

Gorodenkoff/Shutterstock.com

> SEE ALSO:
>
> Appendix C: What are Atoms, Isotopes and Ions?
>
> Appendix D: How to Use the Periodic Table of the Elements
>
> Campbell Biology - Chapter 2, sections 2.3 and 2.4, and Chapter 3, section 3.1
>
> JOVE (video) - Understanding Concentration and Measuring Volumes
>
> Others:

Important Terms to Know

Acid
Anabolic
Anion
Base
Catabolic
Cation
Cellular respiration
Compound
Coefficient
Covalent bond
Decomposition
Dilution
Displacement
Electrolyte

Electronegativity
Endergonic
Exergonic
Hydrogen Bond
Hydrometer
Ionic Bond
Mixture
Molarity
Mole
Neutralize
Non-polar
Octet Rule
pH
pH scale

Polar
Polyatomic ion
Product
Reactant
Reducing agent
Salt
Solute
Solution
Solvent
Specific gravity
Synthesis
Valence

Pre-lab Preparation and Background Information

You should be familiar with the concepts of atoms, subatomic particles, and the periodic table of the elements from your high-school biology or chemistry courses. The following material assumes you have some background in chemistry from high school. If you need to refresh your understanding of these concepts, please refer to **Appendix C: What are Atoms, Isotopes and Ions?** and **Appendix D: How to Use the Periodic Table of the Elements** before continuing.

Atom Stability and the Octet Rule

Any atoms with partially filled energy levels are inherently unstable by themselves. These atoms will react with other atoms to try to establish a full outer energy level. Atoms can do this by giving, receiving, or sharing electrons. When they do this, a chemical bond is formed. And generally speaking, atoms do this according to the "rule of 8," or octet rule.

The **octet rule** states that most atoms will make bonds with one or more other atoms in order to *fill their outer energy level* or to *get eight electrons in their outer energy level*. When different elements interact with each other in order to fill their outer energy level, they form a **compound**. For example, fluorine has seven valence electrons. (It has nine total electrons, but two are in the first energy level.) It is in period 2, group 17, indicating that it has one less electron than necessary to fill the second energy level. (See Figure 2.1.) Refer to Appendix D for a larger periodic table. The further down the periodic table one goes, the more complex the arrangement of electrons gets, due to electrons not filling the energy levels in order. However, elements in the same group have the same number of valence electrons and thus need the same number of electrons in order to satisfy the octet rule. Chlorine is below fluorine, and it also has seven valence electrons. Chlorine also needs one more electron to satisfy the octet rule. Recall that the third energy level can hold 18 electrons, but the octet rule states that atoms are also stable with eight valence electrons, so even though the third energy level won't be filled with one more electron, it will have eight, and having eight will fill a sublevel and make it stable.

The group an element is in, therefore, indicates its valence electrons or number of electrons needed to be stable. All noble gas elements in group 8 are already stable. They satisfy the octet rule with the electrons they have. All elements in group 17 lack one electron to satisfy the octet rule. All elements in group 16 lack two electrons to satisfy the octet rule. All elements in group 1 have one electron in their outermost energy level and want to get rid of it to satisfy the octet rule by emptying out the outermost energy level.

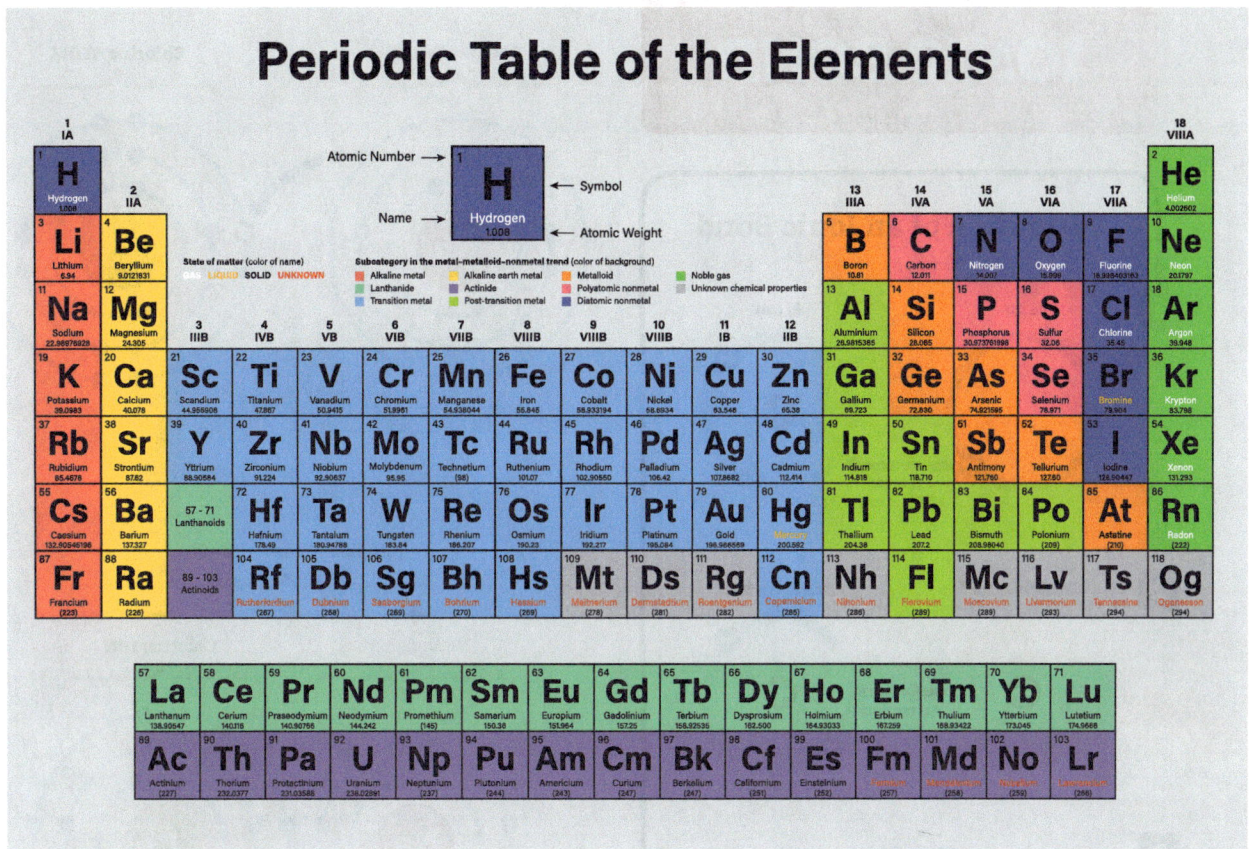

Figure 2.1 A simplified periodic table of elements. See Appendix D for a larger, more detailed version that is easier to read. The appendix also provides more information about how to interpret and use the periodic table including an explanation of the "groups" and "periods" of the periodic table.

Ionic Bonds

Ionic bonds occur between atoms that transfer electrons from one to the other. For example, sodium (Na) is in the first column of the periodic table. A sodium atom has one valence electron, and it would be more stable if it could get rid of it. Chlorine (Cl) is in the 17th column; it has seven valence electrons and would be more stable if it had one more electron. If sodium and chlorine are combined, a sodium atom will give its outermost electron to a chlorine atom (Figure 2.2). This allows both atoms to satisfy the octet rule and become more stable. Consider the effect on the atoms. Sodium now has one less electron, giving it more protons than electrons. Originally, the sodium atom had 11 protons and 11 electrons, as indicated by its atomic number. Since it lost an electron, it now has more positive protons (11) than negative electrons (10), giving it a positive charge of +1.

Chlorine is also affected. The chlorine atom originally had 17 protons and 17 electrons, but if it gains an additional electron (18), the negative electrons outnumber the positive protons (17) by one, giving it a -1 charge. Atoms with charges due to the loss or gain of electrons are called *ions*. Positively charged ions are specifically called **cations**, and negatively charged ions are called **anions**. Anions and cations are oppositely charged and therefore attracted to each other. This attraction forms an ionic bond.

Ionic bonds vary in strength. They can be strong when dry, such as a "rock" of salt (NaCl). Ionic compounds often form geometric patterns or crystals when in solid form. Ionic compounds often dissociate (separate) into ions in water, such as the salt in seawater. **Electrolytes** (dissolved ions) play crucial roles in the physiology of organisms.

If you study the periodic table, it is easy to predict which elements will form ionic bonds: those elements on the far left want to get rid of their valence electrons, and those on the far right want to gain electrons to fill their valence energy level.

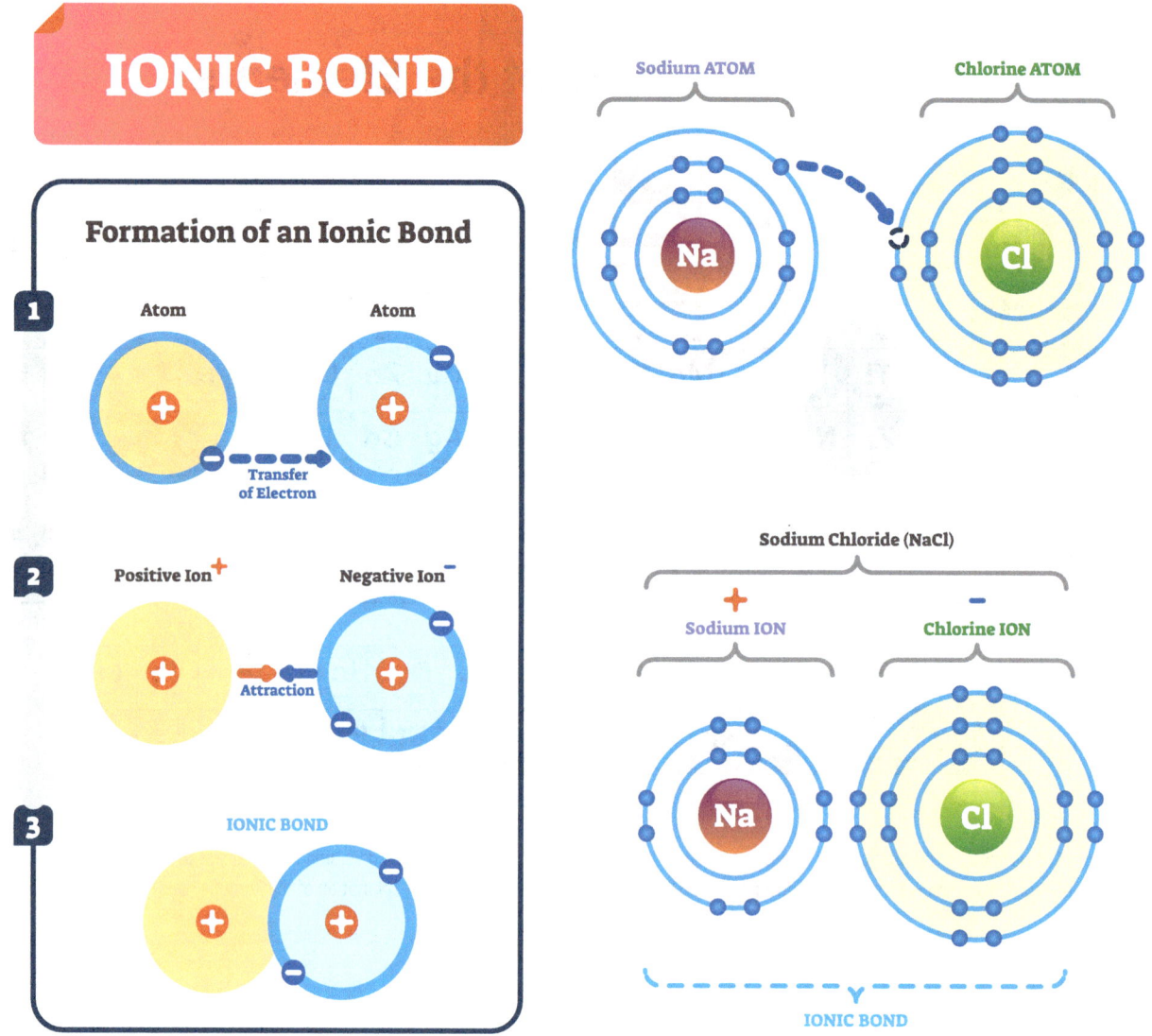

Figure 2.2 Ionic bonding of sodium and chlorine. Note that as the sodium atom loses an electron, it becomes a positively charged cation, but more stable. As the chlorine atom receives the electron, it becomes a negatively charged anion, but more stable. Oppositely charged ions attract each other.

Metals (left) readily bond to non-metals (right) to form ionic compounds. Ionic compounds are not considered true *molecules* because they often dissociate in water and in certain other liquids.

Covalent Bonds (Single, Double, Triple)

Non-metals can also bond with other non-metals, including other atoms of the same element. Yet this seems impossible when we consider that elements on the right side of the periodic table tend to want to receive electrons. If two non-metals—for example two chlorine atoms—each want one electron, can they bond together? The answer is yes, if they both *share* an electron with the other.

Sharing implies that the item being shared is available to both parties using it. When both atoms each share an electron with the other atom, the electrons are available to both atoms. If one atom shares an electron with another atom, the other atom will also share an electron, so the number of electrons being shared is always paired—one electron from each atom.

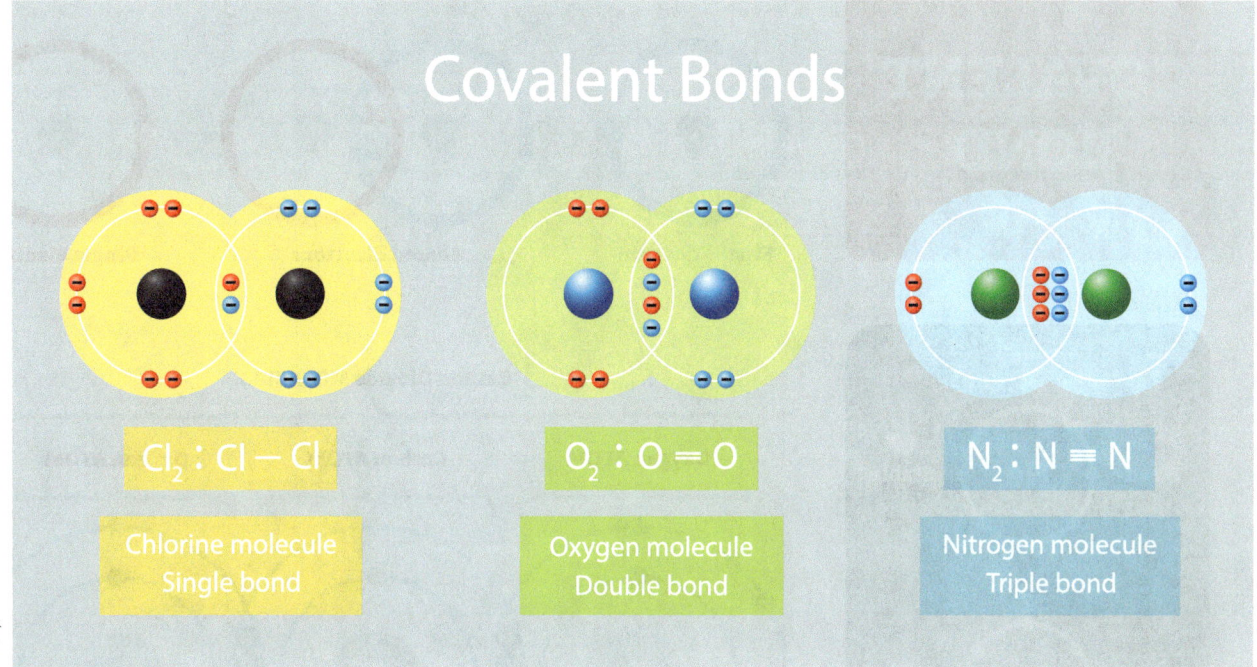

Figure 2.3 Covalent bonds occur when atoms fill their valence energy level by sharing electrons. Note that the electrons shown in the overlapping areas of the circles count for the valence shells of both atoms.

This allows both atoms to increase their number of valence electrons by one. Consider the chlorine atoms in Figure 2.3. The left chlorine shares one red electron with the right chlorine, allowing the right chlorine to fill its valence energy level. The right chlorine shares one blue electron in return, allowing the left chlorine to fill its outer energy level. When each atom shares one electron, the result is a **single covalent bond**.

However, atoms can share more than one electron. In the case of oxygen, each oxygen atom has six valence electrons and needs two more. If two oxygen atoms share two electrons with the other, this fills both valence energy levels. This is a **double covalent bond**. As shown with nitrogen, some atoms such as nitrogen can each share three electrons with each other in order to satisfy the octet rule. This creates a **triple covalent bond**. Covalent bonds can form between any non-metal atoms that can share electrons to satisfy the octet rule. See Figure 2.4.

Polar Covalent Bonds

Covalent bonds get a bit more complex, however, because some atoms are more "greedy" than others when sharing. When covalently bonded atoms do not share electrons equally, this is called a **polar covalent bond**. The desire of an atom to get electrons to satisfy the octet rule is called its **electronegativity**. The simplified rule for determining electronegativity is that those atoms closest to the upper right of the periodic table (excluding the noble gases) tend to have the greatest electronegativity, while those in the lower right corner of the periodic table have the lowest electronegativity. Oxygen is one of the most common elements involved in biochemical reactions, and it is highly electronegative.

When two atoms with different electronegativity covalently bond with each other, the more electronegative atom tends to "pull harder," thus holding on to the shared electron a little more than the other atom. As a result, the atom with the higher electronegativity value gets a *slightly* negative charge and the less electronegative atom gets a *slightly* positive charge. This is not the same as the charge produced by atoms transferring electrons as seen in ionic bonds. The notation for polar charges is rarely given, but when needed it can be represented by a $\partial-$ or $\partial+$ (positive or negative Greek delta) to distinguish that polar compound from a non-polar covalent compound or an ionic compound. Water is a polar covalent

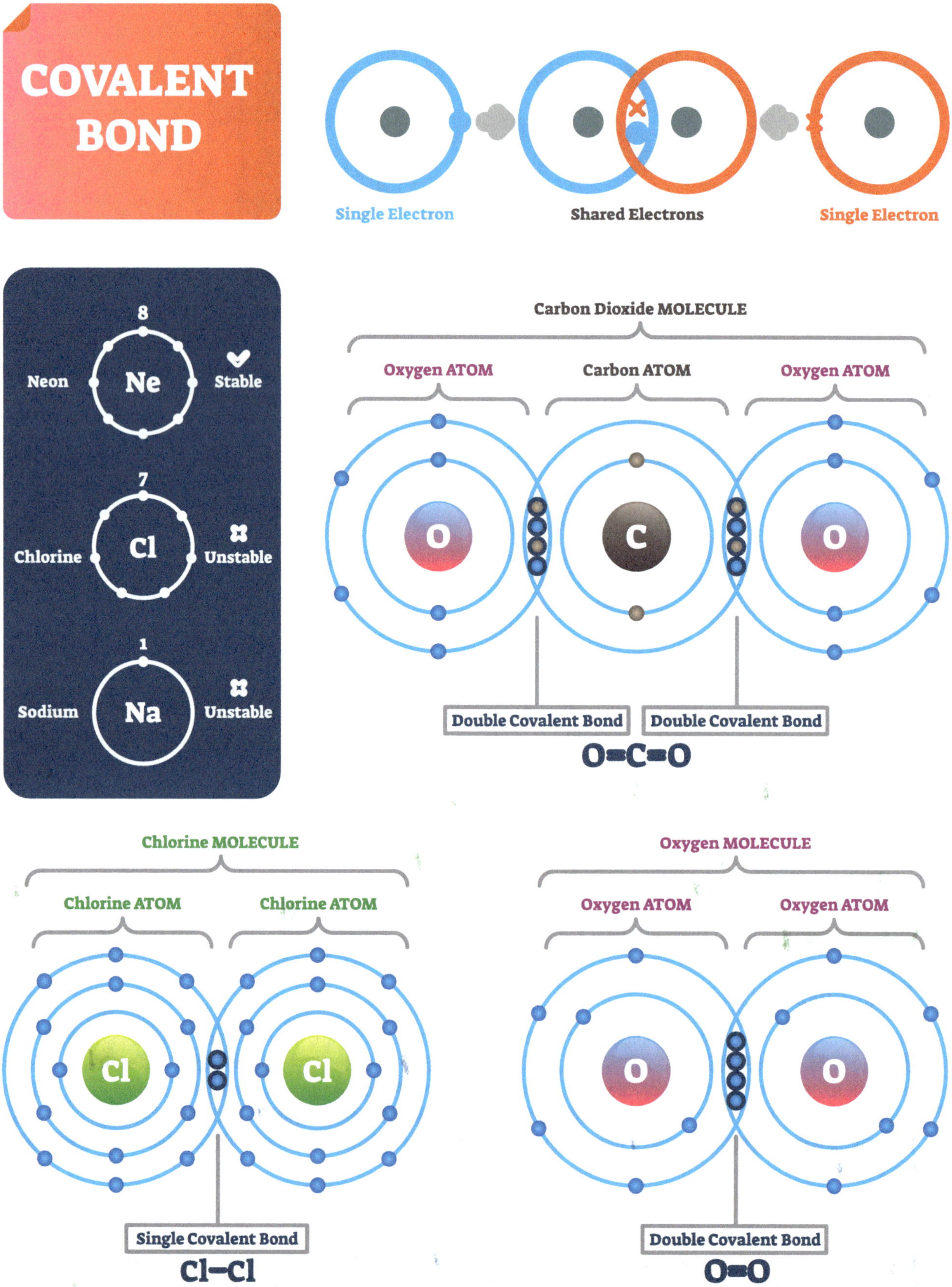

COVALENT BOND

Single Electron — **Shared Electrons** — **Single Electron**

Neon	Ne	8 — Stable
Chlorine	Cl	7 — Unstable
Sodium	Na	1 — Unstable

Carbon Dioxide MOLECULE
Oxygen ATOM — Carbon ATOM — Oxygen ATOM

Double Covalent Bond | Double Covalent Bond

O=C=O

Chlorine MOLECULE
Chlorine ATOM — Chlorine ATOM

Single Covalent Bond

Cl–Cl

Oxygen MOLECULE
Oxygen ATOM — Oxygen ATOM

Double Covalent Bond

O=O

Figure 2.4 These are examples of how atoms can share electrons in order to fill their valence energy level. As long as the octet rule is satisfied, atoms of different elements can form covalent bonds. As shown above, carbon has four valence electrons and oxygen has six. But if one carbon atom shares two of its electrons with one oxygen atom and two more electrons with another oxygen atom, and each oxygen atom shares two electrons with carbon, all three atoms will satisfy the octet rule. This forms carbon dioxide (CO_2) by covalent double bonds.

compound because oxygen is highly electronegative (or "greedy") and doesn't share the electron equally with the hydrogen atoms (Figure 2.5).

Oxygen bonded to itself in a double covalent bond forms a **non-polar** molecule because the two atoms have the same electronegativity. But if an oxygen atom was bonded to a tellurium atom in a double bond, the electronegativity difference between them would result in a polar molecule. All molecules formed by one atom bonded to another atom of the same element are non-polar. Their equal electronegative "pulls" balance out.

However, not all covalent molecules containing different elements are polar. The geometry of the atom determines its polarity. For example, carbon dioxide is not polar because the carbon electrons are pulled on by two oxygen atoms equally from both sides (Figure 2.6).

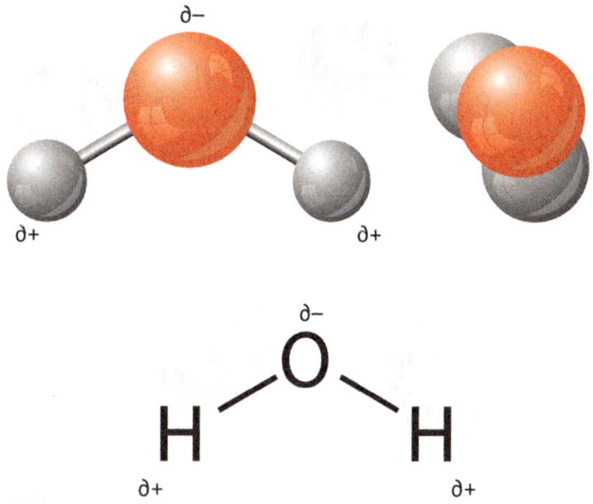

chromatos/Shutterstock.com. Adapted by Kendall Hunt Publishing Company.

Figure 2.5 Water (H_2O) is represented here by a ball and stick representation, and in a structural diagram. The oxygen atom (red) is shown connected to two hydrogen atoms by sticks that represent the covalent bonds. In the structural diagram, single lines represent single covalent bonds, two lines represent double bonds, and three lines represent triple covalent bonds. When needed, polar compounds will have slight polar charges indicated by the delta symbols as shown. Oxygen is slightly negative because the negative electrons are spending more time around the oxygen atom.

Hydrogen Bonds

In chemistry, we never deal with single atoms or molecules; we deal with containers full of molecules. When molecules are polar they interact. For example, water is polar. When water molecules are swirling around in a container, the positive part of a water molecule is weakly attracted to the negative part of another water molecule, causing a slight pull between them (Figure 2.7). This is called a hydrogen bond. **Hydrogen bonds** are weak attractive forces between the oppositely charged parts of *any two or more polar molecules*. Hydrogen bonds hold DNA molecules together, determine the shapes of proteins, and influence the shape and behavior of many molecules involved in cellular activities. One of the most important is water. The polar nature of water and its ability to make hydrogen bonds are critical for living things.

The hydrogen bonds that occur between water molecules account for many of the interesting properties we can observe. For example, one can fill a glass with water *above the rim*. It appears as though there is a thin "skin" on the surface of the water holding it back. This is due to surface tension created by the hydrogen bonds between water molecules at the surface of the water. Water bugs or water spiders appear to run on the surface of ponds and lakes. These insects do not

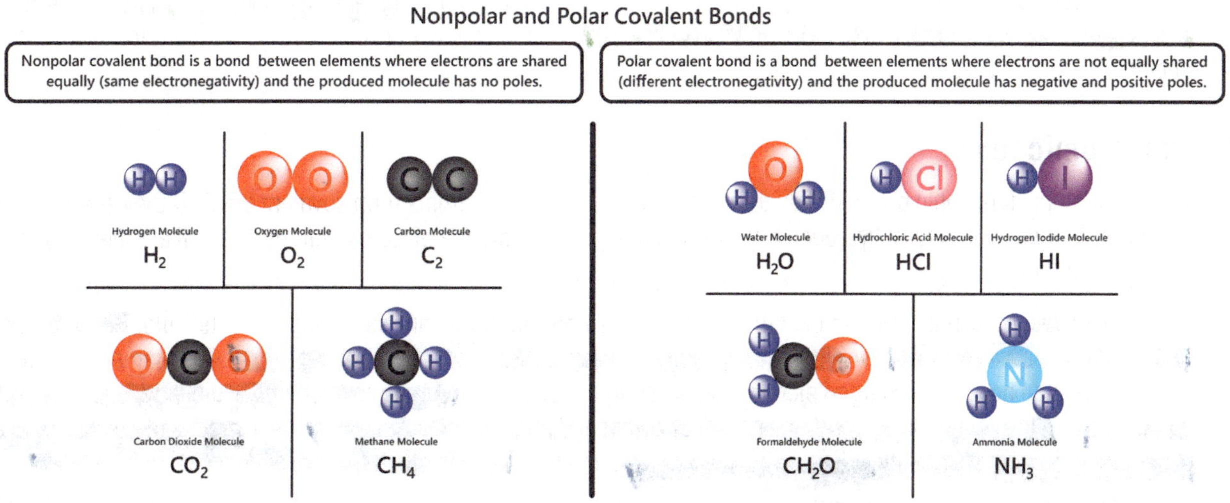

udaix/Shutterstock.com

Figure 2.6 Comparing non-polar and polar covalent bonds. When the electronegative pull is distributed equally among the atoms, the molecule is non-polar. When elements with differing electronegativity share electrons unequally, the molecule is polar. It is not important for you to be able to determine *if* a molecule is polar or not in this course, but it is important to know *why* a molecule may be polar and how that affects its properties.

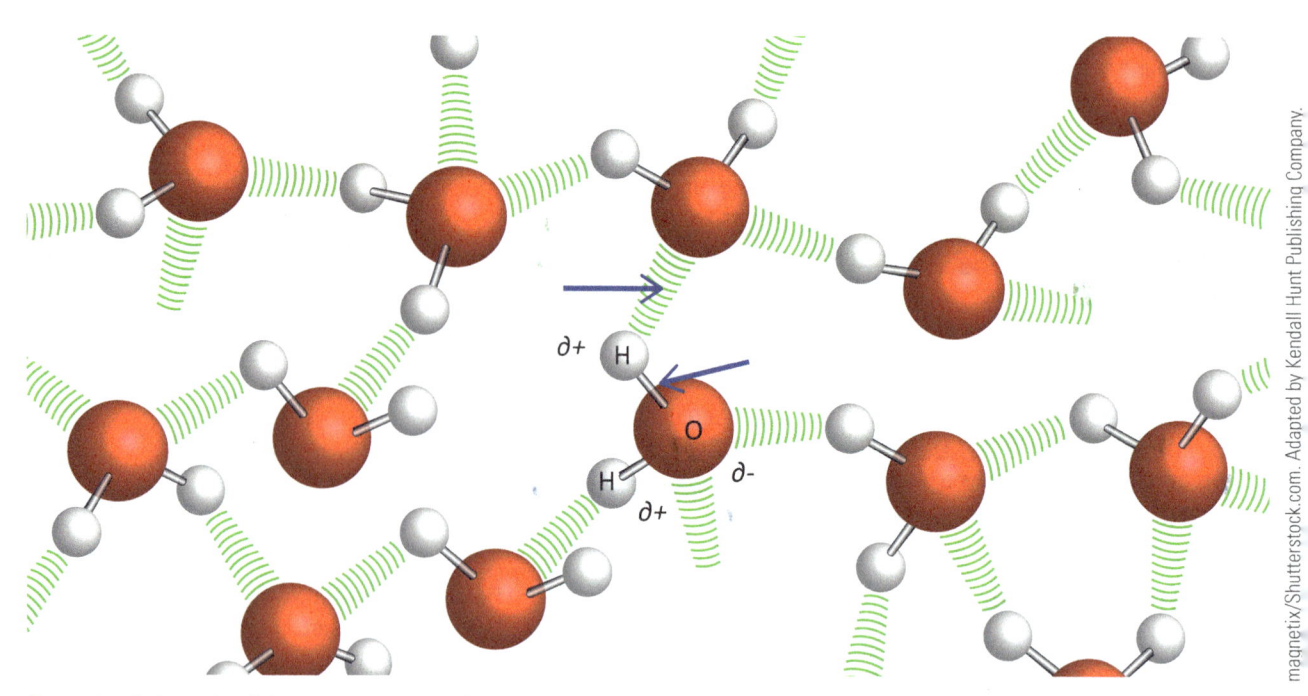

Figure 2.7 Hydrogen bonds between water molecules. The polar covalent bonds in one water molecule (H_2O) are very strong. The hydrogen bonds between water molecules (shown in green) are weak attractive forces that cause water to exhibit many properties important for life.

float. They are light enough to walk on the "skin-like" surface of the water without breaking the hydrogen bonds between the water molecules. If you look closely, they do actually dent the surface of the water much like a person walking across a trampoline.

Although the surface tension of water is significant, it also explains the beading action of water on some surfaces. If a surface is not polar, and water is spilled on it, the water molecules are attracted to each other more than to the non-polar surface, causing the water to bead. Coating a car with a non-polar layer of wax causes water to bead on it. The polar nature of water also causes water to cling to fabrics or materials that are polar.

One of the most important characteristics of water, also due to its polar nature, is its ability to dissolve ionic substances. Table salt, or sodium chloride (Na^+Cl^-), is an ionic compound that is held together in crystal form by the attraction of the positive cation, Na^+, with the negative anion, Cl^-. But when these crystals are added to water, the polar charges on the water molecules compete with the oppositely charged ions in the crystal, causing the ions to become separated and surrounded with water molecules (called a hydration shell). This makes water an important medium for many dissolved chemicals in living tissues.

Polyatomic Ions

Individual atoms form ionic bonds if they transfer electrons. Individual atoms form covalent bonds if they share electrons. Polar covalent *molecules* form hydrogen bonds with other polar covalent molecules. But there is still another way in which atoms can interact.

Covalent *molecules* can exist in an unstable form in which not all atoms are satisfying the octet rule. When a covalent molecule gives away electrons, it becomes a positively charged cation and a **reducing agent**. The element or compound that receives the negatively charged electron has its charge "reduced." When a covalent molecule lacks enough electrons to be stable, it takes them from another element or compound and becomes an anion. These groups of *covalently bonded atoms with an ionic charge* are called **polyatomic ions**, and they behave much like an ion in a chemical reaction.

Hydroxide (OH^-) and hydronium (H_3O^+) are common and important polyatomic ions. These ions form when a water molecule spontaneously splits into a hydrogen ion (H^+) and a hydroxide ion (OH^-). The hydrogen ion will then attach to another water

Autoionization of water

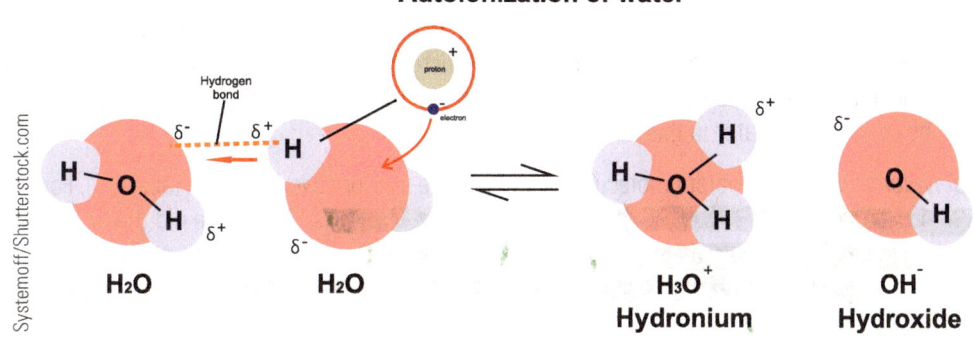

Figure 2.8 Water ionized into H⁺ and OH⁻. The single hydrogen has no electron, and therefore it has a positive charge. Note that the two *uncharged* polar covalent water molecule have become two polyatomic ions. Hydrogen ions are inherently unstable and bond to water or other compounds.

molecule, forming hydronium H_3O^+ (Figure 2.8). In a solution that contains only water, the proportions of hydroxide ions and hydronium (or hydrogen) ions is always equal.

Other common polyatomic ions are listed in Table 2.1. Note that like hydroxide and hydronium, the atoms in any polyatomic ion are covalently bonded together, but the group of atoms has a positive or negative charge due to missing or extra electrons. This makes them able to form ionic bonds. (Note that most of these compounds include highly electronegative oxygen.)

Interpreting Chemical Formulas

Elements are abbreviated by their symbols. The number of each type of atom in a compound is given in a *subscript* number next to the atom symbol. If there is only one of a particular type of atom, it has no number given. So the familiar chemical formula for water, H_2O, indicates that a water molecule has two hydrogen atoms bonded to an oxygen atom.

The chemical formula for the simple sugar glucose, $C_6H_{12}O_6$, describes the molecular structure as being made of 6 carbon atoms, 12 hydrogen atoms, and 6 oxygen atoms. The chemical formula for table salt, Na^+Cl^-, describes the structure of this compound as one sodium atom and one chlorine atom. However, the charges written next to the symbols in *superscript* indicate that this is an *ionic compound*, made of molecules or atoms that have lost or gained electrons. The chemical formula for ionic compounds does not always include the charges, so it is important to check to see if the compound is a metal bonded with a non-metal, or a non-metal bonded with a non-metal. Metals with non-metals, like sodium and chlorine, tend to form ionic compounds. Non-metals with non-metals, like carbon and oxygen, tend to form covalent compounds.

Chemical reactions occur when molecules are combined. The notation for chemical reactions is normally written with **reactants**, or the starting ingredients, on the left side of the equation and **products**, or the end results, on the right side of the equation. Between the reactants and products is an arrow, indicating the chemical change. For example, consider the following chemical formula that represents the reaction for the digestion of glucose in cellular respiration:

$$C_6H_{12}O_6 + 6\ O_2 \longrightarrow 6\ H_2O + 6\ CO_2 + \text{released energy}$$

In this reaction, glucose and oxygen, the reactants, form water and carbon dioxide, the products. The numbers in front of the chemical formulas, like the six in front of O_2, indicate the number of molecules of O_2 needed for this reaction to occur, much like the number of teaspoons of baking soda might be given in a recipe. These numbers are **coefficients**. Notice that the number of each kind of atom on each side of the arrow is exactly the same. No atoms can be lost or gained in a chemical reaction. They just get rearranged.

Table 2.1 Some common polyatomic ions

ammonium $NH_4^{(+1)}$
acetate $C_2H_3O_2^{(-1)}$
carbonate $CO_3^{(-2)}$
chromate $CrO_4^{(-2)}$
hydrogen carbonate $HCO_3^{(-1)}$
hydronium $H_3O^{(+1)}$
hydroxide $OH^{(-1)}$
nitrate $NO_3^{(-1)}$
peroxide $O_2^{(-2)}$
phosphate $PO_4^{(-3)}$
sulfate $SO_4^{(-2)}$
sulfite $SO_3^{(-3)}$

Determining Stable Compounds Containing Polyatomic Ions

A combination of atoms can be stable only if they satisfy the octet rule. Those that do not satisfy the octet rule are inherently unstable and will try to react with other compounds to increase their stability. Some reactions require energy (such as heat) while other reactions release energy (as heat and sometimes as light). The energy involved in breaking and making bonds will not be covered until later in this course.

Recall that *polyatomic ions* are covalent molecules with a charge that behave like ions (Table 2.1). For example, peroxide, $O_2^{(-2)}$, is a pair of covalently bonded oxygen atoms with two extra electrons. But where did the electrons come from? Peroxide must have received them from an element, such as magnesium (in group 2 on the periodic table) that can willingly give up two electrons. In a dry state, this would be an ionic compound MgO_2, or magnesium peroxide. In water, the ions will separate into Mg^{+2} and O_2^{-2}.

What if we substituted lithium for magnesium? Lithium (group 1) just wants to lose one electron. So we would need two lithium atoms to satisfy the electron needs of one peroxide. The dry chemical formula would be Li_2O_2, and again, in water, the ions would separate into Li^+ and O_2^{-2}, but there would be twice as many lithium ions as peroxide ions in the solution.

Can we combine two oppositely charged polyatomic ions, like ammonium, NH_4^{+1} and sulfate $SO_4^{(-2)}$? Yes! But we would need two ammonium ions per sulfate. In order to write this kind of formula we need to put the polyatomic ion in parentheses, and then indicate the number of them with a subscript: $(NH_4)_2SO_4$, or ammonium sulfate. In this case the 2 *outside the parentheses* applies to all atoms in the parentheses. This compound contains two nitrogen and eight hydrogen atoms. But rather than writing it $N_2H_8SO_4$, it should be written in such a way that the polyatomic ions are identified within parentheses. The positive part of a compound (in this case, ammonium) should be written before the negative part (sulfate).

Synthesis Reactions

When compounds are combined, they tend to follow certain patterns. In **synthesis reactions**, two or more *reactants* bond together to form a *product*: $A + B \longrightarrow AB$.

For example, combining oxygen and hydrogen to form water: $O_2 + 2H_2 \longrightarrow 2H_2O$

Remember that oxygen is not stable by itself, and we never deal with one atom at a time, so a container of oxygen (such as a pressurized oxygen tank) will be full of oxygen atoms that have bonded to each other in order to be stable. The same is true of the hydrogen atoms. Once they are combined into the same container, they are free to bond to each other in whatever way allows them to satisfy the octet rule. But if they are already stable, will they?

Is oxygen willing to let go of its partner to bond to hydrogen? How strong are the oxygen atoms bonded together? And the same question arises for hydrogen: Are these hydrogen atoms willing to split up in order to bond to oxygen? The answer to this question is beyond the scope of this exercise, and it requires knowing more about the electronegativity of these atoms. However, almost all reactions that are synthesis reactions require some form of energy (such as heat or light) to be added to them in order to build the new bond between the reactants. Reactions that require energy in order for them to occur are called **endergonic** reactions.

Cooking is much like chemistry. In order to make certain products in the kitchen, like a cake, the ingredients need to be heated. If they are not heated, the cake batter remains a soupy mixture. So without additional energy, adding oxygen gas to hydrogen gas will result in a mixture of oxygen and hydrogen gas, not water. If only a small amount of energy is provided, only some of the reactants may chemically combine to form the product, leaving the rest as a mixture of reactants.

Any reaction in an organism in which new, larger molecules are built is called **anabolic** reaction. These reactions are critical for cellular growth, repair, and the synthesis of compounds needed by the body. These reactions always require an input of energy. Energy is therefore stored in these larger molecules as a potential energy source for future chemical reactions.

Decomposition Reactions

A chemical reaction in which large molecules are broken down into smaller ones is called a **decomposition** reaction. These reactions are the reverse of synthesis reactions: $AB \longrightarrow A + B$. These reactions release energy and are **exergonic**.

Any reaction in an organism in which large molecules are decomposed is called a **catabolic** reaction. For example, the digestion of glucose:

$$C_6H_{12}O_6 + 6\ O_2 \longrightarrow 6\ H_2O + 6\ CO_2 + \text{released energy}$$

This reaction summarizes **cellular respiration**, the process by which individual cells obtain energy for metabolism by breaking down simple sugars in food. Some bonds are being broken, but other bonds are being formed. More bonds are broken than formed, as seen by the coefficients. There are seven reactant molecules (one glucose molecule and six oxygen gas molecules) and twelve product molecules (six water and six carbon dioxide molecules). The overall net energy of this reaction is the sum of the endergonic and exergonic energy processes in the reaction, and since more bonds are broken than formed, this is an exergonic reaction.

Displacement Reactions

Many reactions in organisms are a combination of breaking and building molecules, and these are called **displacement reactions**. Consider the generic reaction: AB + C $\longrightarrow$ AC + B. In this reaction, a bond between A and B is broken, but a new bond between A and C is formed. The number of molecules in the reactants and products hasn't changed, so it is not a synthesis or decomposition reaction. Reactions like these are called **single displacement** reactions.

Consider the reaction: AB + CD $\longrightarrow$ AD + BD. Here two bonds are broken and two new bonds are formed. Reactions like these are called **double displacement** reactions. Again, the number of reactant and product molecules has not changed. Without knowing more about the electronegativity of the atoms involved, there is no way to know if it is endergonic or exergonic, and the terms *catabolic* and *anabolic* do not apply.

Lab Skill 2.1 Understanding the Names of Chemical Compounds

Determining which atoms go with which is a fundamental part of understanding chemistry, but there is also the need to *name* these compounds according to a consistent and logical set of rules. The rules are simple, but they are slightly different for ionic compounds than they are for covalent compounds. It is also important to be able to interpret the names of chemicals in order to know exactly what they are and how they will react with other compounds.

Naming and Interpreting Names of Ionic Compounds

The name of an ionic compound is simply the name of the positive element followed by the name of the negative element, but the ending of the negative part should be changed to *ide*. The endings of polyatomic ion names do not get changed to *ide*.

Examples

- Sodium and chlorine form *sodium chloride*.
- Barium and phosphate form *barium phosphate*. — Polyatomic ion
- Ammonium and fluorine become *ammonium fluoride*.
- *Lithium bromide* is one lithium (group 1) and one bromine (group 7), or Li^+Br^-.
- *Magnesium fluoride* is a magnesium (group 2) and two fluorides (group 7), or $Mg^{+1}F_2^{-1}$.
- *Ammonium phosphate* contains three ammonium polyatomic ions (+1 charge) and one phosphate polyatomic ion (−3 charge), or $(NH_4^{+1})_3PO_4^{-3}$

Naming and Interpreting Names of Covalent Compounds

Covalent compounds are named much like ionic compounds, but one extra step is added. A prefix is used to indicate the number of that kind of atom in the molecule. If there is only one atom of a particular element in the molecule, no prefix is usually given. The prefixes for the number of atoms are as follows: 2—*di*, 3—*tri*, 4—*tetra*, 5—*penta*, 6—*hexa*, 7—*hepta*, 8—*octa*, 9—*nona*, and 10—*deca*. Prefixes are not used with compounds containing polyatomic ions because they are ionic, not covalent.

Examples

- One carbon atom covalently bonded to two oxygen atoms is carbon *dioxide*.
- Two hydrogen atoms covalently bonded to one oxygen is *di*hydrogen ox*ide* (H_2O, water).
- *Carbon tetrachloride* is one carbon covalently bonded to four chlorine atoms, CCl_4.

Recall that it is easy to identify a compound as ionic or covalent. If the compound is made of a metal and a non-metal, it is an ionic compound. If the compound is made of non-metals, it is a covalent compound. If it contains polyatomic ions, the atoms in the polyatomic ions are covalently bonded, but the group acts as an ion.

Lab Skill 2.2 Creating Solutions of Specific Concentration

In living organisms, chemical reactions occur in cells or in fluid outside the cells. This means that the chemical reactions tend to occur in water, and the reactants are dissolved in water. **Solutions** are mixtures in which chemicals are dissolved in a liquid (usually water). **Mixtures** are combinations of two or more kinds of molecules that do not react. The dissolved chemical is called the **solute**, and the liquid in which it is dissolved is called the **solvent**. So, a solution of salt water would contain dissolved sodium and chlorine ion *solutes* in a *solvent* of water. Remember that ionic compounds dissociate, or break into ions, in water.

Concentration as Percent

One method of measuring concentration is by percent. A 1% glucose solution would be 1 gram of glucose (the solute) added to 99 grams of water (the solvent).

Concentration as Molarity

A **mole**, in chemistry, is a number that indicates a specific quantity. A dozen always means twelve, whether talking about donuts or eggs. Likewise, a mole always equals 6.022×10^{23}. This is also called the Avogadro Constant. This is a really big number, so we use the term *mole* to represent it. (If unfamiliar with scientific notation, see Appendix B.)

It is impossible to count out atoms for a solution, but when considering reactions where one atom or molecule needs to react with another, it is important to ensure there are appropriate amounts of those reactants available. Moles allow us to determine this.

The weight of one mole of atoms (or of 6.022×10^{23} atoms) has a mass in grams of that atom's *atomic weight* on the periodic table. This is called its *molar mass*.

If dealing with molecules, a mole of molecules is the combined atomic weight of the atoms in the molecule, expressed in grams. This is also the molar mass of that compound.

- Therefore, *one mole* of carbon (C), which equals 6.022×10^{23} atoms of carbon, has a molar mass of *12.011 grams*. (See the atomic mass of carbon on the periodic table.)
- *One mole* of sodium (Na) has a molar mass of *22.990 g*.
- One mole of sodium chloride (NaCl), or 6.022×10^{23} *molecules* of sodium chloride, has a molar mass equal to the combined molar masses of sodium and chlorine (22.990 + 35.453) or 58.443 g NaCl.

Molarity is a measure of moles of atoms in one liter of water. A one molar (1 M) solution of NaCl therefore would be 58.443 grams of NaCl in a beaker that is then filled to exactly one liter of water. A 2 M (two molar) solution would have twice the amount of sodium chloride per liter of water. A 0.5 M solution would have half the amount of sodium chloride per liter of water. Therefore, a solution of 0.001 M would indicate that an amount equal to the molar mass x 0.001 was added to a liter of water.

$$\frac{1 \text{ mole NaCl}}{1000 \text{ mL } H_2O} = 1 \text{ molar (M)}$$

standard for Molar

Lab Skill 2.3 Creating Dilutions of a Specific Volume

Many reagents used in biology laboratories are sold in concentrated form. If they contain a label marked 2X, 5X, or 10X, this indicates that the solution should be diluted to a proper concentration before use.

A 2X solution needs to have the volume doubled. A 10X solution needs to have the volume increased tenfold. Therefore, you must add enough distilled water to increase the volume that amount. For example, if you have a 100 ml bottle of 2X PBS buffer (a common buffer used in biology experiments), and you need to dilute it to 1X, you would add the 100 ml of 2X concentrated buffer to 100 ml of distilled water to double the volume to 200 ml. The result is 200 ml of a 1X solution.

Likewise, If you have 100 ml of 10X buffer and need to dilute it to 1X, you would need to add 900 ml of water to increase the volume by tenfold to 1000 ml. But what if you do not need that much buffer, or do not want to dilute it all at once? This is where an equation comes in handy.

$$C_1V_1 = C_2V_2$$

C_1 is the starting concentration.
V_1 is the starting volume.

C_2 is the ending concentration.
V_2 is the ending volume.

If you only need 200 ml of the 10X buffer diluted to 1X, you would plug in the numbers and solve for the unknown. In this case what we do *not* know is how much of the original concentrated buffer to use (V_1).

Plug in the numbers: $(10X) \times (V_1) = (1X) \times (200 \text{ ml})$

Solve: $V_1 = (1 \times 200)/10 = 20$ ml.

So 20 ml of 10X buffer should be added to 180 ml of water to create 200 ml of 1X buffer.

The same principle applies to molar solutions.

If you needed 150 ml of 0.001 M hydrochloric acid, but your stock supply is a bottle of 0.1 M hydrochloric acid, how would you make it?

Plug in the numbers: $(0.1 \text{ M}) \times (V_1) = (0.001 \text{ M}) \times (150 \text{ ml})$

Solve: $V_1 = (0.001 \times 150)/0.1 = 1.5$ ml.

So 1.5 ml of 0.1 M hydrochloric acid should be added to 148.5 ml of distilled water to create 150 ml of .001 M hydrochloric acid.

Lab Skill 2.4 Measure Solute Concentration with a Hydrometer

The concentration of a solution can also be determined by comparing the density of that solution to pure water.

A **hydrometer** is a simple device that floats in a solution to determine its density or concentration. Hydrometers are used to analyze many solutions, including measuring the concentration of wastes in urine, the alcohol content of alcoholic beverages like wine and beer, the salt content of seawater and aquarium water, the concentration of chlorine or other chemicals in a swimming pool, and the purity of gasoline.

Specific gravity is the ratio of the weight of a solution compared to water, so pure water has a specific gravity of 1.000. If substances denser than water are added to water, the specific gravity increases. If substances less dense than water are added, the specific gravity decreases.

How to Use a Hydrometer

A hydrometer is a weighted glass bulb with a stem (Figure 2.9). It rests in a tube as shown.

Pour the solution into the tube until the hydrometer is suspended in the liquid and not touching the bottom.

Spin the hydrometer gently to dislodge any bubbles. Ensure the hydrometer is flat on a steady surface. View at eye level where the stem emerges from the liquid.

Read the stem by looking carefully at where the bottom of the meniscus of the solution is level with the markings on the stem.

The numbers on the stem *increase* in value as they go *down* the stem. Near the top of the stem is usually marked with 1.000. The numbers on the stem usually increase in units of 0.010 as they go down the length of the stem. See Figure 2.10. The tiny marks between numbers on this hydrometer indicate a difference of 0.002.

In Figure 2.10 the hydrometer indicates a specific gravity of 1.016.

Figure 2.9 A hydrometer.

Figure 2.10 Reading a hydrometer

Lab Skill 2.5 Calculating pH from Molarity or Molarity from pH

An **acid** is a specific type of compound that dissociates into hydrogen (H^+) ions and an anion in water. For example, HCl, hydrogen chloride (or *hydrochloric acid*), readily dissociates into H^+ and Cl^- in water. (Weak acids do not completely dissociate in water.) The easy rule of thumb is that any compound having a chemical formula that begins with H is likely an acid.

A **base** is a specific type of compound that dissociates into a cation and hydroxide (OH^-) ions in water. For example, NaOH, sodium hydroxide, dissociates into Na^+ and OH^- ions in water. Generally, any compound having a chemical formula that ends with OH is likely a base.

Acids and bases are important in many ways in living things. Humans, for example, use hydrochloric acid in the stomach to help digest food. Acid rain, caused by pollutants released into the air, falls as an acidic solution strong enough to poison lakes and kill plants.

The **pH scale** describes the strength of an acid or base. The scale is based on the concentration of H^+ ions and ranges from 0 to 14. The closer to zero the pH, the stronger the acid. The closer to 14 the pH, the stronger the base. A solution with a pH of exactly 7 is considered neutral and is neither acidic nor basic. See Figure 2.11 for common solutions and their pH levels.

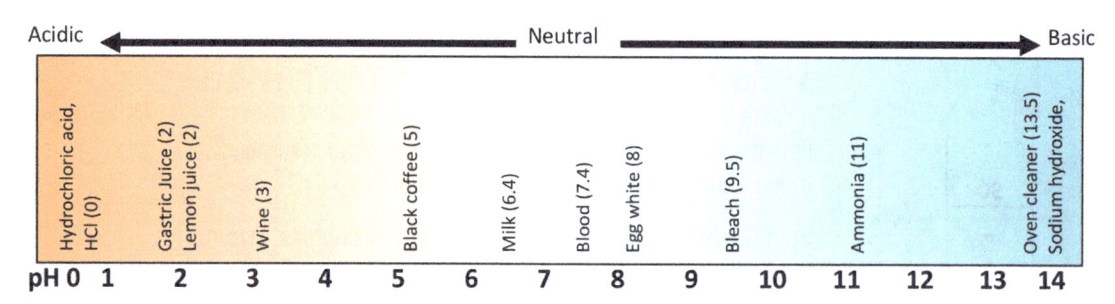

Figure 2.11 The pH scale.

The Relationship between pH and Molarity

The pH of a solution is a measure of the hydrogen ions in that solution. Recall that pure water can ionize into H_3O^+ (or H^+) and OH^- ions, but that there is an equal number of each type. When another compound, like hydrochloric acid (HCl), is added to water, the H^+ ions dissociate from the Cl^- ions, increasing the relative concentration of H^+ ions.

$pH = -\log [H^+]$ where H^+ is the molarity of hydronium (or hydrogen) ions in the solution.

A 0.0001 M solution of HCl can be written using scientific notation as 1×10^{-4} M, in which case the pH is simply equal to the exponent, in this case, 4. $pH = -\log [0.0001] = 4$. Note that a change in pH from one number to the next, say from a pH of 5 to a pH of 4, is an increase in hydrogen ions. Lowering the pH by 1 lowers the negative exponent, which increases the hydrogen ion concentration ten times. So a change in pH by two is a change in hydrogen ion concentration by a factor of *100*, and so forth. Likewise, increasing the pH by 1 reduces the concentration of hydrogen ions by a factor of 10. Thus, every change in the pH value by 1 is actually a ten-fold change in hydrogen ion concentration.

When the molarity of the solution is not a power of 10, the formula must be used. For example, the pH of a solution of 0.004 M HCl has a $pH = -\log (0.004) = 2.398$. If the solution is a strong acid but produces two H^+ ions in water, the molarity must account for this. For example, a 0.0001 M solution of sulfuric acid (H_2SO_4) will produce *two* H^+ ions per mole of sulfuric acid. So the $[H^+]$ must be doubled to 0.0002. Therefore the pH of a 0.0001 M solution of H_2SO_4 is $-\log [0.0002] = 3.699$. (This specifically applies to acids that completely dissociate in water.)

If a solution is a base, the pOH can be calculated the same way as pH. Then $pH = 14 - pOH$. For example, the pOH of a 0.001 M solution of sodium hydroxide (NaOH) is 3. The pH is then $14 - 3$, or 11. (Again, this applies to bases that completely dissociate in water.)

Molarity of acids and bases that completely dissociate in water can be calculated from their pH. If a solution has a pH of 9.21, the molarity of H^+ ions can be calculated as shown:

Switching the equation around: [H⁺] = the inverse log of (−pH) *Note that the inverse log button on a scientific calculator is usually indicated by 10^x*

Therefore, the molarity of the solution is inverse log (−9.21) = 6.16595×10^{-10}

See *Appendix B: Scientific Notation* for help with working with numbers in scientific notation if necessary.

Neutralization of Acids and Bases

When acids and bases are combined, they **neutralize** each other, producing a salt (ions) and water. Antacids are basic compounds that a person can take to relieve the indigestion caused by an overproduction of stomach acid. These neutralization reactions are important for keeping the pH of body fluids near a neutral level so that they do not damage tissues.

Hydrochloric acid and sodium hydroxide are the strongest acid and base respectively. However, if combined in proportional amounts, they produce a neutral saltwater solution. For *complete* neutralization, (a pH of 7), the final concentration of hydrogen ions and hydroxide ions in the final mixture must be exactly equal.

$$HCl + NaOH \longrightarrow NaCl + H2O$$

Note that the H+ ions and OH- ions combine to form water. NaCl is sodium chloride, or common table salt. In fact, any acid combined with any base in a neutralizing reaction will result in a salt and water. A **salt** is generally considered any ionic compound other than an acid or base. Electrolytes are dissolved salts such as sodium (Na⁺), potassium (K⁺), and chloride (Cl⁻).

Problem to Solve 2.1 Can I Make 500 ml of a 0.154 M Saline Solution?

Saline is a widely used solution in biology and in medicine. It is a solution of NaCl and water that matches the salt concentration of most terrestrial animal body fluids. Saline is used to flush wounds during surgery, to help replace fluids in patients who are severely dehydrated, and to keep burns or inflamed tissues moist. Saline is also used to wash contact lenses, in nasal sprays, and in eye drops.

Saline is typically described as a 0.9% solution of NaCl (sodium chloride, or table salt).

A fellow biologist is doing research on autonomic nervous system activation in reptiles. He is planning to do open brain surgery on a lizard. You have been tasked with preparing a 500 ml bottle of saline for use in this project. If the concentration is not correct, the experiment will fail before it begins. Too much or too little salt in the solution will cause the brain sensors to give unreliable results. The researcher also asked you to work out the molarity of the saline solution so he can record it in his notes.

You have been provided NaCl in solid form, a mass scale, distilled water, a hydrometer, and lab glassware. How will you create a **500 ml solution of 0.9% NaCl?**

Write your name, date, and protocol on the notebook page. Some of the calculations have been set up for you. Complete the rest.

Helpful reminders:

The % concentration refers to grams of solute per 100 grams of solvent. So a 0.9% NaCl solution actually means 0.9 grams of salt per 100 g water.

0.9g / 100ml
NaCl   H2O

As you learned in the previous lab, 1 g of water = 1 ml of water.

Conversion factors are equal relationships between two types of variables. A common conversion factor we use all of the time is 12 inches to one foot. In biochemistry, we often must convert from moles to grams or grams to moles. For example, 22.990g Na = 1 mole of Na. **Before starting, determine the mass in grams of 1 mole of NaCl.** This is your conversion factor.

58 g

If you know the number of grams of NaCl is needed for your solution, convert grams to moles to get the molarity. See analysis for how to this use conversion factor.

The standard specific gravity of saline is 1.0048.

Na^{11}_{23} NaCl molecular mass = 58

1 mole NaCl = 6.022×10²³ molecules Cl^{17}_{35} (mole) =

Standardized way of comparing molecules.

= .522M × 500ml

Date: 08/30/2021
Researcher name: Bethany Williams

Title
Make a saline solution and check it for accuracy

Purpose
Goal: Make exactly 500 ml of saline, and use a hydrometer to verify that it is accurate. Determine the molarity of saline.

Protocol

Materials:
- NaCl (crystals)
- Mass balance
- Weigh boat
- Glassware
- Hydrometer
- Distilled water

Protocol:

Data:

Predicted solution density:	1.0048
Actual density from hydrometer reading:	1.001
Percent difference (actual-predicted/predicted):	.3%

1. First knowing that 0.9% NaCl = 0.9 g NaCl / 100 g H2O, determine how many grams of NaCl per 1000 ml H2O.
2. Then using the conversion factor of 1 mole NaCl / 58g NaCl, determine the Molarity of NaCl solution.
3. Place a weigh boat on a scale, zero it out and using a scoop, place 4.5g of NaCl on the weigh boat.
4. Using a graduated cylinder, measure 500 ml of distilled H2O. Put it in a beaker.
5. Add the 4.5g of NaCl to 500 ml and stir until NaCl is fully dissolved.
6. Pour enough of the NaCl solution into the hydrometer to make it float.
7. Then by reading the hydrometer, determine the specific gravity.

Continued from previous page.

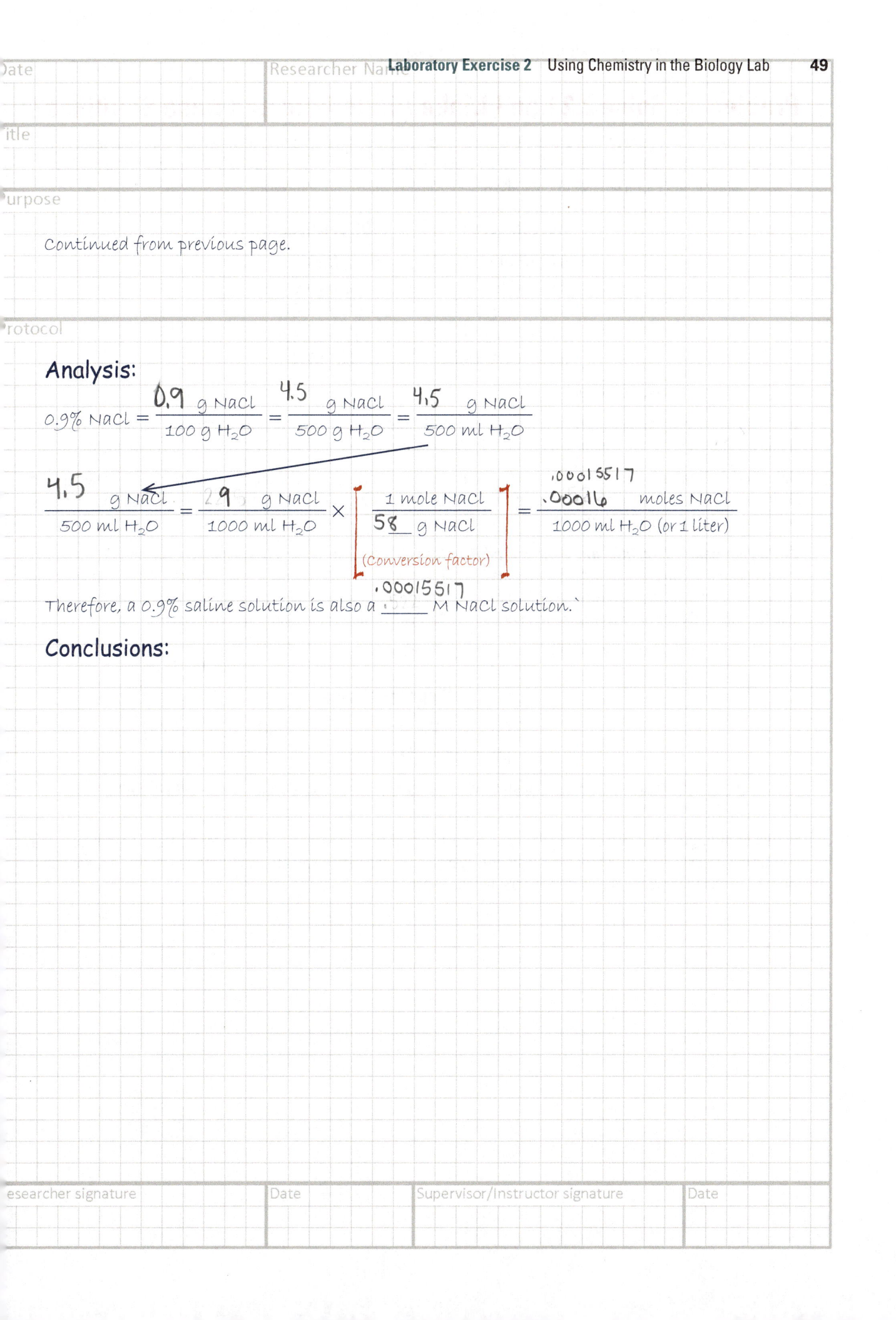

Analysis:

$$0.9\% \text{ NaCl} = \frac{0.9 \text{ g NaCl}}{100 \text{ g H}_2\text{O}} = \frac{4.5 \text{ g NaCl}}{500 \text{ g H}_2\text{O}} = \frac{4.5 \text{ g NaCl}}{500 \text{ ml H}_2\text{O}}$$

$$\frac{4.5 \text{ g NaCl}}{500 \text{ ml H}_2\text{O}} = \frac{9 \text{ g NaCl}}{1000 \text{ ml H}_2\text{O}} \times \left[\frac{1 \text{ mole NaCl}}{58 \text{ g NaCl}} \right] = \frac{.00015517 \; .00016 \text{ moles NaCl}}{1000 \text{ ml H}_2\text{O (or 1 liter)}}$$

(Conversion factor)

Therefore, a 0.9% saline solution is also a .00015517 M NaCl solution.

Conclusions:

Problem to Solve 2.2 Can I Make Dilute Acid and Base Solutions?

Imagine you are part of a team researching the effect of pH on the survival rates of intestinal bacteria. Your responsibility in the project is to prepare solutions of hydrochloric acid (HCl) and sodium hydroxide (NaOH). The project requires you to prepare approximately 1 ml of each of the following concentrations of HCl and NaOH.

HCl solutions	NaOH solutions
0.01 M HCl	0.01 M NaOH
0.001 M HCl	0.001 M NaOH
0.0001 M HCl	0.0001 M NaOH
0.00001 M HCl	0.00001 M NaOH
0.000001 M HCl	0.000001 M NaOH

Distilled water will serve as the control.

What equipment will you need to prepare these solutions? **Dilutions** are less concentrated forms of a solution. How will you obtain accurate and precise dilutions of HCl and NaOH? How can you test the solutions to see if you made them correctly?

Universal indicator is a liquid that changes colors at different pH values. The approximate pH can be determined by adding a few drops of Universal Indicator to a solution and comparing the resulting color to a color chart. (You may also be provided a pH probe or test strips.)

Write your name, date, title, and goal in the lab notebook page. Since this lab problem is about setting up solutions for an experiment, you are not testing a hypothesis here. But you are learning how to make dilute solutions and measure pH. Write this goal in the *Purpose* area of the notebook. Then write any additional materials needed and the steps you will take to prepare these solutions in the *Protocol* section. This will require some planning and thinking!

Some data tables have been provided to help you on the following page.

Helpful reminders

$$C_1V_1 = C_2V_2 \qquad pOH = -\log [OH\text{-}]$$
$$pH = -\log [H+] \qquad pH = 14 - pOH$$

You can create one dilution, then use that dilution to make the next. This is called a *serial dilution* since you are repeating the dilution over and over.

Bonus challenge: A fellow biologist asked you to make 5 ml of a 0.003 M NaOH solution for a different experiment they are doing. Can you? How? What should the pH be? Try making it, and test the pH after you do to see if you did it accurately. (Be sure to record this work somewhere in your notebook pages.)

Title:

Goal: Dilute 0.1 HCl & 0.1 NaOH and determine the PH of all dilutions.

Materials:
Safety glasses and gloves
stock solution of 0.01 M HCl ✓
stock solution of 0.01 M NaOH ✓
Beaker of distilled water ✓

5 HCl 1 distilled 5 NaOH
✓ 11 test tubes and a test tube rack
✓ Adjustable micropipette and tips
✓ Universal indicator (in dropper bottle)
✓ pH probe and test strips (optional)

use new pippette tip each time.

Protocol:
(Write out the steps you will follow.)
1. Set the micropippette to 100 μL
2. extract 100 μl from 0.1M HCl and place in the 0.01 HCl test tube.
3. Using larger micropipette set to 900 μL.
4. Place 900 mL of distilled water into the 0.01 HCl test tube
5. Then extract 100 μL from the 0.01 HCl and place in the second test tube.
6. extract 900 μL H_2O and add to the second test tube to create the 0.001 HCl.
7. extract 100 μL of 0.001 HCl and place in 3rd test tube
8. extract 900 μL of distilled water and add to third test tube
9. extract 100 μL of 0.0001 HCl and add to 4th tube.
10. extract 900 μL of distilled H_2O and add to fourth tube.
11. extract 100 μL of 0.00001 HCl and add to 5th tube
12. extract 900 μL of distilled H_2O and add to fifth tube.
13. add a few drops of universal indicator into each tube.
14. from the color of the solution, determine the PH
15. place 100 μL of 0.1 NaOH into first tube.
16. extract 900 μL of distilled H_2O into the tube
17. extract 100 μL of the 0.01 NaOH and place in 2nd tube
18. extract 900 μL of distilled H_2O and add to 2nd tube
19. extract 100 μL of 0.001 NaOH and add to 3rd tube
20. extract 900 μL of H_2O and add to 3rd tube
21. extract 100 μL of 0.0001 NaOH and add to 4th tube
22. extract 900 μL of H_2O and add to 4th tube.
23. extract 100 μL of 0.00001 NaOH and add to 5th tube
24. extract 900 μL of H_2O & add to 5th tube.
25. add couple drops of indicator and determine PH from color

Continued from previous page.

Data and Observations:

Test tube	Solution and Concentration	Predicted pH	Color after adding Universal Indicator	Actual pH
1	0.01 M HCl		Bright reddish	4
2	0.001 M HCl		reddish orange	5
3	0.0001 M HCl		Yellow	6
4	0.00001 M HCl		yellow darker	6
5	0.000001 M HCl		darker yellow almost green	6
6	Distilled water		Yellow, should be green	7
7	0.000001 M NaOH			
8	0.00001 M NaOH			
9	0.0001 M NaOH			
10	0.001 M NaOH			
11	0.01 M NaOH			

Continued form previous page.

Conclusions:
(Did the actual results match the predicted?)

Name: _____   Date: _____

POST-LAB QUIZ 2: What have you learned?

1. Which group on the periodic table contains elements that readily donate an electron?

2. Clearly define the following types of chemical bonds:
   a. ionic bond
   b. covalent bond
   c. hydrogen bond

3. Give the correct name for the following compounds:
   a. $Ca(OH)_2$
   b. $BaSO_4$
   c. SO_2

4. Give the correct chemical formula for the following compounds.
   a. Magnesium oxide
   b. Bromine monochloride
   c. Ammonium hydroxide

5. Show how to make 1 liter of a 0.001 M solution of NaCl.

6. Predict if the following elements will bond. If so, give the chemical formula and name.
   a. Sodium and oxygen
   b. Chlorine and another chlorine
   c. Sodium and another sodium
   d. Oxygen and fluorine
   e. Oxygen and helium
   f. Calcium and bromine

7. If you had one liter of a 0.001 solution of NaCl, how could you obtain 300 ml of 0.0001 solution of NaCl?

8. This frozen juice concentrate instructs the user to mix the contents of this can with three cans of water. How would you accurately describe the concentration of the undiluted juice in this can?

9. What is the pH of a 0.05 M solution of H_2SO_4? (H_2SO_4 completely dissociates in water.)

10. What is the molarity of an HCl solution with a pH of 5? (HCl completely dissociates in water.)

Laboratory Exercise 3

Biochemistry: Chemical Compounds of Living Organisms

Concepts to Understand

After completing this lab, the student should be able to:

1. Explain the difference between temperature and heat.
2. Explain the Kinetic Theory of Matter.
3. Describe the four types of macromolecules and their component parts.
4. Explain how enzymes facilitate reactions.
5. Explain the effect of temperature on enzyme activity.
6. Explain the role of pancreatic enzymes in digestion.

Skills to Learn

After completing this lab, the student should be able to:

1. Use indicators to detect the presence of specific compounds.
2. Manage the temperature of solutions and compounds in the lab.

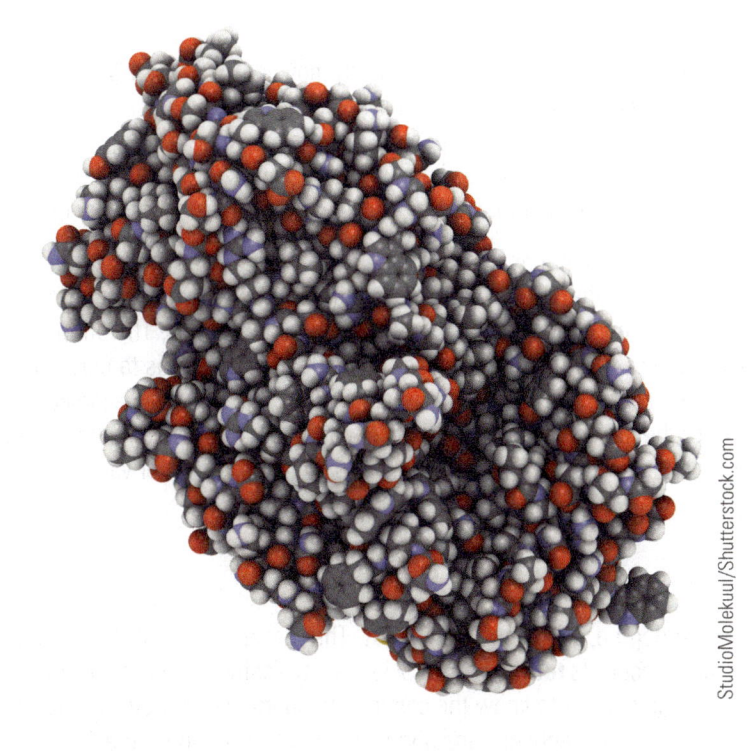

SEE ALSO:

Campbell Biology - Chapter 5, Sections 5.1 – 5.4 and Chapter 8 section 8.4.

JoVE (video) – Regulating Temperature in the Lab: Applying Heat

Others:

Important Terms to Know

- Absolute zero
- Activation energy
- Active site
- Amino acid
- Amylase
- Benedict's Sol.
- Brownian mvmt.
- Carbohydrate
- Dehydration
- Denature
- Disaccharide
- DNA
- Ectothermic
- Endothermic
- Enzyme
- Ester bond
- Fatty acid
- Genetics
- Glycerol
- Glycosidic bond
- Heat
- Hydrolysis
- Incubator
- Iodine
- Kinetic Theory
- Lipase
- Lipid
- Litmus
- Macromolecule
- Monomer
- Monosaccharide
- Metabolism
- Nitrogen base
- Nuclease
- Nucleic Acid
- Nucleotide
- Organic chemistry
- Pancreatin
- Peptidase
- Peptide
- Peptide bond
- Phosphodiester bond
- Phospholipid
- Polymer
- Polysaccharide
- Protease
- Protein
- Proteomics
- RNA
- Substrate
- Temperature
- Triglyceride
- Water bath

Pre-lab Preparation and Background Information

Kinetic Theory of Matter

All matter in the universe is made of atoms and molecules. These molecules are constantly moving. **Brownian Movement** is the inherent movement of all molecules. This movement occurs at such a microscopic level that we cannot usually observe it directly. However, if you use a microscope to observe tiny particles of ink suspended in water you can see that those tiny particles will move and jiggle around regardless of how long you let the slide sit undisturbed. This is not only true of inanimate objects, it is true of the molecules of living things. This molecular movement drives much of the biochemistry of life.

The **Kinetic Theory of Matter** states that matter is solid, liquid, or gas, based on the degree and speed of the molecular movement of that substance. Solid objects contain molecules that vibrate in place. Liquids contain molecules that move faster and slide past each other. Gases contain molecules that move very fast and bounce around freely. The difference between these states of matter is simply the speed and degree of the movement of the molecules. Water is familiar to us in all three states: solid, liquid, and gas. If water is frozen as ice, the molecules are vibrating in place. As these molecules are heated, the heat energy increases the movement of the molecules, causing the ice to melt and become a liquid. When liquid water is heated further, the molecules move even faster to the point of leaping into the air as steam or vapor. When cooled again, the process is reversed.

Temperature and Heat

Since all objects contain molecules that move to some degree, all objects have heat. The measurement of the *average* kinetic energy (energy of movement) of a molecule in an object is its **temperature**. Scientists typically measure temperature in Celsius (C) or Kelvin (K) rather than Fahrenheit (F). It is important to know the common temperatures in Celsius, such as the freezing point and boiling points of water (0° C and 100° C respectively) and normal human body temperature (37° C) as shown in Figure 3.1. The *overall* energy of an object based on the *sum* of the kinetic energy of all of the molecules is that object's **heat**. The heat of an object depends on its temperature and mass. **Absolute zero** ($-273°$ C or 0 K) is defined as

TEMPERATURE CONVERSION

Fahrenheit
$°F = 1.8 × °C + 32$

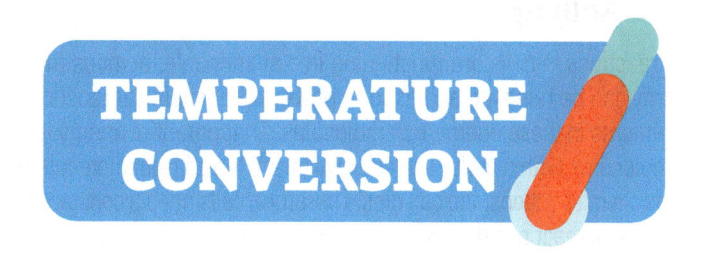

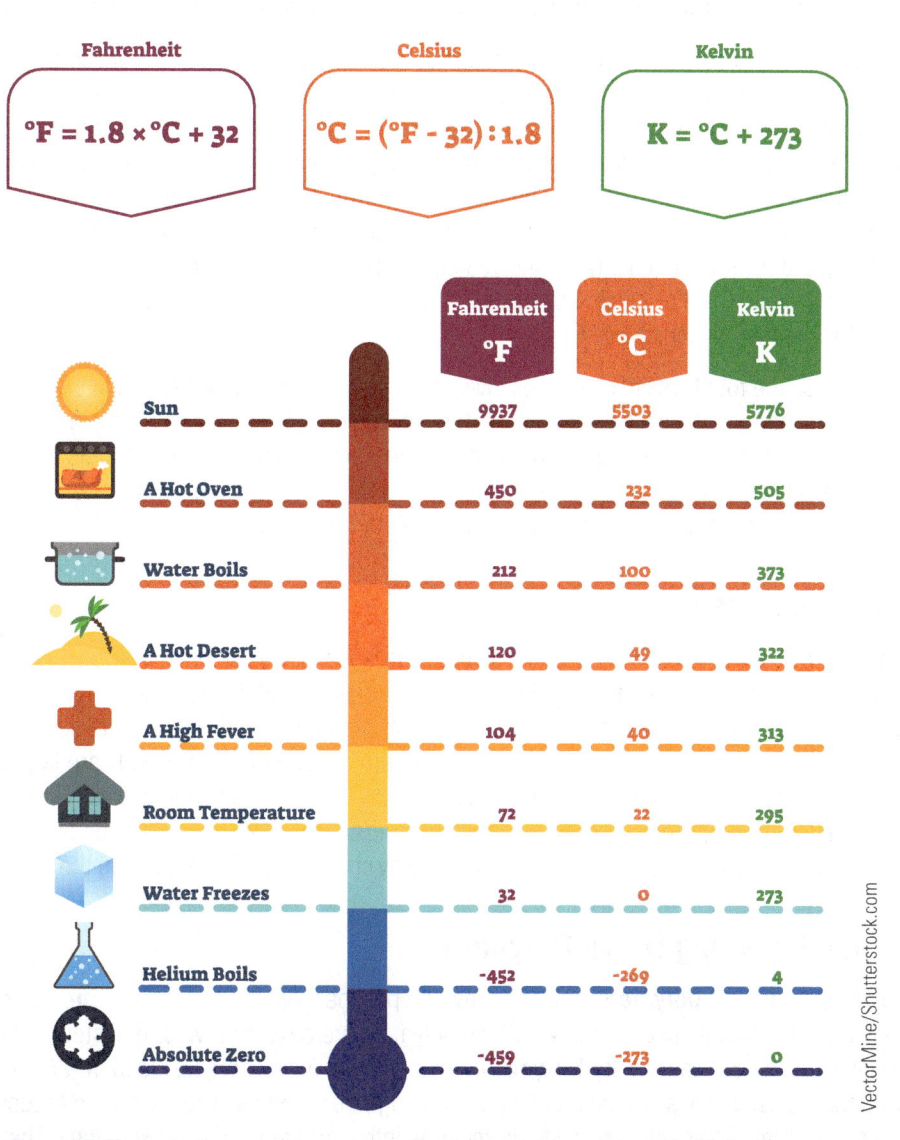

Celsius
$°C = (°F - 32) ÷ 1.8$

Kelvin
$K = °C + 273$

	Fahrenheit °F	Celsius °C	Kelvin K
Sun	9937	5503	5776
A Hot Oven	450	232	505
Water Boils	212	100	373
A Hot Desert	120	49	322
A High Fever	104	40	313
Room Temperature	72	22	295
Water Freezes	32	0	273
Helium Boils	-452	-269	4
Absolute Zero	-459	-273	0

Figure 3.1 Thermometers measure *average molecular movement*, or temperature. Scientists use Celsius and Kelvin, not Fahrenheit in scientific research and writing.

the point at which molecular movement completely stops. This condition does not occur naturally anywhere in the universe. This means that even an ice cube has heat. (It just doesn't have as much as we do, so it *feels* cold.)

Chemical reactions within living organisms must occur at the right rates. Increasing the temperature of the intracellular molecules speeds up molecular movement, increasing the rate of chemical reactions. Decreasing the temperature slows molecular movement and chemical reactions. An organism's **metabolism** refers to the total chemical activity of the organism. This metabolism is directly affected by temperature. Some animals, like amphibians and reptiles, are **ectothermic**. Their metabolic rate is directly proportional to their environmental temperature. Humans, other mammals, and birds are **endothermic**. Their metabolism requires a consistent internal body temperature within a certain range to function properly, so they expend energy to generate their own heat.

Control of Metabolic Activity

Life occurs when chemical reactions in cells are functioning in a sustainable, homeostatic balance. Cells take in energy-rich molecules so that that energy can be released and used to power other chemical reactions in the cell. Cells break down the covalent bonds between elements in these energy-rich molecules. That potential energy is transferred to other chemical reactions and energy-carrying molecules within cells. The energy is transferred from molecule to molecule as cells carry out their biochemical processes. If this chemistry stops, the cell dies. Therefore, cells need a constant supply of energy-rich molecules and all the ingredients for the chemical reactions that must occur in the cells. The study of biology is the study of how cells, tissues, organs, organ systems, organisms, and even populations and communities work toward keeping the chemistry of life going!

It is not enough that cells continuously perform chemical reactions. They must perform the right chemical reactions at the right time. Sometimes, when energy demands are low or energy-rich fuel molecules are plentiful, the cell may need to make energy-storing molecules, like fat (triglycerides) or glycogen. Sometimes when energy-rich molecules are scarce, such as between meals or during vigorous activity, those energy-storing molecules are broken down and used. Sometimes cells need to produce certain compounds, like hormones, neurotransmitters, or stomach acid, in order to perform a specific function. Sometimes cells need to divide and make more cells, or grow larger. All of biochemistry is a complex balance of *anabolic* (building) and *endergonic* (energy-storing) reactions with catabolic (breaking-down) and exergonic (energy-releasing) reactions. All cells do not do all of these reactions continuously and at the same time. These reactions must be controlled and regulated to maintain homeostasis.

This lab will focus on the four types of molecules involved in many of the chemical reactions in living things. Some of these molecules serve as energy-rich fuel, or as a form or stored energy. Some are structural and form much of the physical mass of an organism. Some are functional, serving as chemical messengers or signals. And others store information within their physical structure, much like information can be stored in a series of block letters. Some molecules serve as the regulators of chemical reactions, and others act as on-off switches to start or stop a reaction.

The Macromolecules

Organic chemistry is the study of carbon-based compounds, specifically compounds made of primarily C–C or C–H bonds. These molecules tend to be the structural and functional molecules in all living things. In fact, about 96% of the mass of human tissue is made of carbon, hydrogen, oxygen, and nitrogen atoms. These atoms are combined into four major types of compounds found in all cells, tissues, organs, and organisms. These four compounds are large **macromolecules** that are made of subunit parts. Each of the subunits is composed of carbon-based molecules. The four major organic macromolecules are *carbohydrates, proteins, lipids*, and *nucleic acids*. You will not be able to understand biology without a firm understanding of what these molecules are and how they play critical roles in maintaining the biochemical activity of a cell.

Building and Breaking Down Polymers

The four macromolecules are **polymers** (poly = many). All polymers are built from many **monomers** (mono = one), much like a chain is built from many repeating links. When polymers are broken down or digested, the reaction that breaks the bonds between monomers is called a **hydrolysis** (*hydro* = water; *lysis* = break) reaction. When a polymer is broken into smaller parts, the exposed bonds that held the monomers together must have something to occupy that open bond. Recall from Lab Exercise 2 that water spontaneously dissociates into hydrogen and hydroxide ions. The unstable hydrogen ions will attach to another compound or a water molecule, forming H_3O^+. Since these activities happen continuously in water, hydrogen and hydroxide ions are always available. In hydrolysis H^+ and OH^- from a split water molecule attach to the recently broken sections of the polymer (Figure 3.2).

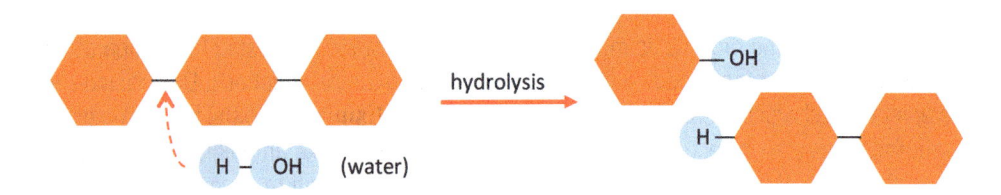

Figure 3.2 Simplified hydrolysis reaction breaking down a polymer.

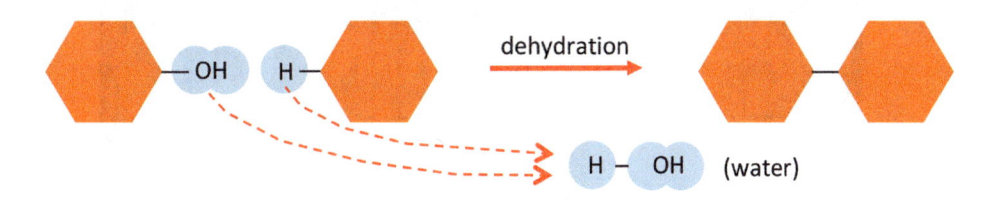

Figure 3.3 Simplified dehydration reaction, linking monomers.

Dehydration reactions occur when polymers are built from monomers. As the name implies, H^+ and OH^- ions are removed, forming a water molecule. The monomers are *dehydrated* (water is removed) and a larger molecule is *synthesized*. It is the reverse of the reaction illustrated in Figure 3.3.

Recall that decomposition reactions (including hydrolysis reactions) are *catabolic, exergonic*, and release chemical energy by breaking covalent bonds. Synthesis reactions (including dehydration reactions) are *anabolic, endergonic*, and store chemical energy in the newly formed covalent bonds. The synthesis and decomposition of organic macromolecules is a major aspect of cellular biochemistry.

Carbohydrates

Carbohydrates are molecules formed from carbon, hydrogen, and oxygen in a 1:2:1 ratio. The monomer of a carbohydrate polymer is a simple sugar such as glucose, $C_6H_{12}O_6$. (Note the 1:2:1 ratio of carbon to hydrogen to oxygen.) Simple carbohydrate monomers are called **monosaccharides**. Two linked monosaccharide molecules are a **disaccharide**. See Figure 3.4. Simple monosaccharides and disaccharides taste sweet. Simple sugars typically have names that end in "-ose."

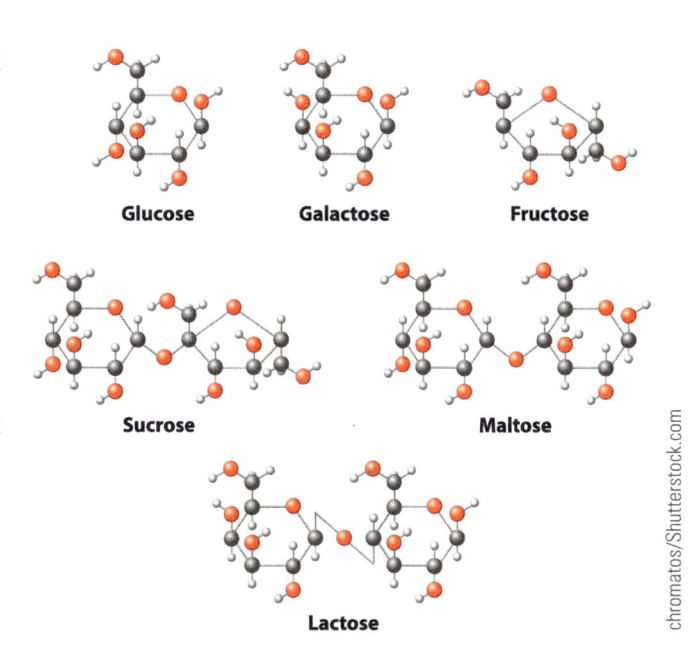

Figure 3.4 Common monosaccharides and disaccharides. Glucose, galactose, and fructose are monosaccharides. Sucrose is formed when glucose and fructose are joined together. Maltose is made from two glucose molecules, and lactose is made from a glucose and a galactose.

Longer chains of many monosaccharides are called **polysaccharides**. Monosaccharides, such as glucose, are generally used as fuel in cellular respiration. More complex polysaccharides, such as starch and glycogen, are a form of stored fuel. Polysaccharides like starch do not taste sweet, but they still contain lots of stored energy. In plants, polysaccharides like cellulose also serve an important structural role.

Lipids

Lipids are complex molecules that also contain carbon, hydrogen, and oxygen but not in a 1:2:1 ratio. They are not true polymers since they do not occur in long chains. Lipids are made of repeated subunit parts. Common lipids are composed of **glycerol** and up to three **fatty acid** chains. **Triglycerides**, or the lipids that make up animal fat, are composed of a glycerol and three fatty acid chains. Triglycerides store energy in animal tissues, and in many animals they also serve as protection and insulation. *Saturated fats* are triglycerides that have no double bonds between the carbon atoms of the fatty acid chains. They tend to be solid at room temperature and are difficult to digest. *Unsaturated fats*, normally referred to as *oils*, are triglycerides that contain carbon double bonds, are liquid at room temperature, and are easier to digest. Both are common in foods, but saturated triglycerides are more likely to contribute to cardiovascular disease. Some lipids include additional groups of atoms. **Phospholipids**, for example, are composed of two fatty acid chains, a glycerol molecule, and a phosphate group (Figure 3.5). These important molecules are a fundamental structural component of all cell membranes. Other lipids, which lack any fatty acid chains, include *steroids* and *cholesterol*.

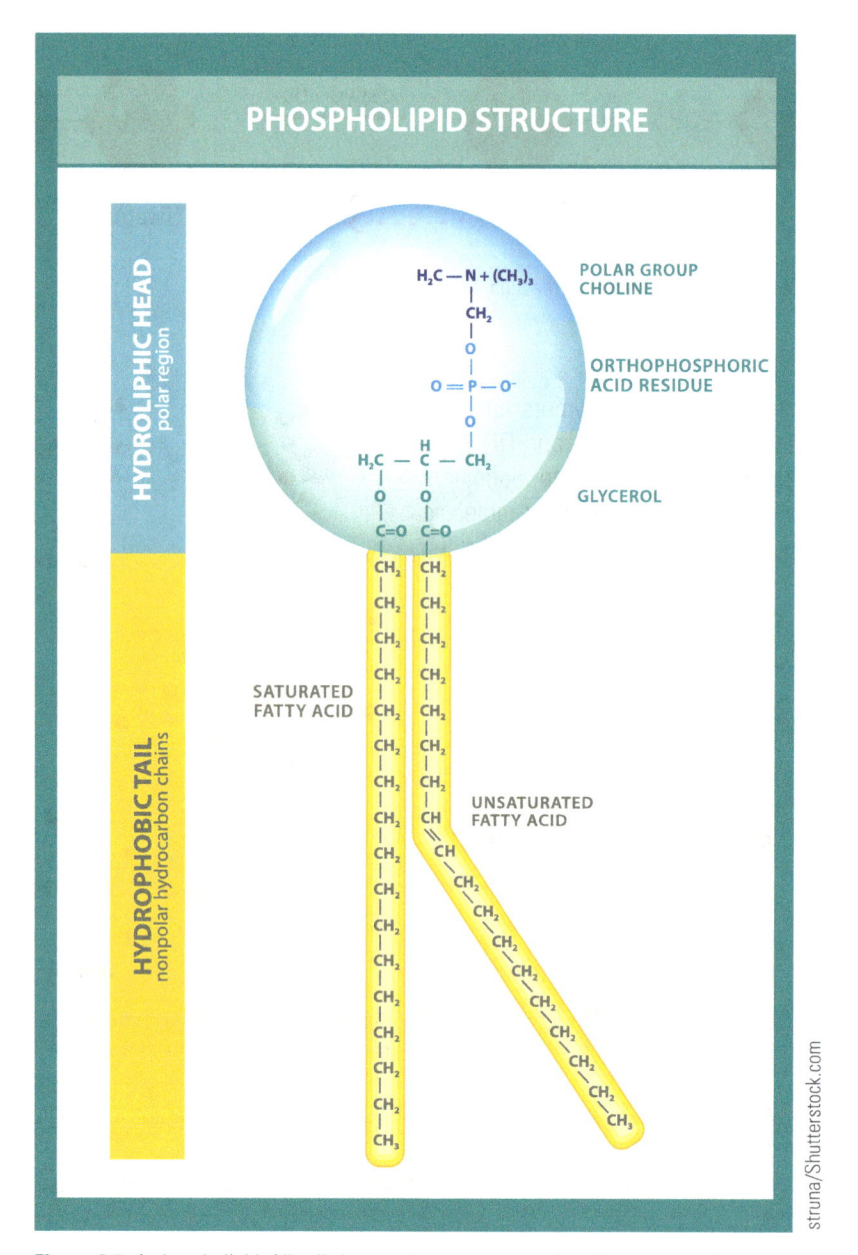

Figure 3.5 A phospholipid. All cells have cell membranes made of these molecules.

Nucleic Acids

Nucleic acids are large, complex molecules composed of **nucleotide** monomers. One nucleotide is composed of a 5-carbon monosaccharide (deoxyribose or ribose), a phosphate group, and one of four ringed carbon-nitrogen structures called a **nitrogen base** (Figure 3.6). Nucleotides are composed of the elements carbon, hydrogen, oxygen, nitrogen, and phosphorus. Examples of nucleic acids include the genetic molecules **DNA** (deoxyribonucleic acid) and **RNA** (ribonucleic acid). DNA and RNA molecules have a "backbone" made of linked sugar and phosphate subgroups. The order of the nitrogen base pairs determines the genetic code (Figure 3.7). The four nitrogen bases in DNA pair up according to how they form hydrogen bonds with each other: thymine with adenine, and guanine with cytosine. The order of these molecules provides the code for all of the genetic activities of the cell and allows for the inheritance of genetic traits from parent organisms to their offspring. The DNA molecules in all of the cells of one organism are generally identical (with the exception being sperm or egg cells). The DNA molecules in closely related organisms are very similar. **Genetics** is the branch of biology that specializes in the study of nucleic acids. Later lab exercises will cover more on the structure and function of DNA.

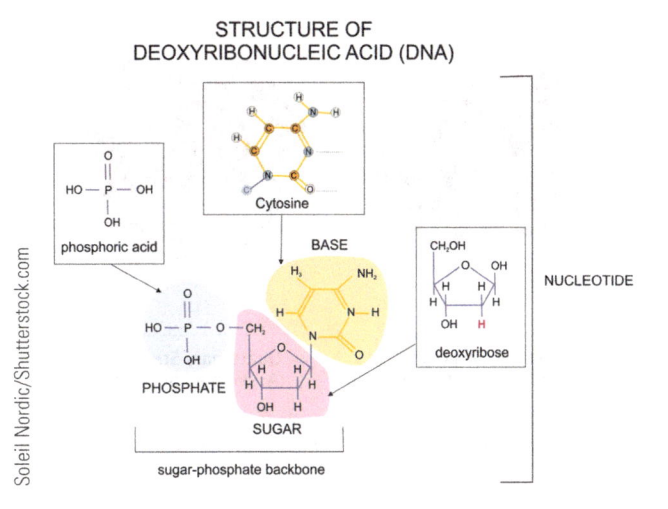

Figure 3.6 Chemical structure of a nucleotide.

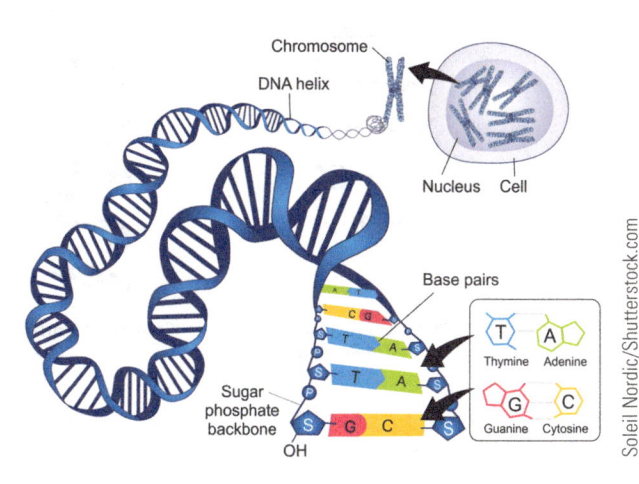

Figure 3.7 Nucleic acid showing sugar-phosphate backbone (blue) and paired nitrogen bases (abbreviated with T, A, G and C.).

Proteins

Proteins are macromolecules composed of chains of 20 different types of **amino acid** monomers. Proteins fundamentally contain carbon, hydrogen, oxygen, and nitrogen, although other types of elements may be found in certain amino acids. Amino acids always contain an amino group, $-NH_2$ and a carboxyl group $-COOH$, plus other groups of atoms (indicated as the R side chain) that vary in size and composition (Figure 3.8). Proteins are extremely important structural and functional molecules in a cell. Many complex cell structures are composed of proteins.

We often think of meats as protein-rich foods. Meat is composed of muscle fibers. The structural components of muscles that allow them to contract are cells full of *myofilaments*. Myofilaments are proteins. Other proteins such as collagen provide structure throughout body, including bones and cartilage. Many hormones and neurotransmitters are proteins. Much of biochemistry, cellular biology, and medicine depends on the understanding the complex roles of proteins in living cells. The study of protein structure and function is called **proteomics**.

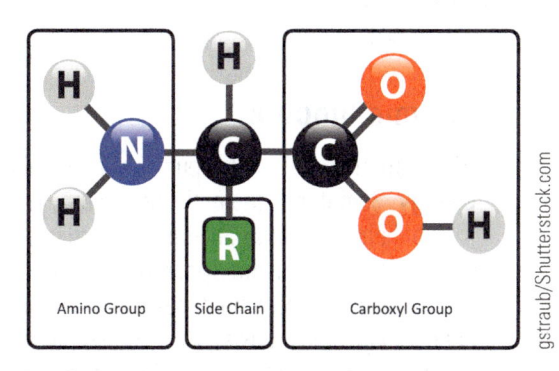

Figure 3.8 General structure of an amino acid. The R side chains vary among the 20 different types of amino acids.

A protein's shape determines its function. The overall shape of a protein is determined by its sequence of amino acids. Some amino acids are charged, and others are polar. Those with charges may be attracted to others with opposite charges. Hydrogen bonds can form between polar amino acids. *Disulfide bridges* can form between cysteine amino acids. Heat (movement of molecules) and pH (excess H+ ions) can disrupt these weak attractions that hold the protein together. This can **denature** the protein (destroy its three-dimensional structure).

There are 4 levels of protein structure. The *primary* level is simply the sequence of amino acids in the protein. The *secondary* structure refers to the shapes formed in parts of the amino acid strand, which are typically *alpha-helices* (spirals) or *beta-pleated sheets* (zig-zag patterns). The *tertiary* (third) level of structure is the overall folding of the molecule, and the *quaternary* structure refers to the result when proteins physically fit together to form an even more complex structure (Figure 3.9).

Enzymes are molecules that do not directly participate in metabolic chemical reactions, but they help facilitate the reaction. Therefore, the presence of an enzyme in a cell "turns on" or activates a chemical reaction, and the absence of that enzyme "turns off" or slows/stops the reaction. Enzymes are only able to facilitate reactions in which the **substrates** (reactants) fit the **active site** of the enzyme. This allows a certain enzyme to have control over a specific reaction.

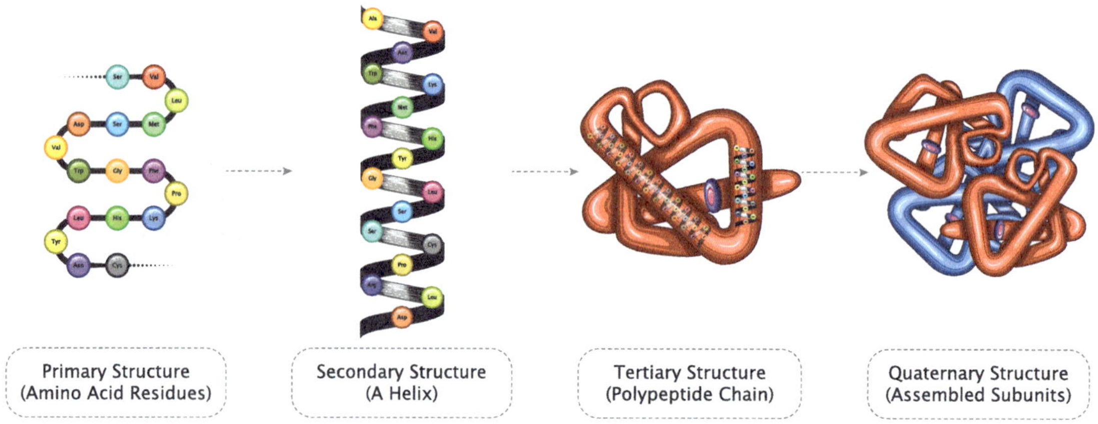

N.Vinoth Narasingam/Shutterstock.com

Figure 3.9 The structural levels of proteins. Note that "Polypeptide Chain" in this figure refers to one protein, and "Assembled Subunits" refers to multiple proteins linked together.

Therefore, the cell's production of various enzymes controls many of the metabolic activities of the cell and, ultimately, the metabolic activities of a complex organism. Enzymes can facilitate synthesis reactions and decomposition (or digestion) reactions (Figure 3.10).

Enzyme function

All reactions require an input of energy called **activation energy**. Sometimes the activation energy is provided in the form of heat, such as when we heat food to change the properties of the ingredients from runny cake batter to a cake. In living organisms, body heat may provide much of the needed energy to get reactions started. But some reactions need to occur at certain times and therefore are not triggered by body heat alone. Enzymes lower the activation energy of a reaction. Therefore, if a reaction does not happen with available heat energy, the production of an enzyme may lower the activation energy threshold needed to trigger the reaction (Figure 3.11).

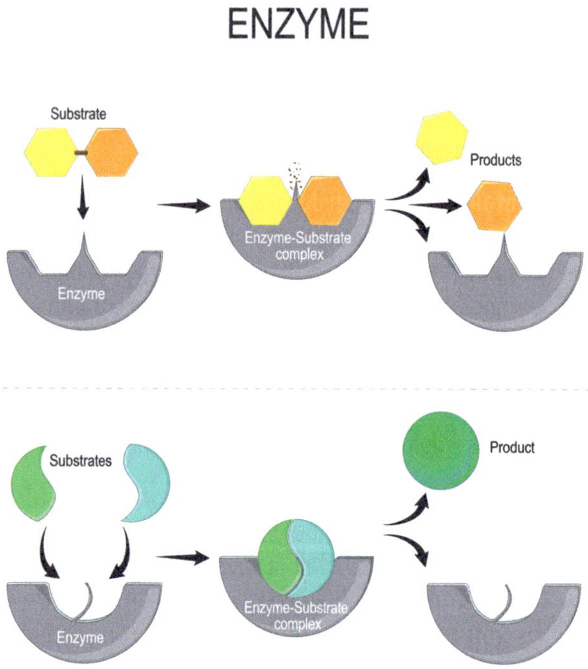

Designua/Shutterstock.com

Figure 3.10 Enzymes facilitate reactions when substrate molecules fit the enzyme's active site. The enzyme may facilitate decomposition or synthesis reactions.

Pancreatic Enzymes

The pancreas is an important organ in the human body. Many higher animals have a pancreas, and it functions in two body systems: digestive and endocrine. The endocrine system deals with hormones, and the pancreas produces two important protein-based hormones, *insulin* and *glucagon*. These will be discussed further in a later lab. Here we are interested in the role of the pancreas in the digestive system.

In humans, the pancreas is located just below the stomach. Food does not go through the pancreas, but the pancreas is critically important for digesting food. The pancreas produces four categories of enzymes: amylases, proteases/peptidases, nucleases, and lipases. **Amylases** (also called *carbohydrases*) are enzymes that break down polysaccharides into mono- and disaccharides. **Proteases** break proteins down into shorter fragments called **peptides**, and **peptidases** break the peptides into amino acids. **Nucleases** break down nucleic acids into nucleotides. **Lipases** break lipids down into glycerol and fatty acids. **Pancreatin** is a solution made from liquified pancreas that contains all of these enzymes. Other organs also produce digestive enzymes. The mouth produces *salivary amylase* in saliva that digests carbohydrates, and the stomach produces *pepsin,* which helps digest proteins.

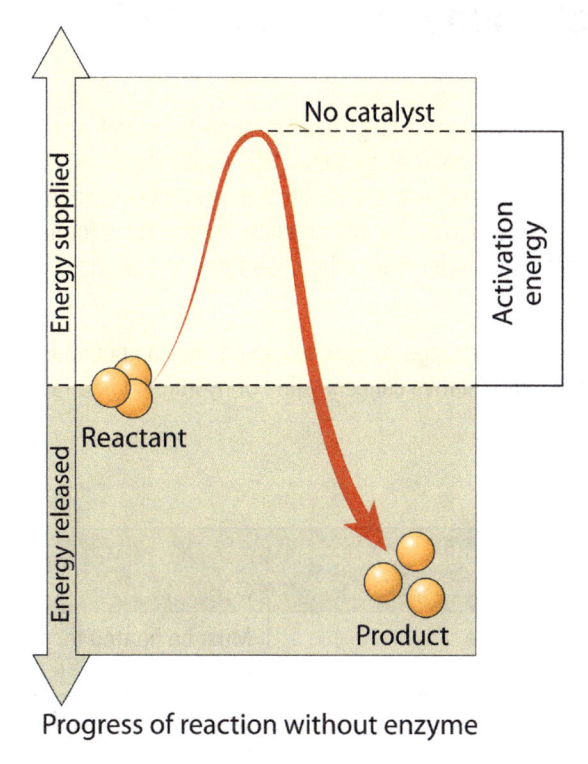

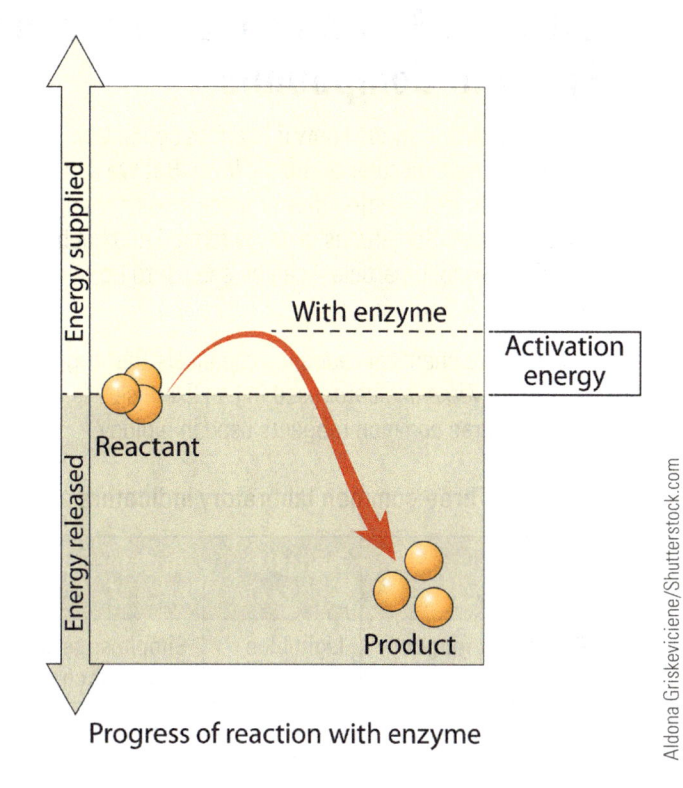

Laboratory Exercise 3 Biochemistry: Chemical Compounds of Living Organisms

Figure 3.11 Effect of enzymes on chemical reactions.

Notice that enzyme names tend to start with the substrate upon which they act, and end with "ase" This makes it easy to identify most enzyme names and helps indicate their function. For example, the term "*amino acetyl tRNA synthase*" looks pretty intimidating until you realize that it is just an enzyme (ends with "ase") that *synthesizes* (combines) an *amino acid* with a nucleotide—a special type of RNA molecule called *tRNA*.

The monomers of each type of macromolecule are held together by covalent bonds. However, the covalent bonds between different types of monomers are given additional special names based on the macromolecule type. The bonds between monosaccharides in a carbohydrate are called **glycosidic bonds**. The bonds between glycerol and fatty acids in a lipid are called **ester bonds**. The bonds between amino acids in a protein are called. **peptide bonds**. And the bonds between nucleotides in a nucleic acid are called **phosphodiester bonds**. Therefore, the four different types of digestive enzymes break the four different types of covalent bonds in macromolecules (Table 3.1). The pancreas secretes these enzymes into the small intestine to help digest food, which contains all of these macromolecules.

Table 3.1 Types of bonds in macromolecules and the enzymes that act upon them

Macromolecule	Type of covalent bond between subunits of macromolecule	Type of enzyme that breaks the bonds between subunits
Carbohydrate	Glycosidic	amylase
Lipid	Ester	lipase
Nucleid Acid	Phosphodiester	nuclease
Protein	Peptide	protease and peptidase

Lab Skill 3.1 Using Indicators to Determine the Presence of Specific Compounds

You may wonder how we know if enzymes actually work. It is impossible to look through a microscope and watch a chemical reaction at the molecular or atomic level. But we can detect changes in a solution as a chemical reaction occurs. In some cases, a chemical reaction may produce a color change. In ectothermic reactions, the reaction may produce a substantial amount of heat. Sometimes, a gas is formed and bubbles can be observed in a solution, or a new compound is formed and a precipitate (solid particles) can be seen settling on the bottom of the solution. But when there are no obvious changes, we may use indicators.

Indicators are chemical *reagents* (chemicals that trigger a reaction) that change color when a specific compound is present. The *universal indicator* used in Lab Exercise 2 is used to detect the relative concentration of hydrogen ions. Table 3.2 describes three common reagents used in biology.

Table 3.2 Three common laboratory indicators and their uses

Indicator/reagent	Normal reagent color	Tests for the presence of	When chemical is present, changes color to	Special considerations
Benedict's reagent	Light blue	Simple sugars (mono- and disaccharides)	Orange	Must be heated to boiling
Blue litmus	Dark purple-blue	Acid	Pink	Color change is difficult to detect without a standard for comparison
Iodine	Orange-brown	Complex carbohydrates (starch)	Purple-black	Provides instant and obvious color change

Using Benedict's reagent

Typically, the solution being tested is placed in a test tube. Benedict's reagent can be added to the test tube using a pipette, wash bottle, or graduated cylinder. Typically, the amount of Benedict's reagent added to the test tube is just enough to turn the solution being tested a light blue color. After adding Benedict's reagent, the mixture should then be gently swirled. There will be no immediate color change other than making the tested solution blue. (A blue color is not a positive result for simple sugars.)

To allow Benedict's reagent to work, the mixture of the test solution and Benedict's reagent must be heated in a beaker about half-full of boiling water for approximately one minute. The test tube containing the mixture of test solution and Benedict's reagent should be carefully lowered into the beaker of boiling water so that it does not spill. Use test tube clamps to avoid burns.

After about one minute, remove the test tube from the beaker and observe. If the liquid is still blue, the test is negative, and there are no simple sugars present in the solution. If the liquid in the tube is orange, the test is positive. If there is a slight greenish color, continue heating for one more minute. A green color may be produced when a very small amount of Benedict's reagent changes color and mixes with the rest of the blue reagent. After use, discard the Benedict's reagent mixture as directed by your instructor (Figure 3.12).

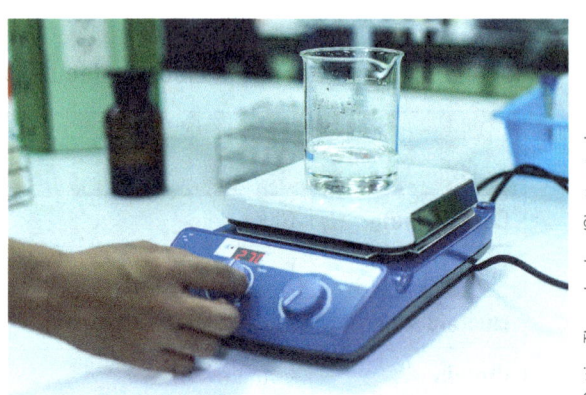

Figure 3.12 Use caution when heating liquids on a hotplate in the lab. Many hotplates have magnetic stirrers in the burner to help ensure even heating.

Using blue litmus

Blue litmus is much like the universal indicator used in Lab 2. It is usually stored in a bottle and accessed with a graduated cylinder or beaker. Enough blue litmus should be added to a test solution to see a purple-blue (negative) or pink (positive) color. Blue litmus changes from a dark purple-blue color to a pink color depending on the concentration of acid present. The change from purple to pink may be so minor that the solution only looks slightly lighter or a shade pinker than it originally did. Since the color change can be subtle, it is helpful to have a control tube for comparison. Blue litmus does not have to be heated to work. After using, discard the blue litmus mixture as directed by your instructor. When not in use, blue litmus should be kept in the lab refrigerator.

Using iodine

Iodine, a rusty orange-brown liquid, is usually stored in a bottle or in small dropper bottles. Very little iodine is needed to determine the presence of starch. Add a few drops to the solution being tested and swirl gently. Typically, a few drops of iodine will either turn the liquid orange-brown (a negative result) or very dark purple-black (a positive result). Iodine works almost instantly and does not need to be heated to work. After use, discard the iodine mixture as directed by your instructor.

Using a test strip

In some cases, simple chemical tests can be made with indicator strips. The test strip may have multiple testing compounds on it. The test strip shown in Figure 3.13 is the type used to look for compounds that may indicate disease when present in urine. Note the color chart. The test strip must be oriented correctly before comparing the strip to the chart or bottle.

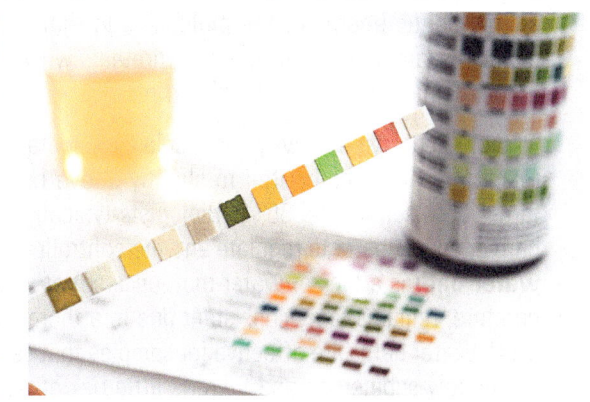

Figure 3.13 A test strip indicator.

Lab Skill 3.2 Managing Heat in the Laboratory

It is often necessary to manipulate the temperature of solutions in the laboratory. Solutions can be heated quickly on a hotplate. However, in some cases, the hotplate may get too hot, or may not regulate the temperature as well as a different type of lab equipment.

Recall that enzymes are proteins that must have a specific shape to work properly. Enzymes are found in living cells. Cells have evolved to produce enzymes that tend to work best near or slightly above their homeostatic temperature range. Human body temperature is approximately 37° C. The rate of enzyme-catalyzed reactions increases as the temperature increases—but only to a point. Recall that temperature and heat are measures of the kinetic energy of the molecules in that object. If the molecules are moving too violently, proteins will denature and other chemical bonds will be disrupted.

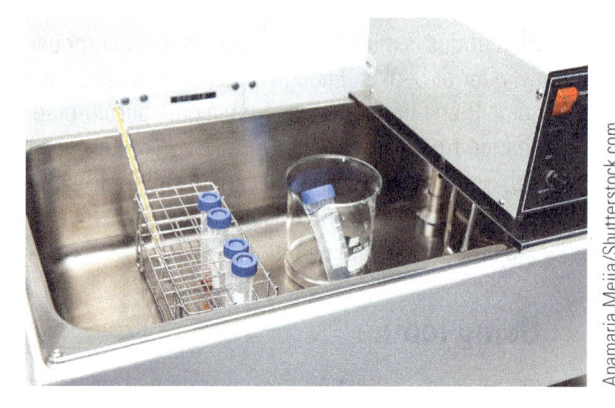

Figure 3.14 A laboratory hot water bath.

To promote enzyme activity, most solutions containing enzyme reactions should be heated to slightly above body temperature—between 40 and 55° C. Biologists typically use a water bath or an incubator to regulate enzyme-controlled reactions. A **water bath** is a tank of water that, once set at a certain temperature, remains fairly consistent due to a thermostat (Figure 3.14). Water baths are effective for samples that can be almost completely submerged and there is little risk of water getting into the sample or the sample leaking out. However, it may take time to heat the water in a water bath if it is not already at the desired temperature.

Figure 3.15 A laboratory incubator.

An **incubator** works much like an oven. It heats the air around a sample, which evenly heats the sample throughout (Figure 3.15). There is no risk of the sample getting contaminated by being submerged in water. However, an incubator loses some heat each time the door is opened, making temperature fluctuations a bigger concern. Incubators are often used for growing cultures of bacteria in petri dishes that cannot be placed in a water bath.

Many biological experiments require manipulating the temperature of a reaction, specimen, or culture. It is helpful to note the different methods used for managing temperatures, and the strengths and weaknesses of these different approaches.

Problem to Solve 3.1 Can I Observe Enzymes Digesting Macromolecules?

This laboratory activity will look at three major concepts simultaneously: the composition of macromolecules, the activity of enzymes, and the effect of temperature on enzymes.

Protein Digestion

Protein digestion will be shown as a class demonstration. Proteins take a longer time to digest, so this part of the experiment was prepared in advance. Observe a boiled egg in a beaker of water, and contrast it with the boiled egg in a beaker of pancreatin. Note the difference on your notebook page in the Data section. What must be in the pancreatin to cause this result?

The remaining part of this lab will require setting up multiple tubes, placing them in the water bath or incubator, using indicators to test for substances, and observing the results. It is important that you follow the protocol carefully. Make sure all glassware is clean when you start and that you wash beakers and graduated cylinders after each use to avoid cross-contamination.

Hypothesis

Before you begin, work with your lab partners to answer the following five questions.

1. Based on what you have learned in the pre-lab, what will happen if pancreatin is added to a starch solution? Why? How could you find out if you are correct?

2. Based on what you have learned in the pre-lab, what will happen if you mix salivary amylase with starch? Why? How could you find out if you are correct?

3. Based on what you have learned in the pre-lab, what will happen if you mix pancreatin with a lipid-rich solution? Why? How could you find out if you are correct?

4. At what temperature should you run your experiment to test these hypotheses? Why?

5. If pancreatin is boiled, will it still have the same effect on macromolecules? How could you test this?

Laboratory Exercise 3 Biochemistry: Chemical Compounds of Living Organisms

Title

Observing Enzymatic Digestion of Macromolecules

Purpose

Observe enzyme activity, use indicators, and test effect of heat on enzyme activity.
Hypothesis:_____
Predictions:_____

Materials (per lab group):

6-10 pieces of freshly popped popcorn
1 small graduated cylinder
3 small beakers (100-150 ml)
1 medium beaker (250-300 ml)
15 test tubes
Test tube rack
Marker and tape to label test tubes
Hotplate
Bottle of blue litmus
Bottle of Benedict's reagent

Bottle of iodine
Bottle of distilled water
Water bath set at 45° C
1 liter flask of 5% pancreatin (for class)
1 liter flask of 1% starch sol. (for class)
Quart container of cream (for class)
Powdered dextrose (for class)
Powdered starch (for class)
Dropper bottle of .01 M HCl (for class)

Protocol:

Prep solutions

1. Label small beakers "starch," "cream," and "pancreatin."

2. Pour approximately 20 ml starch into the starch beaker.

3. Pour approximately 10 ml pancreatin into the pancreatin beaker.

4. Pour approximately 20 ml of cream into the cream beaker.

5. Fill the medium beaker half full of water and set it on the hotplate. Turn on the hotplate to high. Once the water comes to a boil, turn it down to just barely boiling.

6. Use the tape and marker to label the top of one test tube "BP."

Continued from previous page.

Protocol

7. Use clean graduated cylinder to transfer 5 ml pancreatin to the BP test tube.

8. Place the BP tube into the beaker of boiling water for at least 5 minutes. Continue preparing other solutions as it heats. (Turn down but do not turn off hotplate when finished heating. You will use the boiling water again.)

9. Use clean graduated cylinder to add 20 ml blue litmus to the beaker containing cream. Stir or swirl until thoroughly mixed.

Prepare test tubes

10. Label 8 test tubes with group number and letters A-H. (Label at top as shown—make sure tape goes all the way around top of tube.)

11. Crumble two pieces of popcorn and add the pieces to test tube A.

12. Add 2 ml distilled water to test tube A.

13. Select one person from group to chew (but not liquefy) 3-4 pieces of popcorn and spit the pieces into clean test tube B. They must add about 2 ml saliva - enough until tube B contains the same amount of liquid as tube A. The goal is to make tubes A and B match as closely as possible.

14. Use cleaned graduated cylinder to transfer 5 ml of starch solution to each test tube labeled C, D, and E.

15. Use cleaned graduated cylinder to transfer 10 ml starch/blue litmus mixture to each test tube labeled F, G, and H.

Title

Purpose

Continued from previous page.

Protocol

Add pancreatin and water to tubes (See table in Data section.)

16. Use cleaned graduated cylinder to transfer 2 ml pancreatin from beaker to tube C (which already contains 5 ml starch).

17. Add 2 ml pancreatin from beaker to tube F (which already contains 10 ml cream/blue litmus mixture).

18. Add 2 ml of <u>boiled</u> pancreatin from the test tube labeled BP to test tube D.

19. Add 2 ml of <u>boiled</u> pancreatin from the test tube labeled BP to test tube G.

20. Add 2 ml distilled water to each of the tubes labeled E and H.

Place all tubes labeled with group number and A through H into the test tube racks in the 45° C water bath. (Ensure they are clearly labeled with group number. Make sure labels do not get wet or the ink and/or tape may come off. Incubate 45 minutes.

Practice with indicators

21. While waiting, label 6 test tubes S1, S2, D1, D2, B1, and B2.

22. Add a small amount of powdered starch to tubes S1 and S2 and add 5 ml distilled water to both tubes. Cover the tops and shake tubes to mix.

23. Add a small amount of dextrose to tubes D1 and D2 and add 5 ml distilled water to both tubes. Cover tops and shake tubes to mix.

24. Add 5 ml blue litmus to tubes B1 and B2.

Continued from previous page.

Read the following and record your predictions before doing steps 25-28.

25. Add 5 ml Benedict's reagent to Tubes S1 and D1 and then heat them for one minute in the beaker containing boiling water.

26. Add a few drops of iodine to tubes S2 and D2.

27. Add a few drops of 0.01 M hydrochloric acid to tube B2.

28. Add a few drops of iodine to a piece of popcorn.

Tube	Contents	Predicted result	Actual result
S1	Starch, water and Benedict's reagent		
S2	Starch, water, and iodine		
D1	Dextrose, water, and Benedict's reagent		
D2	Dextrose, water, and iodine		
B1	Blue litmus		
B2	Blue litmus and HCl		
--	Popcorn and iodine		

Purpose

Continued from previous page.

Protocol

Finish experiment

29. After test tubes A-H have incubated in the water bath for 45 minutes, remove them and return them to your test tube rack.

30. Turn hot plate back on and bring beaker of water back to a boil.

31. Add 5 ml Benedict's solution to test tubes A-E.

32. Place test tubes A-E in the beaker of boiling water for about 1 minute.

33. Carefully remove test tubes A-E and place them back in test tube rack. Examine tubes. Record your results in the table in the Data and Observations section.

34. Write your summary and conclusions.

35. Clean and return all glassware.

Continued from previous page.

Data and Observations:

Tube	Solution	Molecule of Interest	Treatment	After in water bath for 45 min.	Observations
A	Crumbled popcorn		2 ml water	5 ml Benedict's reagent and heat	
B	Chewed popcorn		2 ml saliva	5 ml Benedict's reagent and heat	
C	5 ml 1% starch solution		2 ml pancreatin	5 ml Benedict's reagent and heat	
D	5 ml 1% starch solution		2 ml boiled pancreatin	5 ml Benedict's reagent and heat	
E	5 ml 1% starch solution		2 ml water	5 ml Benedict's reagent and heat	
F	10 ml cream/blue litmus mixture		2 ml pancreatin	X	
G	10 ml cream/blue litmus mixture		2 ml boiled pancreatin	X	
H	10 ml cream/blue litmus mixture		2 ml water	X	

Continued from previous page.

Summary and Conclusions:

Name: _____   Date: _____

POST-LAB QUIZ 3: What have you learned?

1. What purpose did tubes A, E, and H serve?

2. Why were the test tubes heated in the water bath?

3. Why were test tubes A–E heated in the boiling water after the water bath?

4. Why were tubes F–H not placed in the boiling water?

5. What kind of macromolecule is popcorn made of? How do you know?

6. Tubes A and B both contained popcorn. Did they produce the same results? Why or why not? What enzyme was being used in test tube B?

7. Tubes C and D both contained pancreatin and starch. They were both treated with Benedict's solution. Did they produce the same results? Why or why not?

8. Tubes F and G both contained the cream/blue litmus mixture and pancreatin. Did they produce the same results? Why or why not?

9. Did the *crumbled* popcorn produce a positive result for simple sugars? If so, why?

10. It is impossible to live without a pancreas. Based on the results of this lab alone, what would be the effects of not having a functioning pancreas?

Laboratory Exercise 4

Molecular Biology I: Understanding and Isolating DNA

Concepts to Understand

After completing this lab the student should be able to:

1. Describe the structure and function of DNA.
2. Explain the central dogma of molecular biology.
3. Explain how transcription and translation occur in a cell.
4. Explain the events in each phase of mitosis.

Skills to Learn

After completing this lab the student should be able to:

1. Predict the sequence of amino acids produced by a section of DNA
2. Identify the stages of mitosis in plant and animal cells.
3. Extract DNA from human cells.

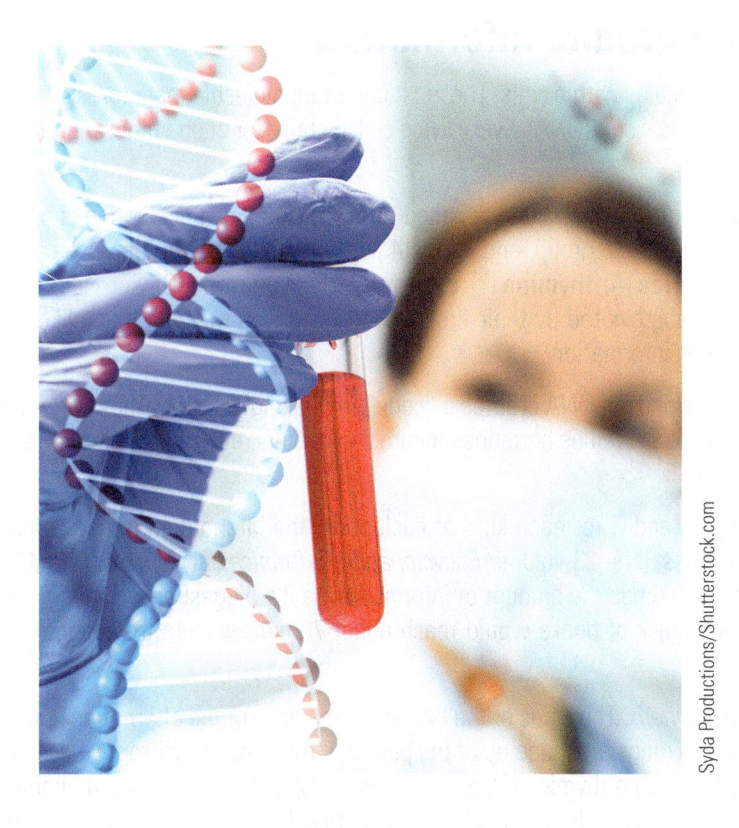

SEE ALSO

Campbell Biology – Chapter 12, section 12.1 – 12.2; Chapter 17, section 17.1

JOVE (video) – An Inroduction to the Centrifuge

YouTube: Mitosis BioFlix

YouTube: From DNA to Protein – 3D

Others:

Important Terms to Know

- Adenine
- Anaphase
- Anticodon
- Cell plate
- Central dogma
- Centriole
- Centrifuge
- Centromere
- Chromatid
- Chromatin
- Chromosome
- Cleavage furrow
- Codon
- Cytokinesis
- Cytosine
- Deoxyribose
- DNA polymerase
- DNA ligase
- G1 and G2 phases
- Gene
- Genome
- Guanine
- Helicase
- Homologous
- Interphase
- Karyokinesis
- Kinetochore
- Lagging strand
- Leading strand
- Metaphase
- Mitosis
- Molecular biology
- Nitrogen base/nucleobase
- Nucleotide
- Okazaki fragments
- Pellet
- Prophase
- Protein synthesis
- Proteomics
- Replication
- Ribose
- Ribonucleic acid (mRNA, rRNA, and tRNA)
- RNA polymerase
- S-phase
- Sense strand (of DNA)
- Spindle Fibers
- Supernatant
- Telophase
- Template (antisense) strand
- Thymine
- Transcription
- Translation
- Uracil

Pre-lab Preparation and Background Information

Molecular biology is the study of the functional molecules within cells. The most important molecule that stores genetic information and directs cellular activity is *deoxyribonucleic acid* (DNA). Every living cell contains (or at one time, contained) DNA.

Nucleic acids are huge "macro"-molecules made up of repeating subunits called **nucleotides**. Each nucleotide is made up of 3 parts — a sugar molecule (**deoxyribose**), a phosphate group, and one of 4 possible **nitrogen bases** (also called **nucleobases**). In DNA the four nitrogen bases are **adenine** (A), **thymine** (T), **cytosine** (C), and **guanine** (G). See Figure 4.1. The *order* of the 4 kinds of nitrogen bases (typically abbreviated by their first letters) stores information, much like the specific order of letters and spaces on this page stores information.

The ordered combination of the 4 nucleotides in the strands of DNA creates a "code" for making proteins. These proteins make up much of the physical structure of an organism, as well as hormones, neurotransmitters, enzymes, pigments, and many other extracellular and subcellular components.

If the information was written down as letters (A,T,C, and G for each kind of nucleotide) the information in one human nucleus would fill about one million encyclopedia pages. (The 23-volume *Encyclopaedia Brittanica* only has about 25,000 pages!) This means every cell in your body contains 40 times the amount of information as the biggest encyclopedia set. If this information were written into book form, the stack of books would reach about 70 meters high (the height of a twenty-story building).

The structure of DNA is like a twisted ladder, or *double helix*. The sides of the ladder are made up of the alternating deoxyribose (a pentose sugar) and phosphate molecules. The rungs are made up of the paired nitrogen bases. Nitrogen bases fit like puzzle pieces so that an adenine (A) always pairs with a thymine (T), and a cytosine (C) always pairs with a guanine (G) due to their ability to form either two or three hydrogen bonds with each other (Figure 4.1).

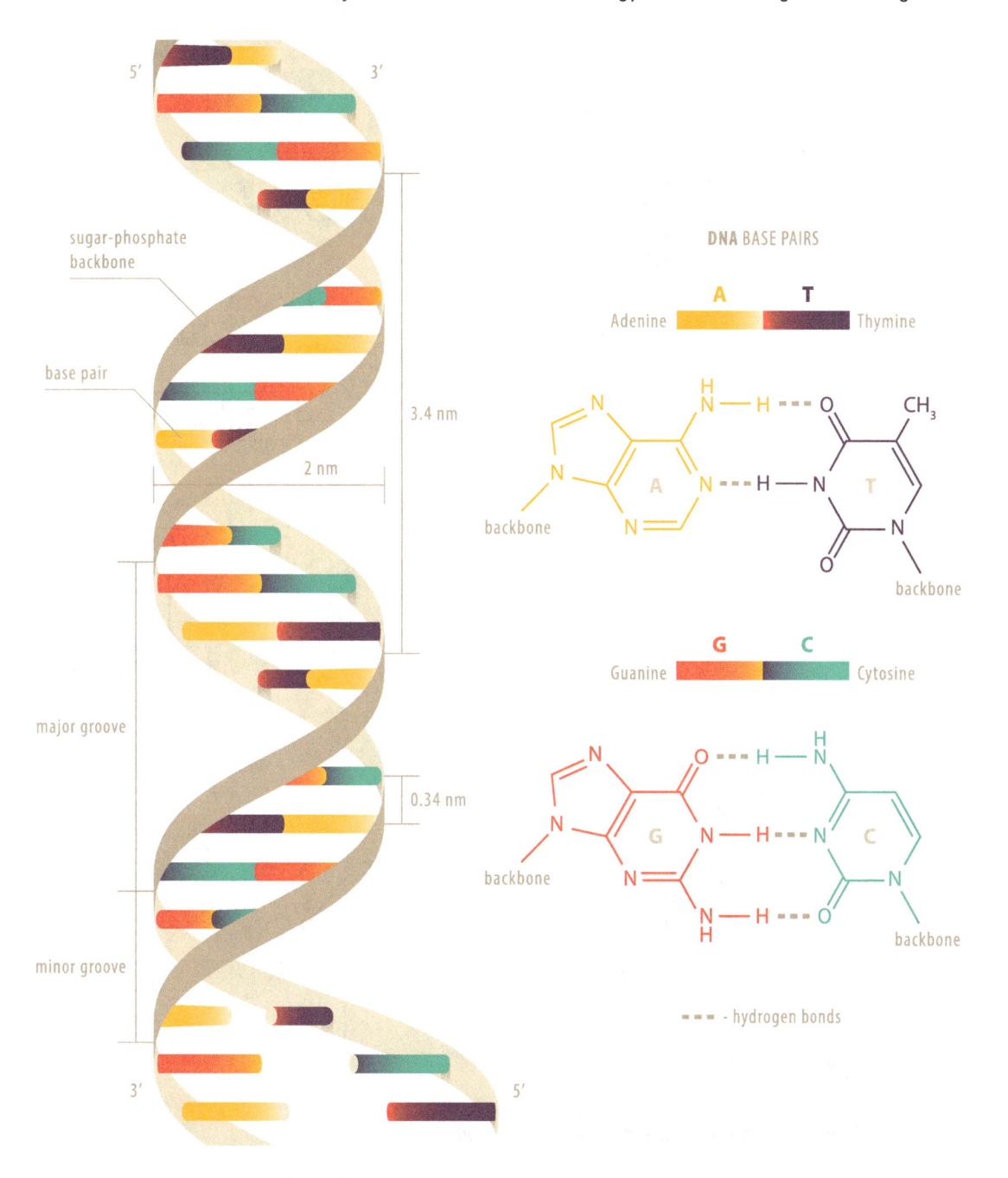

Laboratory Exercise 4 Molecular Biology I: Understanding and Isolating DNA 83

Figure 4.1 DNA structure showing adenine (yellow), thymine (purple), guanine (red), and cytosine (teal).

When a cell is not dividing, the nuclear membrane serves to isolate and protect the DNA. DNA is only visible as **chromatin**, a tangled mass of nucleic acid and proteins inside the nucleus. When cells undergo cell division, the nuclear membrane breaks down and the DNA condenses into forty-six **chromosomes** (See Figure 4.2.) The DNA is normally coiled around special *histone* proteins. If all of the pieces of DNA from the 46 chromosomes in one human cell nucleus were unwrapped and joined end to end, they would form a very thin molecule about two meters long.

The quantity of DNA and the number of chromosomes vary in different species. Humans have 46 chromosomes, or *23 pairs* of chromosomes. Chromosomes occur in pairs because the two chromosomes in a pair came from two parents. Pairs of chromosomes are **homologous**. They look similar and contain similar types of information, but they are not identical.

Chromosomes are often represented as "X-shaped" structures. However, they are not always shaped that way. An X-shaped chromosome is a **replicated** chromosome containing two identical copies of the DNA held together by **kinetochore**

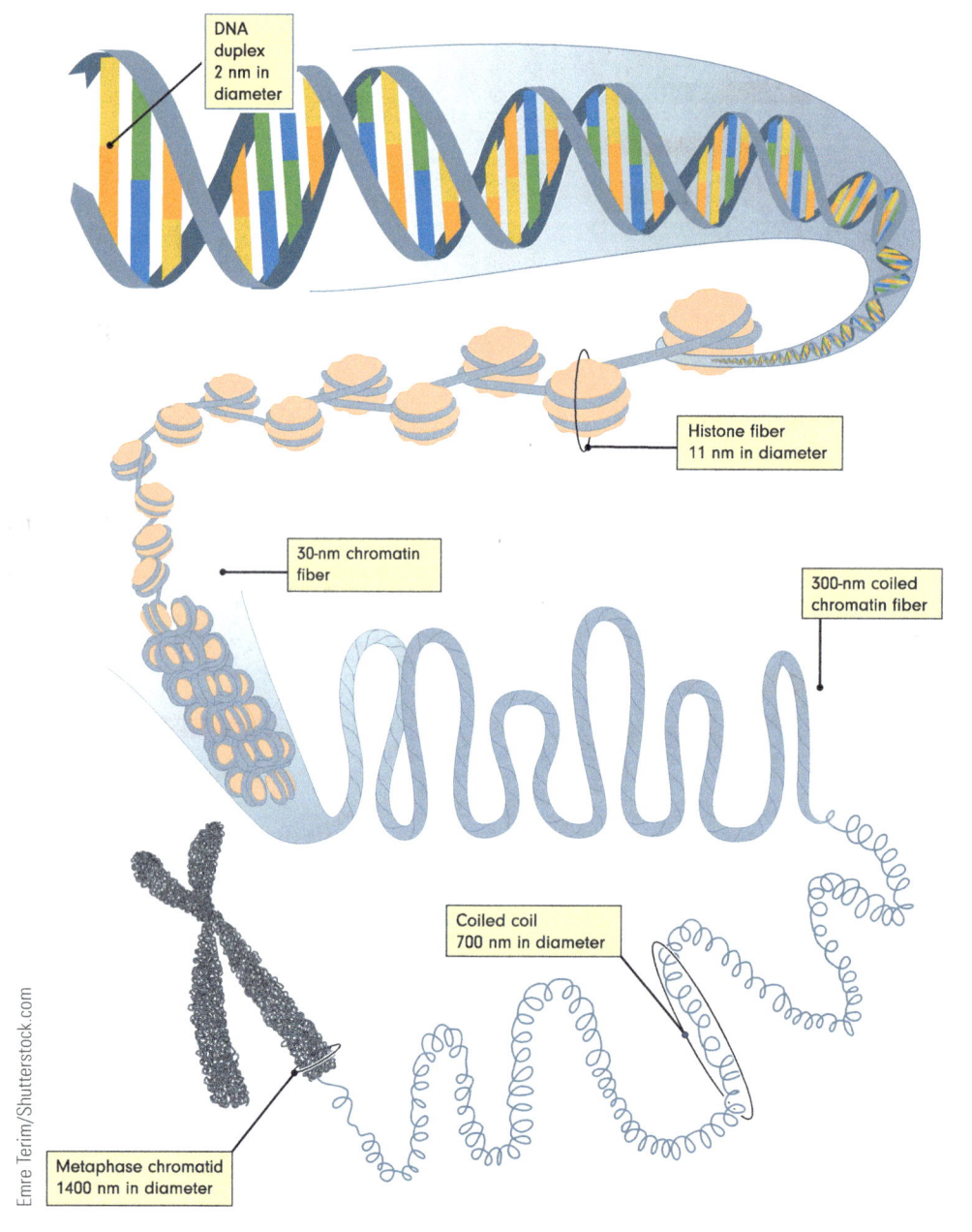

Figure 4.2 Representation of the DNA in a chromosome.

proteins at the **centromere** (Figure 4.3). The two copies are called sister **chromatids**. The term chromatid only applies to replicated chromosomes. A common mistake is confusing *chromatids* with paired *chromosomes*. The left side of Figure 4.3 represents a homologous pair of *unreplicated chromosomes* and *no sister chromatids*. The right side of Figure 4.3 represents a homologous pair of *replicated* chromosomes, each with *two sister chromatids*.

Chromosomes are simply pieces of DNA that very in size from very small to very large. The number of chromosomes should not be equated with the total quantity of DNA in a cell or the complexity of the organism. Humans have 46 chromosomes in their *somatic* (normal body) cells (Figure 4.4), but *Drosophila melanogaster*, the fruit fly, has only eight chromosomes, and a fern has 1260 chromosomes. Chromosomes are just pieces of the **genome** (total DNA content) of a cell.

A **gene** is a section of DNA responsible for a certain protein. Recent studies indicate that the human genome contains approximately 22,000 genes. Each gene is like a recipe for making a body part at the molecular level. For example, a gene for hair color (a genetic trait) is the part of the DNA responsible for creating proteins found in hair cells. If brown proteins,

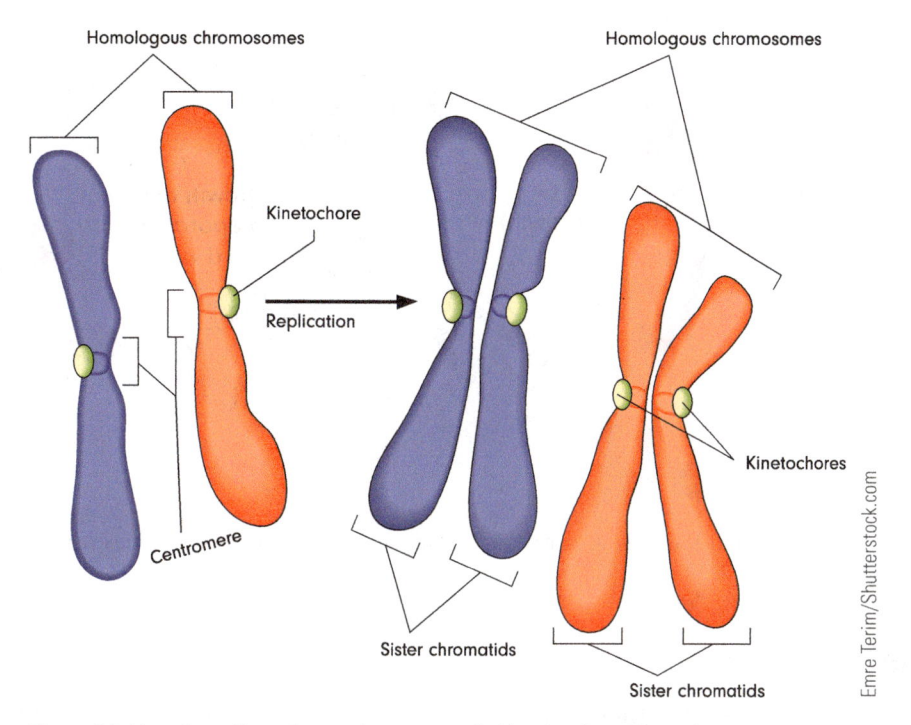

Figure 4.3 Unreplicated homologous chromosomes (left) and replicated homologous chromosomes (right). The term "sister chromatid" only applies to half of a replicated chromosome. It does not apply to unreplicated chromosomes. Chromosomes can only be seen during cell division.

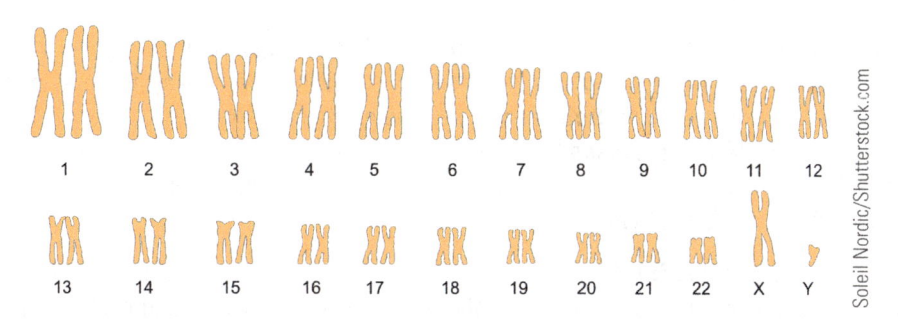

Figure 4.4 The 46 human chromosomes shown as homologous pairs. Note the last pair contains the X and Y chromosomes that determine the sex of the organism.

like melanin, are produced in these cells, the hair appears brown. Other genes may not produce an obvious outward trait. Genes produce enzymes, hormones, clotting factors, antibodies, neurotransmitters, cellular structures, and many other kinds of proteins that influence the traits of the organism.

Mitosis

Mitosis, also called **karyokinesis** (*karyo* = nucleus; *kinetic* = movement), is the process of one nucleus splitting into two identical nuclei. This happens so that the genetic information in the DNA can be passed on to new cells as the organism grows. Though it is a common misconception, mitosis is not the physical separation of one *cell* into two cells, it is the separation of one *nucleus* into two nuclei. The physical separation of the rest of the cell (the cell membrane and organelles) is called **cytokinesis** (*cyto* = cell). Except for sperm and egg cells, every one of the trillions of cells in a human body came from a single fertilized egg, was produced by mitosis, and contains the same genetic code as all the others. Mitosis is part of the *cell cycle* (Figure 4.5). The cell cycle describes the cyclic nature of cells as they divide, grow, and divide again perpetually during the life of an organism.

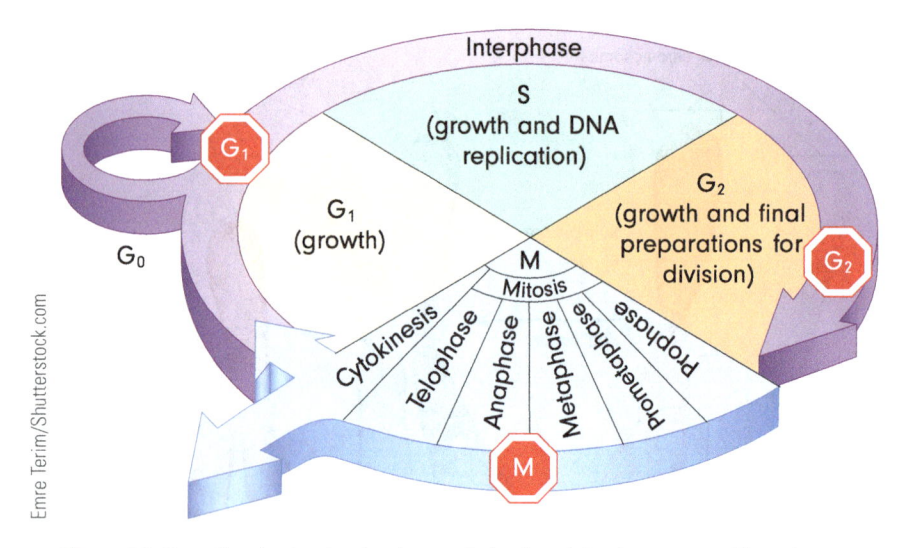

Figure 4.5 The cell cycle showing the phases of mitosis and the three stages of interphase.

Interphase

Cells spend most of their time in **interphase** (*inter* = between; *phase* = mitotic phases). During interphase a cell appears normal, has an intact nucleus, is performing its normal functions, and is not dividing. Cells undergo mitosis when the organism is growing, healing from an injury, or replacing worn out cells. Some parts of an organism are highly mitotic all the time, like the surface of the skin. New skin must replace the dead cells that constantly flake off. Other parts, including most parts of a human brain, never go through mitosis after growing to adult size. Many other body parts only undergo mitosis as needed, such as after an injury.

Interphase is broken into three stages: **Gap 1 (G_1), Synthesis (S),** and **Gap 2 (G_2)**. During the gap (or growth) stages the cell grows larger, synthesizes proteins, and makes any needed organelles. During the synthesis stage the DNA is replicated. DNA replication begins when an enzyme, **helicase**, breaks the hydrogen bonds between the nitrogen bases of a double stranded DNA molecule. **DNA polymerases** (enzymes that build DNA polymers) then match complimentary bases to the unzipped strands, forming two new DNA strands from the original split strand (Figure 4.6). DNA replication occurs during the S phase of interphase, while the DNA is packed into the nucleus.

There is a 3' and 5' directionality to DNA strands based on the chemical structure of the sugar-phosphate backbone Figure4.6). DNA polymerase always builds new DNA strands in the 5' to 3' direction. The strand being formed in the direction the

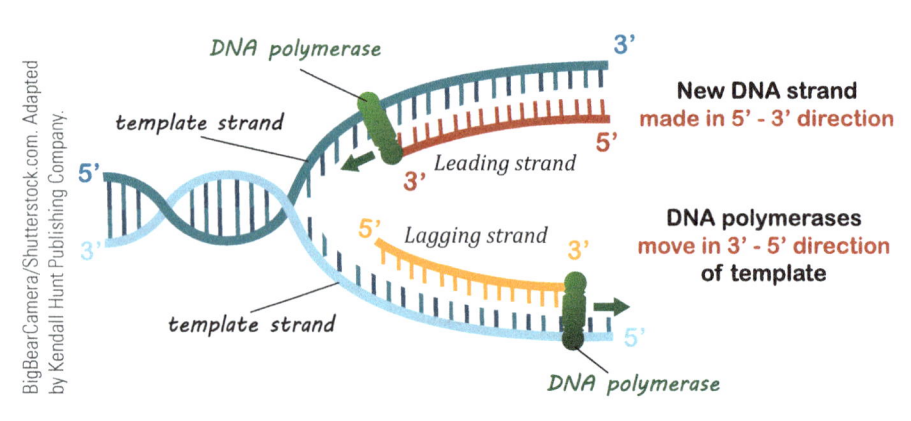

Figure 4.6 DNA Replication occurs when the enzyme helicase (not shown) breaks the hydrogen bonds between DNA strands, and then DNA polymerase rebuilds each strand forming two copies of the original. Note the leading strand is shown in red, and the lagging strand in yellow.

DNA is opening is called the **leading strand**, and the strand being built in the opposite direction is the **lagging strand**. Lagging strands are necessarily formed in pieces. An additional enzyme, called **DNA ligase**, connects these pieces, which are called **Okazaki fragments**.

Mitotic Phases

The phases of mitosis are typically defined as **prophase**, **metaphase, anaphase** and **telophase**. Note that interphase is *not* a phase of mitosis (Figure 4.7). Some texts, also include *prometaphase*, a step between prophase and metaphase as an additional mitotic phase. Consult your textbook for the details for each step. They are *briefly* summarized in table 4.1, but you may be expected to know more.

Mitosis

Interphase Prophase Metaphase Anaphase Telophase

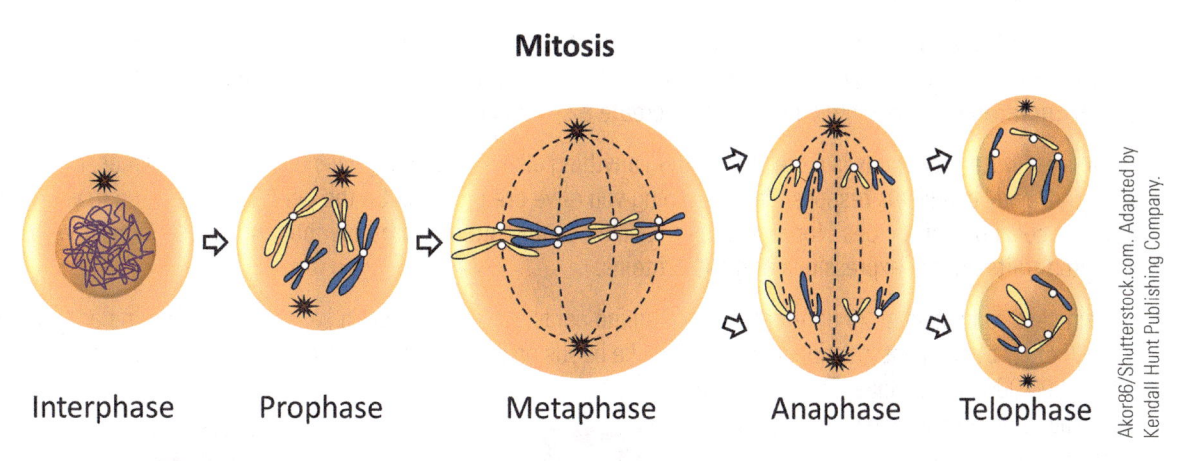

Akor86/Shutterstock.com. Adapted by Kendall Hunt Publishing Company.

Figure 4.7 The stages of interphase and mitosis.

Table 4.1 Brief summary of the events during interphase and the phases of mitosis.

Phase		Major Events
Interphase		Cells carry out normal cell functions. DNA replication occurs. (Not a phase of mitosis!)
Mitosis	Prophase	Nuclear membrane breaks down. Chromatin condenses into chromosomes. Centrioles migrate to opposite poles of the cell.
	Metaphase	Spindle fibers from the centrioles attach to centromeres of the chromosomes, arranging them on the equatorial disk of the cell.
	Anaphase	Spindle fibers pull sister chromatids apart. Each chromatid, once separated, becomes a single chromosome, doubling the number of chromosomes in the cell temporarily. (Though not part of mitosis, cleavage furrow forms; cytokinesis begins.)
	Telophase	Chromosomes clump into two new nuclei. Nuclear membranes form around chromosomes. Chromosomes turn back into chromatin. Cytokinesis continues until daughter cells separate.
Interphase		Cells must grow, synthesize new organelles, and replicate the DNA again.

Mitosis in Plants and Animals

Animal cells lack a rigid cell wall. When animal cells divide, the cells separate by forming a **cleavage furrow**, which looks much like tightening a string around a balloon. During cytokinesis, fibers around the circumference of the cell tighten and pinch the cell apart, eventually producing two "daughter cells" that are each half the size of the original cell. Plant cells are surrounded by rigid cell walls. Plant cells must instead create a new cell wall between the new nuclei being formed in telophase. This new cell wall is called the **cell plate** as it begins to form. It often starts in the center of the cell and grows wider until it separates the cell into two roughly equal parts. Most plant species also lack centrioles. They have a *centrosome* area that produces microtubules that form the **spindle fibers** of the spindle. But the dark spots indicating **centrioles** that are visible at the end of the spindle in animal cells are not present in plant cells.

Protein Synthesis

Protein synthesis is the process by which the information in a DNA molecule is used to synthesize proteins. The **central dogma** of molecular biology states that information in the genes of DNA is **transcribed** (rewritten) into RNA which is then **translated** (changed into a new form) to create a specific protein.

Aside from water, your body is more protein than anything else. Additionally, most hormones, enzymes, neurotransmitters, and other body chemicals responsible for keeping you alive are proteins. So the genetic information in your DNA is really a giant "cookbook of recipes" for making the specific combination of proteins that makes you exactly who you are. Our inherited traits are the expressions of these proteins.

Proteins are composed of 20 different kinds of amino acids. As you learned in Lab Exercise 3, amino acids are connected to each other by covalent (peptide) bonds, much like beads on a string, to form proteins. Proteins may have anywhere from 300 to 300,000 amino acids. The order of the amino acids determines the shape and function of the protein.

Ribonucleic acid (RNA) molecules are also made up of nucleotide subunits similar to DNA nucleotides, with a few significant differences. RNA nucleotides contain **ribose**, a different pentose sugar. RNA nucleotides also lack thymine (T) nitrogen bases, and instead replace it with **uracil** (U). Uracil and ribose molecules are found only in RNA, not DNA. RNA is single stranded, unlike double-stranded DNA. And RNA molecules are much shorter in length than the DNA in chromosomes. See Figure 4.8.

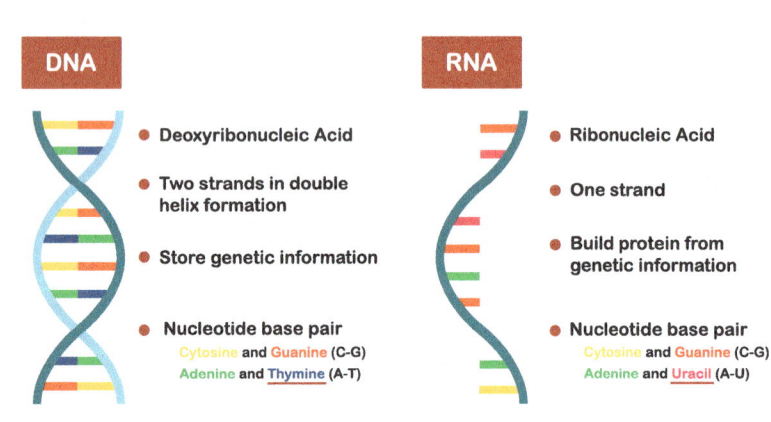

DNA
- Deoxyribonucleic Acid
- Two strands in double helix formation
- Store genetic information
- Nucleotide base pair
  Cytosine and Guanine (C-G)
  Adenine and Thymine (A-T)

RNA
- Ribonucleic Acid
- One strand
- Build protein from genetic information
- Nucleotide base pair
  Cytosine and Guanine (C-G)
  Adenine and Uracil (A-U)

Figure 4.8 Comparison of DNA and RNA.

RNA molecules are crucial for protein synthesis. There are three types of RNA based on what they do:

- **mRNA** acts as a *messenger* that carries genetic information from the cell's nucleus to a ribosome, the organelle that assembles amino acids into protein.
- **tRNA** acts as a *translator* or *transferring* molecule that is pre-attached to a specific amino acid. The intracellular enzyme *aminoacyl tRNA synthase* continuously matches tRNA molecules to specific amino acids.
- **rRNA**, with some proteins, forms the structure of a *ribosome*. It is more important as a structural component than as a functional one.

You can combine similar ingredients in many different ways to make various types of dishes. This is also true for assembling chains of twenty different kinds of amino acids. A cell "reads" the order of the nitrogen bases in its DNA and then creates the exact proteins it needs at the right time. How it does this is amazing and a major aspect of molecular biology research.

Figure 4.9 Transcription occurs in the nucleus as RNA polymerase opens a portion of the DNA double helix, assembles a "complementary" (matching) RNA strand from one side, and then closes the DNA again. The strand of DNA that is used to make mRNA is called the **template** (or **anti-sense**) **strand**. The strand that is not used to make mRNA is called the **sense strand**.

Transcription

Transcription refers to the process of opening the ladder-shaped DNA strand, and "rewriting" the information in a section of the DNA into the form of mRNA. The enzyme **RNA polymerase** does this in the nucleus when the cell needs to generate a certain protein. See Figure 4.9. (Note that "*polymer-ase*" refers to an enzyme that makes a polymer.)

In DNA, nitrogen bases are paired by hydrogen bonds. RNA polymerase breaks those bonds and facilitates the formation of new hydrogen bonds between free-floating RNA nucleotides and the DNA nucleotides of the opened strand. Recall that there are no RNA nucleotides containing thymine. Instead, RNA nucleotides containing the nitrogen base uracil are matched to the adenine nitrogen bases in the DNA strand. Then the RNA is released from the DNA as a new single-stranded nucleic acid molecule. The DNA closes back into the double helix. The mRNA strand is free to diffuse out of the nucleus.

Translation

Translation is the process of matching the tRNA up to the mRNA so that the right amino acids can be combined together to make the right protein. This is done by a ribosome out in the cytoplasm of a cell. (Some ribosomes are floating in the cytoplasm, and some are attached to the *endoplasmic reticulum*. The ribosome reads the mRNA strand 3 nitrogen bases at a time. These 3-base sections of mRNA are called **codons**. The ribosome matches a corresponding 3-base section of tRNA, called an **anticodon**, to each codon.

Recall that amino acids are paired up with pieces of tRNA. If the nitrogen bases of the mRNA codon and tRNA anticodon correspond (by pairing complimentary bases, A with U and C with G, the ribosome joins the new amino acid with previous ones, like adding beads on a string. See Figure 4.10. The

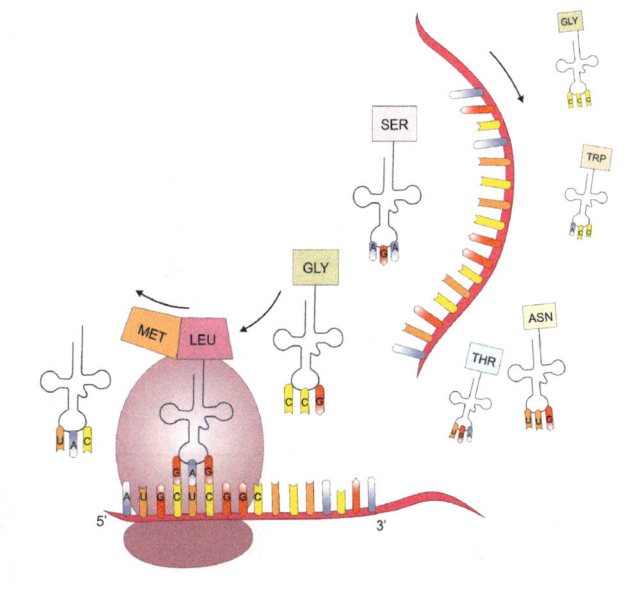

Figure 4.10 Translation of RNA into a protein. The three tRNA nitrogen bases must correspond to the three mRNA nitrogen bases. The ribosome moves down the mRNA strand, assembling the protein as it goes. In this case the amino acid chain is only composed of MET (methionine) and LEU (leucene). The next amino acid will be GLY (glycine) as the CCG anticodon of the tRNA matches with the GGC codon of the mRNA.

polypeptide (unfinished protein) is assembled by the ribosome according to the template provided by the mRNA until it reaches a STOP codon.

There are a number of events that occur during protein synthesis that affect the resulting protein. Between transcription and translation, the mRNA is often "edited" or processed. Intracellular enzymes or the Golgi apparatus may also modify the protein (Figure 4.11). The ways in which cells regulate protein synthesis before transcription, between transcription and translation, and after translation are ongoing areas of research in molecular biology.

Proteomics is a relatively new field in biology primarily concerned with understanding the shape and function of proteins in cells. Websites like the *Basic Local Alignment Search Tool (BLAST)* (https://blast.ncbi.nlm.nih.gov/Blast.cgi) allow researchers to upload gene sequence data and find proteins associated with that sequence.

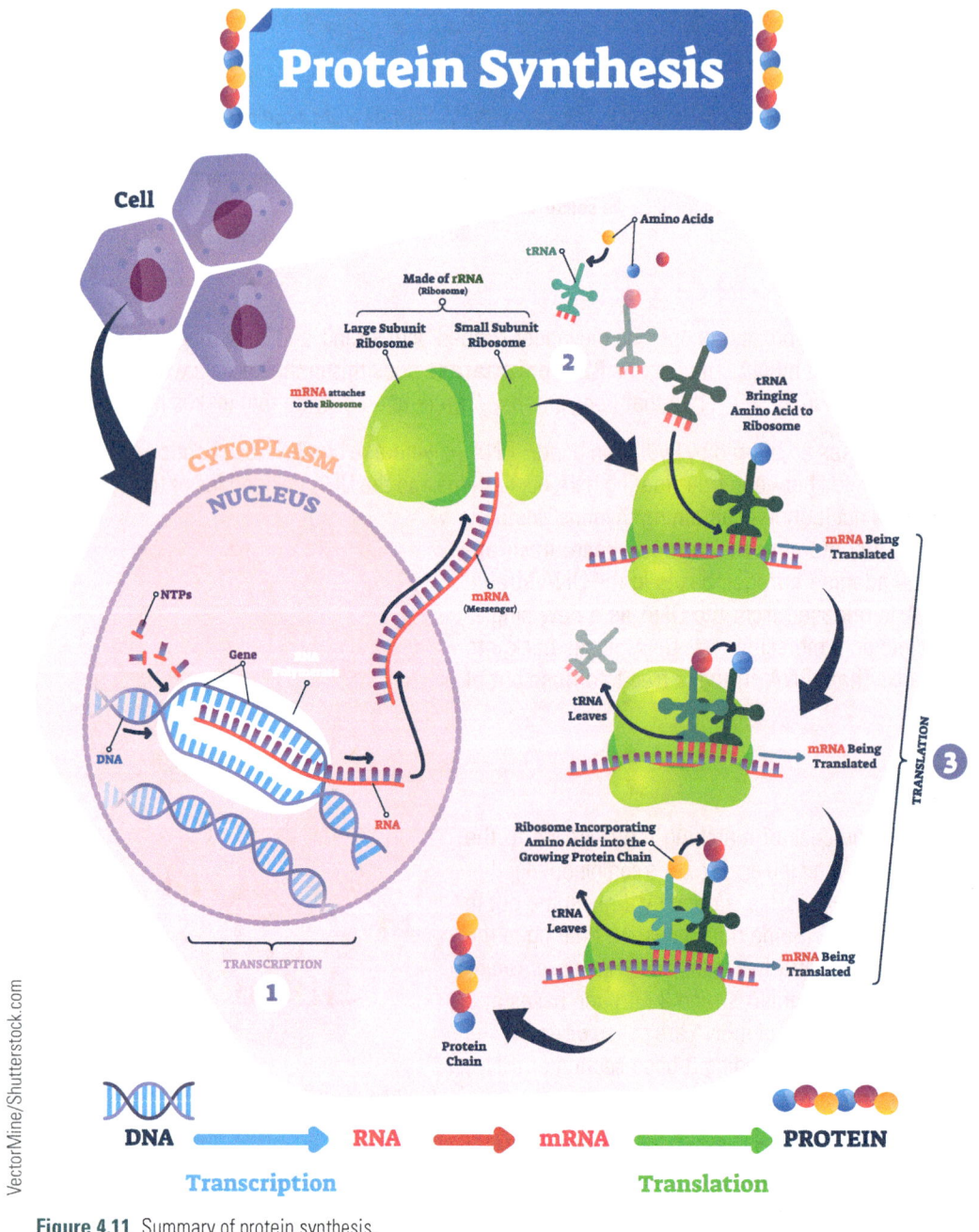

Figure 4.11 Summary of protein synthesis.

Lab Skill 4.1 Using a Benchtop Centrifuge

A **centrifuge** is a piece of lab equipment that can be used to separate the parts of a mixture. Biology laboratories often have one or more centrifuges. They range in size from small, hand-held versions, to bench top-sized models (Figure 4.12) to large floor models. A centrifuge spins the sample so that the bottom of the tube points away from the axis of the spinning motion. This applies centrifugal force to the particles in the mixture. The densest particles are forced outward, toward the bottom of the tube (Figure 4.13). When finished, the mixture will be separated into layers, with the densest materials located at the bottom of the tube, and the least dense at the top.

Steps for using a benchtop centrifuge

4. The sample must be placed in a *centrifuge* tube. Centrifuge tubes typically have a more pointed bottom than a test tube. They come in many different sizes depending on the size of the centrifuge. It is important that the tube properly fit the rotor of the centrifuge so that it doesn't come loose during spinning.
5. When loading the centrifuge, each tube should have another tube opposite it, or the tubes should be as evenly spaced out as possible. If only centrifuging one tube, prepare a tube equally full of water to serve as a balance for your sample tube. If the samples are not balanced, the centrifuge will be unstable and may bounce around or even "leap" off the lab bench.
6. Ensure the samples are snug and appropriately spaced in the centrifuge. If it has an inner lid, make sure it is in place. This lid prevents any liquids or debris from flying out of the centrifuge as it spins.
7. Close the centrifuge and set the speed and timer as required by the protocol. The centrifuge speed is measured by revolutions per minute (RPMs) or as a relative centrifugal force (RCF) which is a factor of gravitational force measured in Gs. In some cases unit used by the centrifuge will be RPM, but in others it may be in G. Some centrifuges can use either unit. Although you will not need it in this lab, the G-force produced by a centrifuge with a specified rotor radius and speed in RPMs can be calculated as follows:
$G = 1.118 * 10^{-5} * r * (RPM)^2$ (Note that r is the radius of the rotor.)
In some cases the centrifuge may only be needed for a few seconds to "spin down" a sample. The "pulse" setting of a centrifuge is used for this.

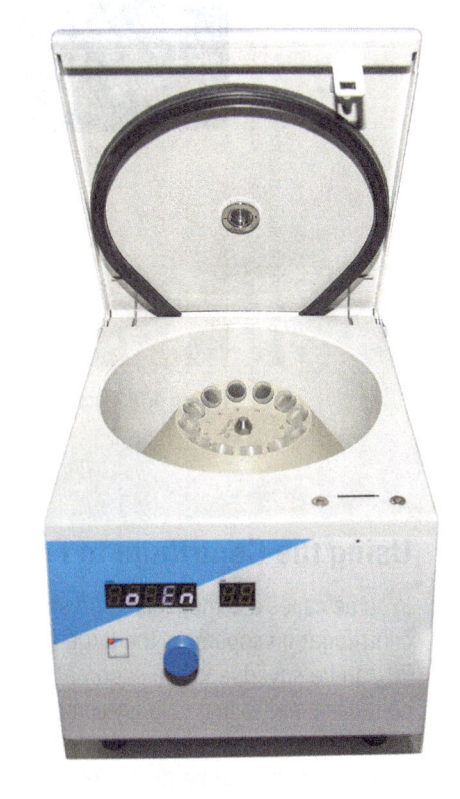

Figure 4.12 A typical benchtop centrifuge.

CENTRIFUGATION

Figure 4.13 The results of centrifugation.

8. Once the sample has been spun at the desired speed and for the necessary length of time, the centrifuge will stop automatically. It may take a few minutes for the rotor to come to rest. At that point the lid will unlock so the sample may be removed.
9. The result should appear as shown in Figure 4.14, where the bottom of the tube contains the **pellet** of dense material, and the upper part contains the **supernatant**, or remaining dissolved materials. The longer a sample is run at high speeds, the more particles collect in the pellet, so if the pellet does not appear substantial enough, you can repeat centrifugation for a longer time or at a higher speed.

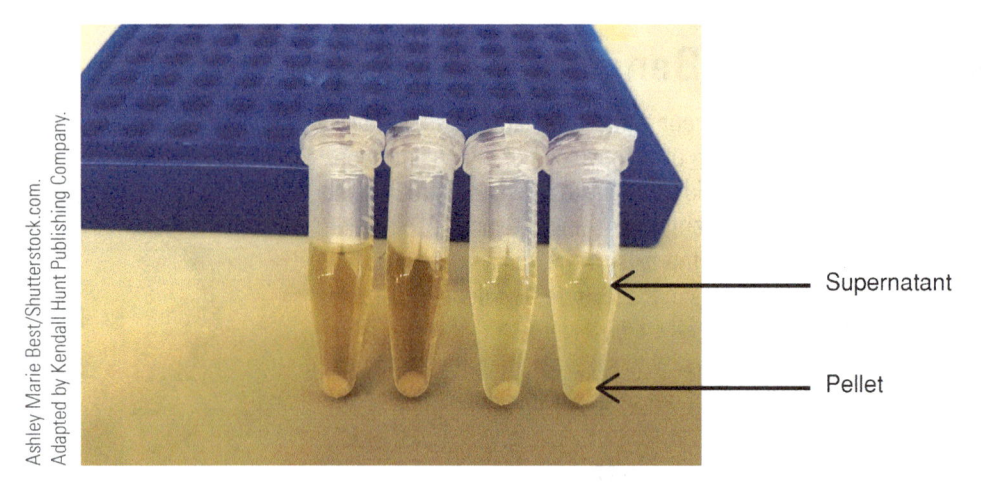

Figure 4.14 Bacteria pellets in centrifuge tubes after centrifugation.

Using the Centrifuge to Fraction a Cell

In some cases a complex mixture such as parts of cells, can be fractioned (divided) by *differential centrifugation*. The first centrifugation separates the large particles from the smaller particles still dissolved in the supernatant. But then the pellet can be saved, and the supernatant removed and centrifuged again. That process can be repeated so that each pellet contains smaller and smaller components of the original mixture, separated by size (Figure 4.15). It is always important to know which part of the sample you wish to keep. In some cases the pellet will contain the desired materials and the supernatant should be discarded. In others the supernatant will contain the desired materials and the pellet should be discarded. In some cases the components in both the supernatant and pellet are needed; they just need to be separated.

DIFFERENTIAL CENTRIFUGATION

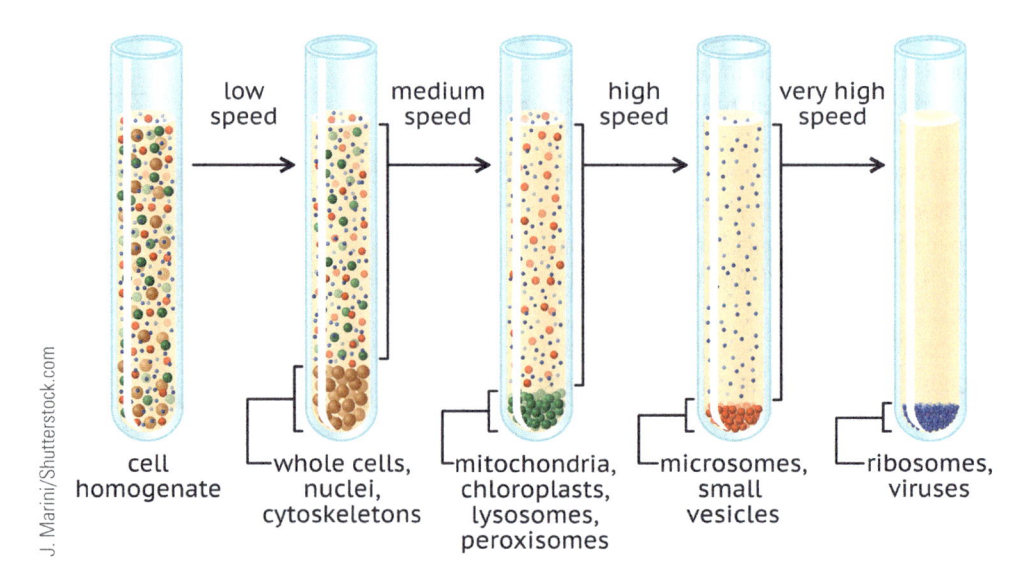

Figure 4.15 Differential centrifugation. The first tube contains blended cells. After centrifugation, the largest structures are in the pellet as shown in the second tube. The supernatant is transferred to a new tube and centrifuged again, resulting in the third tube with midsized components in the pellet. Further centrifugation of the supernatant can separate even smaller components.

Problem to Solve 4.1 Can I Recognize the Phases of Mitosis?

Identifying microscopic images of cells is a common part of biology. Noticing when cells are undergoing mitosis is particularly important when viewing potentially cancerous tissue. Cancer is simply a failure of the cell's control mechanisms to stop mitosis. When this system fails, cells may divide rapidly and form a *tumor* (mass of undifferentiated cells).

You will not be looking for cancer in this lab activity, but you will be looking for cells in each stage of mitosis. Practice with the mitosis models first to ensure you can recognize the characteristics of cells in the different stages of the cell cycle.

Once you are comfortable identifying the stages of mitosis on the models, observe actual animal and plant cells.

The animal cells you will observe are from recently fertilized fish eggs that were rapidly dividing and growing into an embryo. The cluster of cells was frozen, sliced, and stained. Do not confuse the *clusters* of cells with single cells. You will need to observe the clusters of cells on high power to clearly see individual cells. When you do, you will notice that most cells are in interphase. But as you look around on the slide, you should see some cells in various stages of mitosis. The cells have been stained so that the DNA should appear darker than the rest of the slide. You may notice that some cells seem to be empty and have no nucleus. Others may not appear normal or do not appear as expected. Realize that this is because the cells were sliced very thinly, and sometimes this damages the cells or produces cells that appear to be empty. Instead of trying to identify every cell, try to find *ideal* cells that represent each stage of the cell cycle.

The second slide you will observe is of plant cells at the tip of a root. These cells will appear rectangular, as is normal for plant cells. However, some cells will clearly be undergoing mitosis. Growth in plants occurs at the tips of roots and stems. So scan the slide until you find the cells near the tip of the root. Most mitosis will be occurring there.

Date: 9/13/2021

Title
Title: Observing mitosis in plant and animal slides

Purpose
Goal: Be able to correctly identify specific phases of mitosis in prepared slides of animal and plant cells.

Protocol

Materials:
Slide labeled "fish blastula"

Slide labeled "onion (or Allium) root tip"

Microscope

Protocol:

1. Review Lab Skill 5.1 Properly Using a Light Microscope if necessary.

2. Observe the "fish blastula" slide. Work up to the high power objective and carefully scan the cells. Find cells in each stage of the cell cycle and sketch them. Label the sketch and indicate magnification. Do this for interphase and all four phases of mitosis.

3. Repeat the process with the onion (Allium) root tip slide.

Date: 9/13/2021

Laboratory Exercise 4 Molecular Biology I: Understanding and Isolating DNA

Continued from previous page.

Data:

Animal cells: Fish blastula

Phase: __Interphase__
Magnification: __400__

Phase: __Prophase__
Magnification: __400__

Phase: __Metaphase__
Magnification: __400__

Phase: __Anaphase__
Magnification: __400__

Title

Purpose

Continued from previous page.

Protocol

Data Continued:
Animal Cells: Fish blastula

Phase: telephase
Magnification: 400

Plant cells: Onion (Allium) root tip:
Phase: Interphase
Magnification: 400

Data continued:

Plant cells: Onion (Allium) root tip

Phase: Prophase
Magnification: 400

Phase: metaphase
Magnification: 400

Phase: anaphase
Magnification: 400

Phase: telephase
Magnification: 400

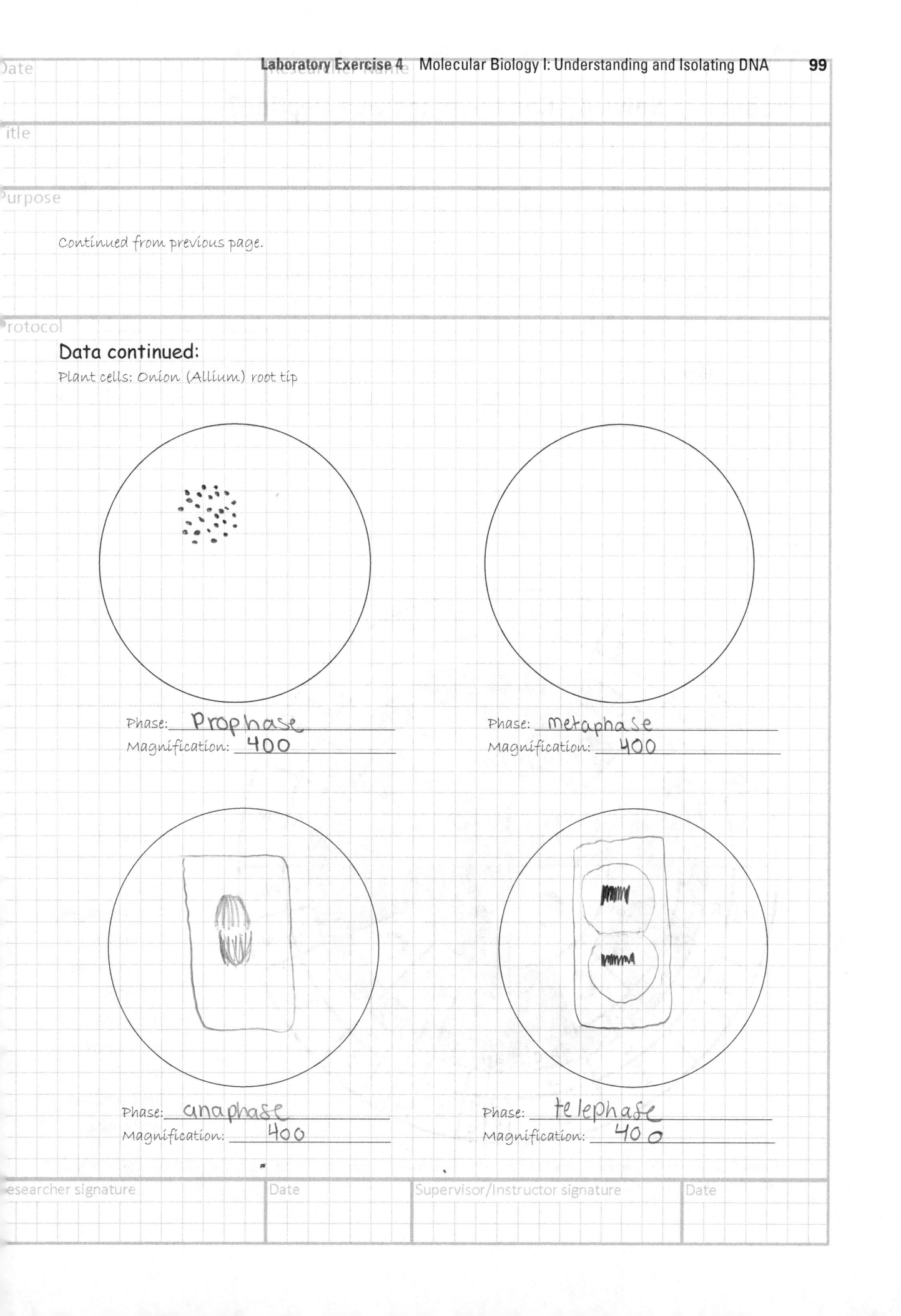

100 Laboratory Exercise 4 Molecular Biology I: Understanding and Isolating DNA

Problem to Solve 4.2 Can I Synthesize a Peptide?

The following protocol asks you and your lab partner to work together to simulate protein synthesis by transcribing and translating a partial section of DNA on paper. A double strand of DNA is below. In this case the top strand of DNA will serve as the *template* strand. The blue lines represent the deoxyribose-phosphate "backbones" of the DNA strand. The letters represent the DNA nitrogen bases. The dots between the letters represent hydrogen bonds. Your task is to decode its peptide.

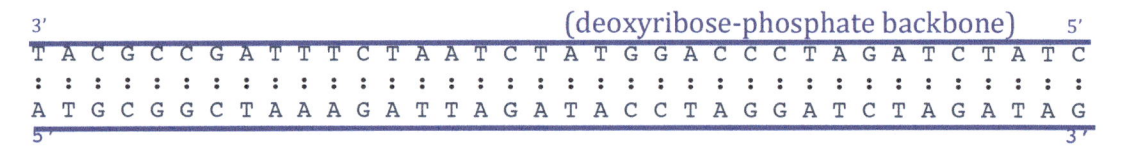

```
3'                          (deoxyribose-phosphate backbone)            5'
T A C G C C G A T T T C T A A T C T A T G G A C C C T A G A T C T A T C
: : : : : : : : : : : : : : : : : : : : : : : : : : : : : : : : : : : :
A T G C G G C T A A A G A T T A G A T A C C T A G G A T C T A G A T A G
5'                                                                      3'
```

Use Figure 4.16 below to translate *codons* into *amino acids*. For example, to translate the codon *AUG*, Find the first base, A, in the center of the chart. Then, move outward (in this case to the left) to the purple U that is adjacent to the big blue A. Then continue to move outward to the orange G that is adjacent to the U. Continue to move outward to identify the amino acid. In this case it is Met (methionine).

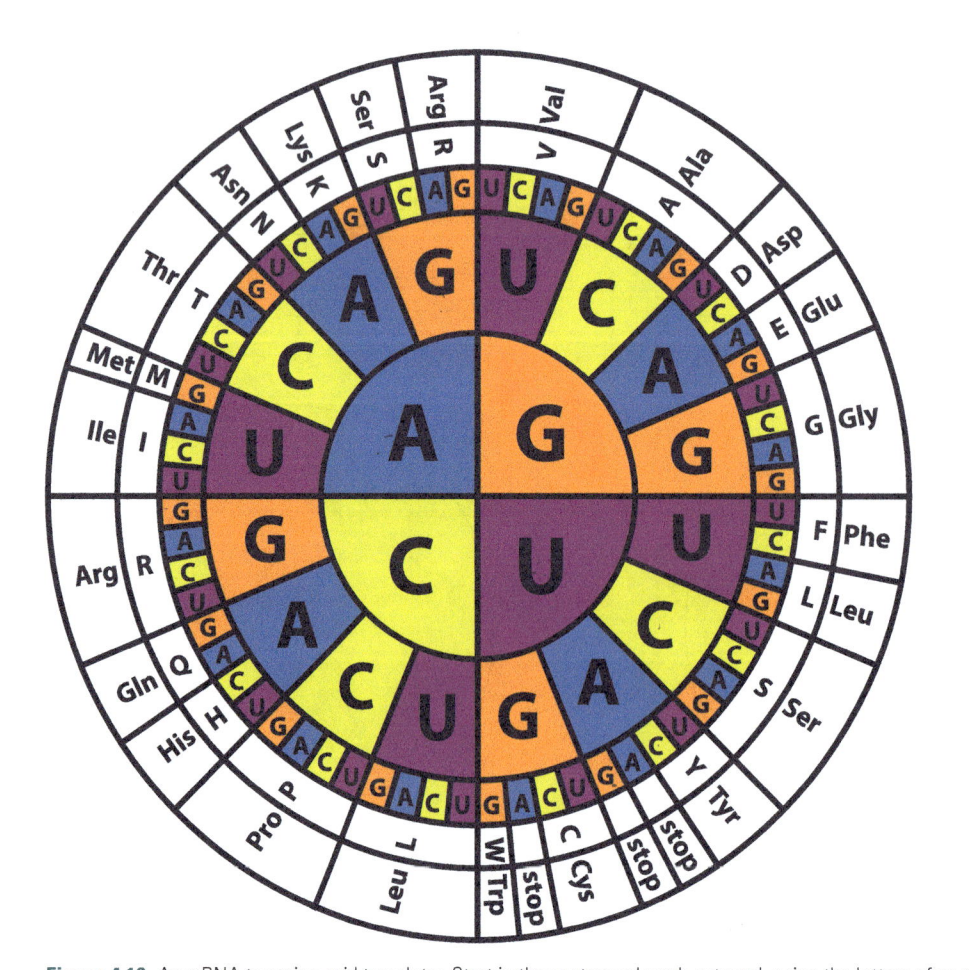

The 20 Amino Acids
Ala: Alanine
Cys: Cysteine
Asp: Aspartic acid
Glu: Glutamic acid
Phe: Phenylalanine
Gly: Glycine
His: Histidine
Ile: Isoleucine
Lys: Lysine
Leu: Leucine
Met: Methionine
Asn: Asparagine
Pro: Proline
Gln: Glutamine
Arg: Arginine
Ser: Serine
Thr: Threonine
Val: Valine
Trp: Tryptophan
Tyr: Tyrosine

Figure 4.16 An mRNA to amino acid translator. Start in the center and work outward, using the letters of an mRNA codon to get to an abbreviation of an amino acid.

Title

Title: Decoding the information in a partial DNA strand

Purpose

Goal: Demonstrate transcription and translation. Determine the correct sequence of amino acids encoded in this section of DNA. Correctly use an mRNA to amino acid translation cart.

Protocol

Materials:
pen or pencil, scrap paper, decoder chart.

Protocol and Data:

1. *Transcribe* the DNA strand. It is provided again here. The red line represents a ribose-phosphate backbone.

in nucleus

```
3'                                    (deoxyribose-phosphate backbone)        5'
T A C G C C G A T T T C T A A T C T A T G G A C C C T A G A T C T A T C
: : : : : : : : : : : : : : : : : : : : : : : : : : : : : : : : : : :
A U G C G G C U A A A G A U U A G A U A C G U G G A U C U A G A U A G
5'                                    (ribose-phosphate backbone)              3'
```

2. *Translate* the codons of this mRNA strand into amino acids. In order to do this, first place a slash after every three nitrogen bases in the mRNA strand. This will separate the mRNA into 3-letter codons.

Use the decoder chart (Figure 4.16) to list the amino acids for these codons on the blanks below. (Use first 3 letters of the amino acid's name to fill in one blank.)

Met Arg Leu Lys Ile Arg Tyr Val Gly Ser Arg Stop

Laboratory Exercise 4 Molecular Biology I: Understanding and Isolating DNA

Continued from previous page.

3. Ribosomes do not use decoder charts to translate mRNA into proteins. Instead they use tRNA molecules. This figure represents the first 3 tRNA molecules used to make the previous peptide. Place the correct amino acids, anticodons, and codons, in the yellow boxes.

Amino acid	Amino acid	Amino acid
Met	Arg	Leu

anticodon	anticodon	anticodon
UAC	GCC	GAU

codon	codon	codon
AUG	CGG	CUA

(Ribose-phosphate backbone)

Continued from previous page.

Conclusion:

Does protein synthesis always work perfectly? Are enzymes reliable? Recall Lab Exercise 3. What might happen if the cells got very hot? We often think that heat kills cells, but <u>why</u>?

Do enzymes always match the nitrogen bases accurately? Imagine RNA polymerase failed to insert one mRNA nitrogen base during transcription. Delete the first C in the mRNA sequence in step 1 of the protocol. (This is called a frame shift deletion.) Ribosomes always match tRNA molecules to mRNA molecules three bases at a time. If one base is "lost" in transcription, the codons to the right of the deletion change. Redefine the codons and decode them. List the new amino acids that will be produced after this deletion.

Aug,

How did the one missing nitrogen base impact the peptide?

What if the C was not deleted but instead replaced with an A? How would that affect the peptide?

Problem to Solve 4.3 Can I Collect and Isolate My DNA?

Molecular biologists work with DNA in many ways. You will be learning more about them in the following lab exercises. Table 4.2 lists some of them. But to manipulate or analyze the DNA, it must be first separated from the rest of the components cell without being destroyed in the process.

Recall that cells are complex structures composed of nucleic acids, phospholipids, proteins, and carbohydrate. Some proteins are enzymes that can damage or hydrolyze the DNA molecules once the DNA is released from the nucleus. DNA molecules are fragile. The technique for separating DNA from other substances in the cell must preserve the integrity of the DNA while effectively disabling destructive enzymes, breaking down other macromolecules, and providing a way to physically isolate the DNA from the other compounds.

Table 4.2 A Few Ways Isolated DNA May be Used

Identification techniques	Restriction Fragment Length Polymorphism (RFLP)	*Restriction enzymes* are used to "cut" the DNA at specific locations, and the fragments are separated by *electrophoresis* (in a gel subjected to an electric field). The pattern produced by the DNA fragments can be compared with other DNA samples. DNA fingerprinting is comparing DNA collected at a crime scene with the DNA of a suspect. (This will be covered in Lab Exercise 7.)
	Paternity Testing	DNA of a male can be cut with restriction enzymes, separated by electrophoresis, and compared to a child's DNA. An analysis of the male's DNA fragments can clearly indicate if he is *not* the father. If the male's DNA patterns are similar to the child's, further analysis can be done to determine the probability that he is the father.
	Ancestry	DNA can be cut with standard restriction enzymes and the fragments compared to a database of standard fragmenting patterns. Different populations of humans around the world have historically maintained certain features in their DNA. The number of these features that show up in a person's DNA can suggest their ancestry.
Sequencing techniques	Testing for Genetic Disorders	Some diseases, such as Hemophilia, result when the code for a necessary protein doesn't work. DNA can be taken from parents to determine the likelihood that they can pass on such a gene to their offspring, and fetuses can be tested to determine if they have a genetic disease even before birth.
	Genomic Sequencing	The DNA of the members of any particular species is very similar. Over the last few decades biologists have sequenced the DNA of many organisms including humans. Databases of the gene sequences are readily available, and the sequencing of isolated DNA has become routine. Biologists can now enter the DNA sequence of a specimen into a database to ensure correct identification of a species.
DNA manipulation	Genetic Modification and CRISPR	Genes are simply fragments of DNA, so if a desired gene of an organism is isolated and cut, a selected fragment containing a desired gene can then be inserted into another cell. The recombined (or recombinant) DNA effectively alters the genetic instructions for that cell. This has been done with many organisms, especially plants, to cause them to express better traits.
	Cloning	DNA can be isolated from one organism and then inserted into a fertilized egg that has had its nucleus removed. The fertilized egg can grow normally, but the organism produced will be a clone of the DNA donor. This has been done with many organisms, including Dolly, the famous cloned sheep. It is not legal to clone humans, but the technology is there.

Most molecular biology experiments require that the DNA of interest be isolated from the rest of the cellular components. The following protocol is one standard method for isolating DNA, but there are many others. It is important to understand the underlying principles involved, shown in Figure 4.17 so that you can effectively isolate DNA regardless of the specific chemicals or steps called for in whichever protocol you decide to use.

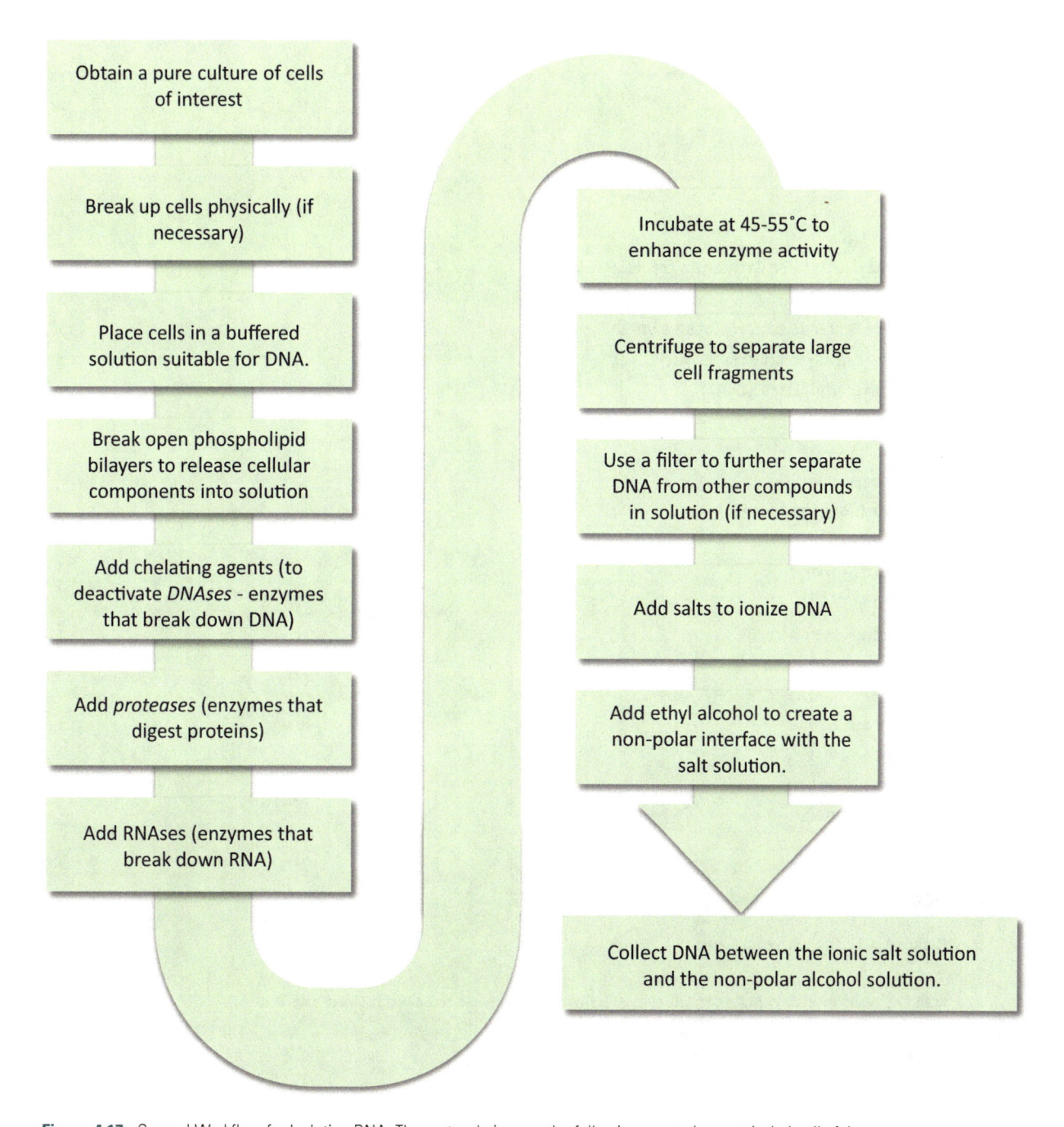

- Obtain a pure culture of cells of interest
- Break up cells physically (if necessary)
- Place cells in a buffered solution suitable for DNA.
- Break open phospholipid bilayers to release cellular components into solution
- Add chelating agents (to deactivate *DNAses* - enzymes that break down DNA)
- Add *proteases* (enzymes that digest proteins)
- Add RNAses (enzymes that break down RNA)
- Incubate at 45-55°C to enhance enzyme activity
- Centrifuge to separate large cell fragments
- Use a filter to further separate DNA from other compounds in solution (if necessary)
- Add salts to ionize DNA
- Add ethyl alcohol to create a non-polar interface with the salt solution.
- Collect DNA between the ionic salt solution and the non-polar alcohol solution.

Figure 4.17 General Workflow for Isolating DNA. The protocol given on the following pages does not include all of these steps.

Title: Isolating DNA from Cellular Components

Goal: To learn the protocol for isolating DNA and to create a sample of my own DNA isolated from my cheek cells.

Materials:
(Per student)
- 1 small test tube
- 1 small centrifuge tube with snap top (1.5 ml)
- 1 microcentrifuge tube with a snap top (0.5 ml)
- 1 package of 2 sterile cotton tipped swabs
- 3 graduated transfer pipettes (3 ml)
- 2 small transfer pipettes (1 ml)
- lanyard

(Per table)
- 1 small beaker
- 1 small graduated cylinder
- 1 test tube rack
- 1 permanent marker
- Labeling tape
- 1 bottle methylene blue

(Per class)
- Lysis buffer
- Protease solution
- NaCl solution
- Ethyl alcohol (kept at -4 °C)
- Water bath at 55 °C
- Microcentrifuge

Protocol:
1. Ensure all glassware is clean.
2. Use the permanent marker to label the 1.5 ml centrifuge tube with your intials.
3. Label the test tube with tape and your initials.

Continued from previous page

4. Use a clean 3 ml transfer pipette to transfer 1 ml of lysis buffer to the 1.5 ml centrifuge tube.

5. Remove one cotton swab from the pack (taking care not to contaminate them) and vigorously swab the inside of your cheek. Then place the swab in the lysis buffer to dislodge the cells. Squeeze the liquid out of the swab by pressing it against the side of the tube, and then discard it in the trash.

6. Repeat the same process with the second swab.

7. Go to your instructor to have three drops of protease added to your centrifuge tube.

8. Cap the tube and invert to mix. The lysis buffer contains sodium dodecyl sulfate, a detergent that disrupts cell membranes, and other buffering salts. Protease breaks up proteins including the histones the DNA is wrapped around. This helps unspool the DNA.

9. Carefully place the labeled centrifuge tube in the 55°C water bath. This helps speed the action of the protease to maximize unspooling.

10. After 10 minutes, remove the centrifuge tube from the water bath.

11. Place the centrifuge tube in the benchtop centrifuge. Centrifuge the sample for 5 minutes at 14,000 RPM. (You can do this with several other students at once. Be sure all centrifuge tubes are evenly spaced and balanced.)

12. After centrifugation is complete, remove the tube and use a clean 3 ml transfer pipette to transfer the supernatant (not the pellet) to your labeled test tube.

13. Go to your instructor to have three drops of NaCl solution added to your test tube. This makes the solution ionic and makes the DNA less soluble.

14. Obtain 1 ml of very cold ethyl alcohol in a small beaker.

15. Very carefully, while holding test tube very still at a 45-degree angle, use a clean 3 ml transfer pipette to slowly drip 1 ml of cold ethyl alcohol down the inside of the test tube. Do this slowly. You should be able to see a layer of alcohol form on top of the salt/DNA solution.

Date

Title

Purpose

Continued from previous page

Protocol

16. Slowly and carefully place the test tube in the test tube rack and allow it to sit for at least 5 minutes. Be careful not to disturb the layered solutions. Alcohol is non-polar, and the DNA will become trapped between the ionic salt solution and the non-polar alcohol.

17. After a few minutes you should see white stringy material forming at the interface of the alcohol and the salt solution.

18. Use a small 1 ml transfer pipette to add one drop of methylene blue to the solution. It will pass through the alcohol and diffuse into the salt solution, but it will also stick to the DNA and stain it dark blue. Allow the test tube to sit undisturbed for at least 3 minutes to ensure the DNA is stained.

19. Use the remaining small 1 ml transfer pipette to extract the DNA. You will need to squeeze the bulb slightly, insert the tip of the pipette as close to the darkest blue area of the tube as possible, and carefully release the pressure on the bulb. This should draw in a small amount of blue DNA.

20. Transfer the blue-stained DNA to the 0.5ml centrifuge tube. You may repeat the previous step if you want to obtain more of your DNA sample.

21. Cap the tube. Add a lanyard if you desire. You have successfully isolated a sample of your DNA and stained it with methylene blue.

Observations and Conclusions:

Name: _____   Date: _____

POST-LAB QUIZ 4: What have you learned?

1. What is the fundamental role of DNA in cells? (Be specific.)

2. If mitosis continually replaces old cells, why do we age? Provide a good hypothesis.

3. What phase of the cell cycle is shown in the cells indicated?

   A:
   B:
   C:
   D:
   E:

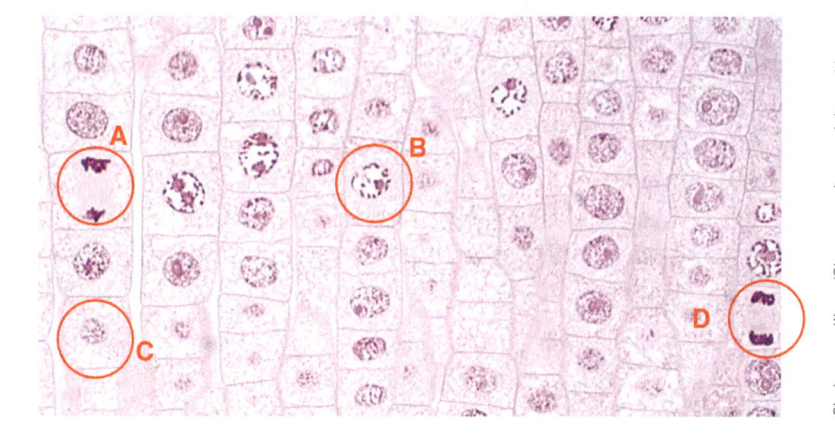

   F:
   G:
   H:

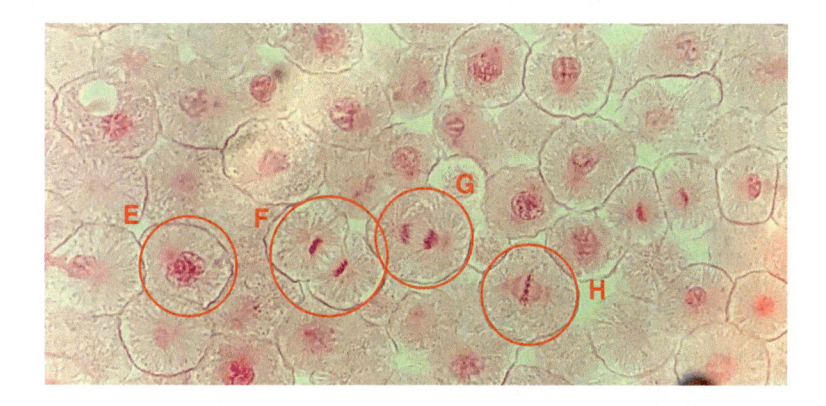

4. Why do you think the brain is virtually amitotic? Provide a good hypothesis.

5. Describe two observable differences in plant mitosis and animal mitosis.

6. *Replicate* the partial DNA strand: $^{(3')}$ T C A T A A C G C G A C $^{(5')}$

7. *Transcribe* and *translate* the partial DNA strand provided in question 6 above. (Use Figure 4.16).

    Amino acids: _____ _____ _____ _____

8. Which would likely have a greater negative consequence for an organism, if *RNA polymerase* made an error during *transcription*, or if *DNA polymerase* made an error during *replication*? Why?

9. When isolating your DNA, what specifically was in the *pellet* after centrifugation? What was in the *supernatant*?

10. In the DNA isolation experiment, the following chemicals were added to the solution. What specific purpose did each serve?
    a.) Lysis buffer
    b.) Protease
    c.) Methylene blue
    d.) NaCl
    e.) Ethyl alcohol

Laboratory Exercise 5

Microscopy: Observing Life at the Cellular Level

Concepts to Understand

After completing this lab, the student should be able to:

1. Name and give the function of the parts of a binocular light microscope.
2. Understand the difference between the major types of prokaryotic and eukaryotic cells.
3. List and give the function of prokaryotic and eukaryotic cell structures.

Skills to Learn

After completing this lab, the student should be able to:

1. Properly use a microscope to view prepared slides.
2. Calculate the magnification of any microscope ocular and objective lens combination.
3. Estimate the size of microscopic objects.
4. Use the oil immersion objective to view bacteria.
5. Make a wet mount slide of plant and animal cells.
6. Visually identify types of cells by division and kingdom.
7. Identify and give the functions of the organelles on plant and animal cell models.

SEE ALSO:

Campbell Biology - Chapter 6, Section 6.1 and 6.2

JoVE (video) - Introduction to Light Microscopy

Others:

Important Terms to Know

Archaea	Eukarya	Organ
Bacillus	Eukaryote	Organelle
Bacteria	Field of view	Peroxisome
Cell wall	Flagellum	Pilus
Centriole	Focus	Plasma membrane
Chloroplast	Golgi apparatus	Plasmid
Chromatin	Kingdom	Plasmodesmata
Cilia	Lysosome	Prokaryote
Coccus	Magnification	Protist
Condenser	Mechanical stage	Resolution
Contrast	Mitochondrion	Ribosome
Cytoplasm	Nucleoid	Spirillum
Cytoskeleton	Nucleolus	Stage
Cytosol	Nucleus	Tissue
Domain	Objective lens	Vacuole
Endoplasmic reticulum (rough and smooth)	Ocular lens	Vesicle
	Oil immersion	

Pre-lab Preparation and Background Information

Life is an emergent property that results from sustained chemical reactions in tiny packages we call cells. All living things are made of cells, and cells only come from other cells. Cells are considered the basic, functional unit of life. You are alive because your cells are alive.

Life requires more than just the ingredients of a cell. Dead bodies are made of cells, but they are not alive any more. So it is important to connect the ideas from the last two lab exercises to this one. Cells are alive when they undergo continuous chemical reactions we collectively describe as the cell's metabolism. Biochemistry and molecular biology are two areas of biology that specialize in studying the complex chemical interactions that occur in cells.

Cells come in a number of forms. There are some similarities among all cells. For example, all cells have a **plasma membrane**, an internal liquid called **cytosol**, genetic material, and **ribosomes**. Bacteria have special features that are not present in other types of cells. Animal and plant cells are both much more complex than bacterial cells, but they also differ from each other. The goal of this lab is to help you recognize different kinds of cells based on their visible characteristics. Your textbook will go into more detail about the functional characteristics of these cells.

The Domains and Kingdoms of Life

All living organisms belong to three domains: **Archaea, Bacteria,** and **Eukarya**. A biological **domain** is a large grouping of organisms that share fundamental characteristics. Domains are then broken into smaller groupings. Biological **kingdoms** are the largest subdivisions of a domain. You are probably familiar with the kingdoms of the domain Eukarya, which are *protists, fungi, plantae,* and *animalia*. The other domains also have kingdoms, but they will not be discussed in this lab exercise.

Additionally, cells of all living things are characterized as prokaryotic or eukaryotic. These terms describe the two fundamental types of cells. **Prokaryotic** cells (pro = before, karyon = nucleus) are simple cells that lack a true nucleus. Both Archaea and Bacteria **domains** (large categories of living things) are made up of prokaryotic organisms. **Eukaryotic** organisms are larger and more complex, with a true nucleus and membrane-bound organelles. All organisms in the domain Eukarya are eukaryotic, including all protists, fungi, plants, and animals.

Domain Archaea

Domain *Archaea* is composed of organisms that can be found on Earth today but that are thought to most resemble the earliest ("archaic") life on the planet. They are found in extreme environments where most other organisms could not survive, such as highly salty, acidic, or hot aquatic environments. They are thought to have existed in these places for over 3 billion years, and that all other life evolved from them. There are multiple kingdoms of Archeaeans. They are always unicellular and prokaryotic, though some are *colonial* (they stick to each other but do not function together). You will not be viewing any Archaea in lab today.

Domain Bacteria

Domain Bacteria includes all of the other prokaryotic organisms. Bacteria are classified in multiple kingdoms. They are very diverse and numerous, but too small to see with the naked eye. Bacteria can be found almost everywhere on the planet. They are incredibly important in biology. They are important medically when they cause disease, but they are also important for their benefits, including decomposing toxins, helping us digest food, and recycling many compounds in the environment. Bacteria may be unicellular or colonial.

Bacteria are simple, but they have some distinct features that can be observed with specialized microscopes (Figure 5.1). Since bacteria are prokaryotic, they lack a true nucleus. However, they still have DNA in a region called the nucleoid, as well as small circles of DNA called plasmids. They contain ribosomes, which are tiny structures that assemble proteins. Some bacteria have extensions such as pili and flagella, and some may have a slime coating or capsule. These structures are too small to see with a normal light microscope.

Bacteria come in three basic shapes: **Coccus** bacteria are spherical, **bacillus** are oval or rod-shaped, and **spirillum** (or spirochete) are spiral-shaped (Figure 5.2). Bacteria that naturally occur in pairs are described with the prefix "diplo-," those in chains with the prefix "strepto-," and those in clusters with the prefix "staphlo-." So a cluster

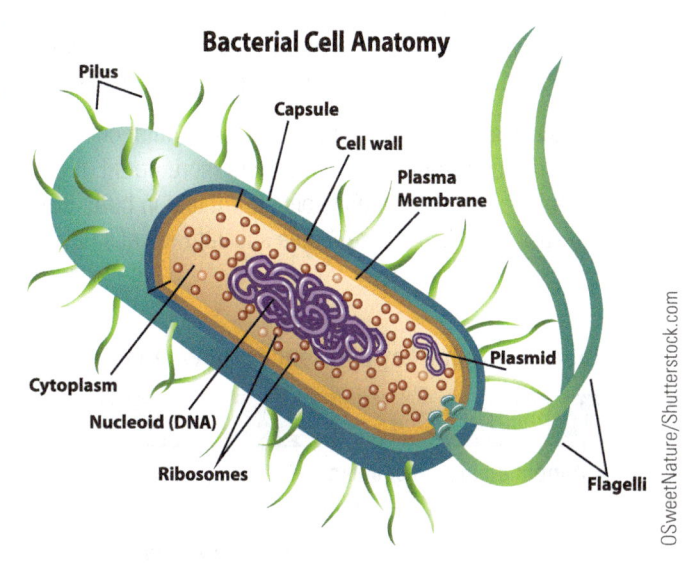

Figure 5.1 Structures that may be present in bacteria. Bacteria vary in shape and structural characteristics. Different types may lack pili, flagella, or a capsule. No prokaryotic cells have a nucleus or membrane-bound organelles.

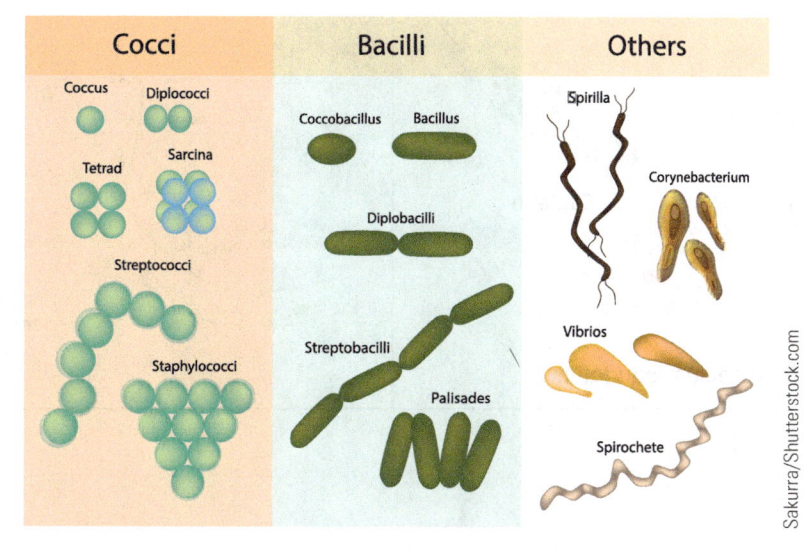

Figure 5.2 Bacterial shapes. The common shapes are coccus (round), bacillus (rod), or spirillum (spiral).

of round bacteria would be visually described as staphylococcus bacteria. You can usually observe the shapes of bacteria with a light microscope.

Domain Eukarya

Eukaryotic cells are typically much larger than prokaryotic cells. Prokaryotic cells are 0.1 to 0.5 µm in size. Eukaryotic cells typically range between 10 and 100 µm, but are sometimes much larger depending on their function. Neurons, for example, can be up to about a meter long.

All multicellular organisms are composed of eukaryotic cells. However, there are some unicellular eukaryotic organisms as well. The general cellular characteristics of organisms in the four kingdoms of domain Eukarya are summarized in Table 5.1.

Eukaryotic organisms may be unicellular, colonial, or multicellular. *Unicellular* organisms are composed of a single cell that acts independently. *Colonial* organisms are composed of cells that tend to cluster together, sometimes in complex ways, but the cells do not cooperate or specialize. *Multicellular* organisms are complex. Groups of similar cells that work together, like the cells in a muscle, constitute a **tissue**. Multiple different tissues that work together form an **organ**, like the stomach. And a multicellular organism is typically made of many organs and organ systems.

Organelles, on the other hand, are subcellular components of a cell. It is important to know their *general* functions at this point, though future labs will go into more detail. It is important to note that some organelles are specific to plant cells, others to animal cells, and some are found in both. Eukaryotic cells are very diverse, and specialized cells may lack all of the structures listed here. Some specialized cells have additional organelles.

Table 5.1 Some key differences between the cells of organisms in the four eukaryotic kingdoms

Kingdom Protista	Kingdom Fungi
Almost all are unicellular.	Almost all fungi are multicellular (yeast being the exception).
Some cells grow in colonies.	Most cells are non-motile.
Some cells are motile and use cilia, flagella, or pseudopodia to move.	Cells typically grow in long chains called hyphae.
Most protists are aquatic.	Cells have cell walls made of chitin.
Some are photosynthetic.	Fungal cells do not do photosynthesis.
Kingdom has a great deal of diversity.	
Amoeba, Euglena, Paramecium caudatum (Kazakova Maryia/Shutterstock.com)	Hyphae, Vacuole, Septum, Cell wall, Nucleus (Designua/Shutterstock.com)

Kingdom Plantae	Kingdom Animalia
All plants are multicellular.	All animals are multicellular.
Most cells are non-motile.	Animal cells do not have cell walls.
Plant cells have rigid, boxy cell walls containing cellulose.	Animal cells are not photosynthetic.
	Some cells (e.g. sperm, white blood cells) are motile.
Cell walls contain plasmodesmata (holes).	Cells are highly specialized for many different functions.

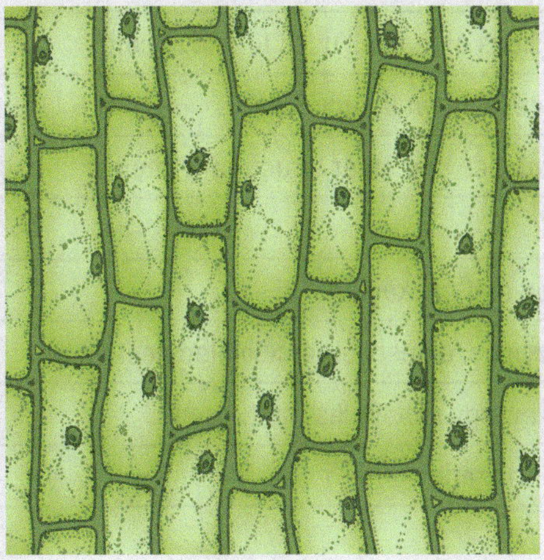

An Introduction to the Biology Lab Microscope

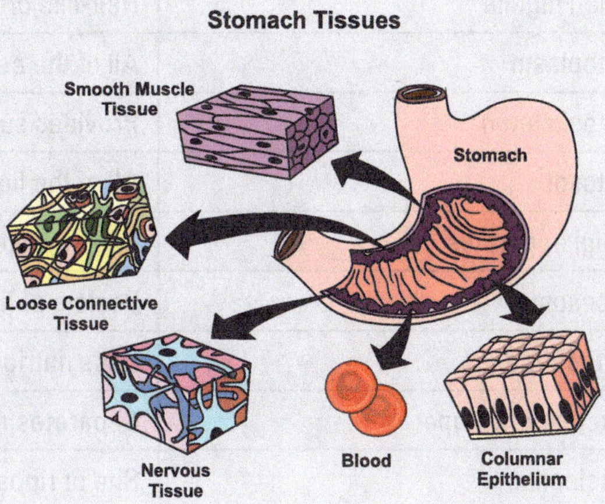

You will be using a binocular compound microscope in this lab. The term "binocular" refers to the fact that there are two ocular lenses. Using both eyes to view a specimen allows for a clearer image and depth perception. The ocular lenses can be moved closer together or farther away to adjust to the width of each user's eyes. It may take some practice, but once you get used to using a binocular microscope, you will appreciate the clarity a binocular scope can provide.

Most cells cannot be seen with the naked eye because they are too small and, in most cases, too transparent. A microscope provides the biologist a way to study cells or tiny specimens by providing three valuable resources: magnification, resolution, and contrast. **Magnification** is defined as how many times larger an image appears. **Resolution**, or resolving power, refers to how well one can see the difference between one object and another one very close to it. The highest resolution of a light microscope tends to be about 0.2 µm. Objects closer than 0.2 µm will appear as one object. **Contrast** refers to how well one can differentiate dark and light areas of the image. Biologists often improve contrast by adding a colored dye to the slide or by adjusting the amount of light going through the slide.

The microscope is an expensive instrument, and when students try to use one without proper training they can easily break it or simply become frustrated with not being able to see specimens clearly. The steps for correctly using a microscope will be described in the lab skills.

Ocular and Objective Lenses

A microscope contains multiple lenses. Ocular lenses typically magnify an image 10X, though others are available. On the nosepiece of the microscope are usually four more objective lenses that increase the magnification further. Table 5.3 lists the common objective lenses found on most biology laboratory microscopes, their magnification, their total magnification, and their field of view. The *total magnification* of a microscope is determined by multiplying the magnification power of the ocular lenses by the objective lens. The **field of view** is the width of the circular area seen through the microscope. (See Figure 5.3.)

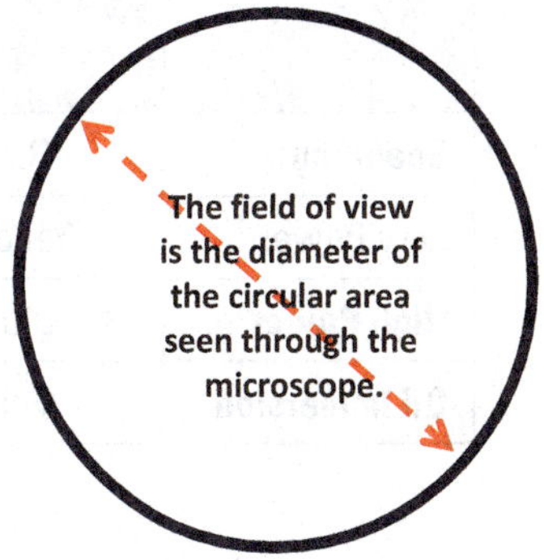

Figure 5.3 Field of View

Table 5.2 Common organelles found in eukaryotic cells and their functions illustrated in Figures 5.4 and 5.5.

Organelle	General function
Cell wall	Outer rigid barrier made of cellulose
Centriole/centrosome	Produce microtubules; involved in cell division
Chloroplast	Does photosynthesis (converts light to energy molecules)
Chromatin	Nuclear material composed of DNA, RNA, and proteins
Cilia/Flagella	Hair-like or tail-like extensions of the cell used for movement
Cytoplasm	All of the contents of the cell except the nucleus
Cytoskeleton	Provides support—microtubules and other structural proteins
Cytosol	All of the liquid in the cell
Golgi apparatus	Site of protein modification and packaging
Lysosome	Vesicle of hydrolyzing enzymes; digests organic molecules
Mitochondrion	Uses nutrients to provide energy molecules for the cell
Nuclear envelope	Separates and protects chromatin
Nucleolus	Site of ribosome synthesis
Nucleus	Houses chromatin
Peroxisome	Detoxifies cell
Plasma membrane	Outer membrane of cell
Plasmodesmata	Holes in cell wall to allow for water to pass from cell to cell
Ribosome	Assembles proteins
Rough endoplasmic reticulum	Site of protein synthesis; contains ribosomes
Smooth endoplasmic reticulum	Produces lipid molecules
Vacuole	Storage area for water in cell
Vesicle	Transports materials

Table 5.3 Common objective lenses and their characteristics

Objective	Color of stripe	Magnification	Total Magnification	Field of View (mm)
Scanning	Red	4X	40X	3.33
Low Power	Yellow	10X	100X	1.50
High Power	Blue	40X	400X	0.33
Oil Immersion	White	100X	1000X	0.15

Laboratory Exercise 5 Microscopy: Observing Life at the Cellular Level

ANATOMY OF A CELL

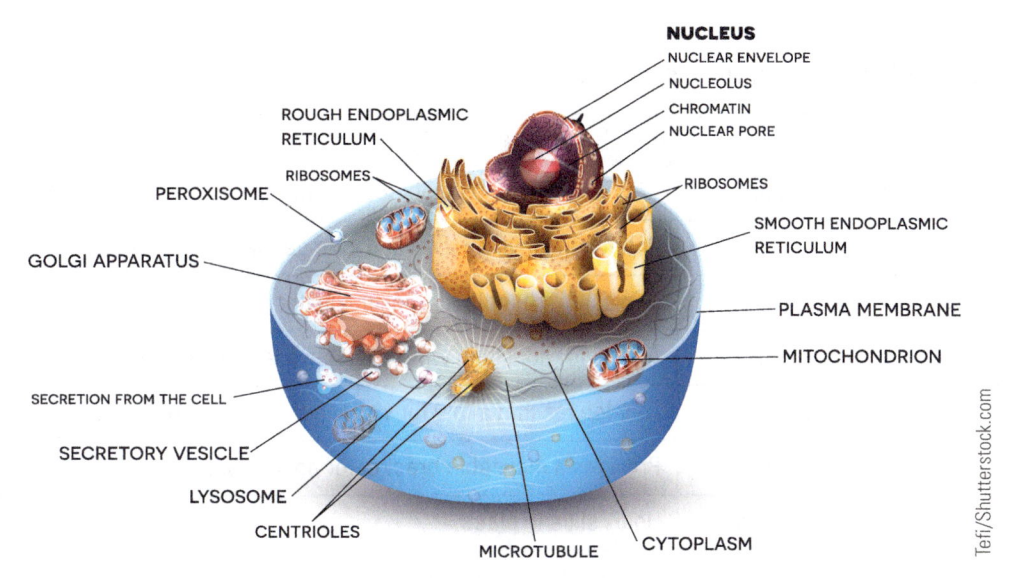

Figure 5.4 Parts of a typical animal cell. Cells with specific functions may be shaped differently and have additional organelles like cilia, flagella, and microvilli. Some animal cells are highly specialized and unique compared to this general representation of a cell.

ANATOMY OF A PLANT CELL

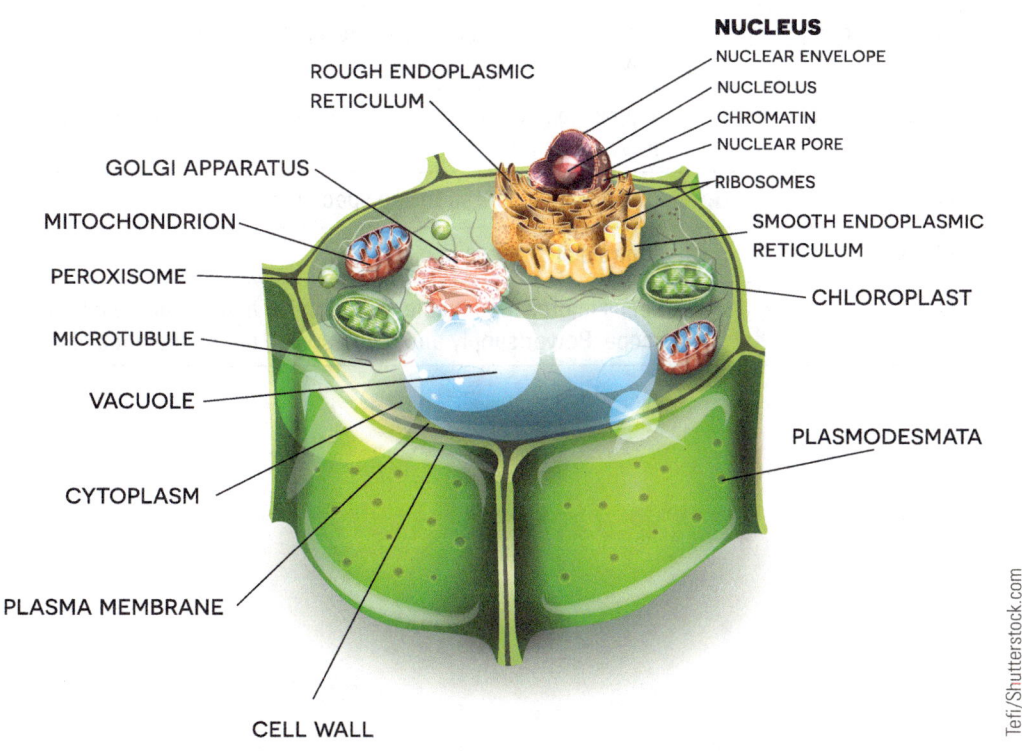

Figure 5.5 Parts of a typical plant cell. Plant cells have organelles not found in animal cells, such as a large, water-filled central vacuole and chloroplasts. Not all plant cells have all structures shown here. Most complex plants lack cilia or flagella, but simple, seedless plants may have some specialized cells with cilia and flagella.

Microscope parts

The table below lists the parts of the microscope and their function. Match the number from the table to the number to the image of the microscope in Figure 5.6.

Table 5.4 Microscope parts and functions

	Name of Part	Function
1	Ocular lens (eyepiece)	*Magnifies* image (with objective lenses); two oculars allow for depth perception.
2	Head	Rotates to move oculars and changes width of oculars.
3	Arm	Serves as a handle for picking up microscope.
4	Rotating nosepiece	Allows for adjusting which objective lens is being used.
5	Objective lenses (4)	Four lenses that allow for viewing the specimen at different *magnifications*.
6	Mechanical stage	Provides a moveable bracket for holding slide.
7	Mechanical stage clip	Holds slide in place.
8	Stage	Surface for placing slide.
9	Mechanical stage adjustment knobs	Moves mechanical stage (and slide) left, right, up and down.
10	Coarse focus	Moves stage up and down in large increments.
11	Fine focus	Moves stage up and down in small increments; helps improve *resolution*.
12	Condenser	Focuses light onto the slide from below.
13	Condenser diaphragm lever	Controls the amount of light passing through the stage to the specimen; helps improve *contrast*.
14	Condenser adjustment knob	Adjusts height of condenser; controls how the light hits the slide and helps improve *resolution*.
15	Lamp	Provides light source to see specimen.
16	Rheostat	Controls brightness of lamp.
17	Base	Keeps microscope stable, with arm serving as handle for picking up the microscope. Power supply and on/off switch are on the back of the base.

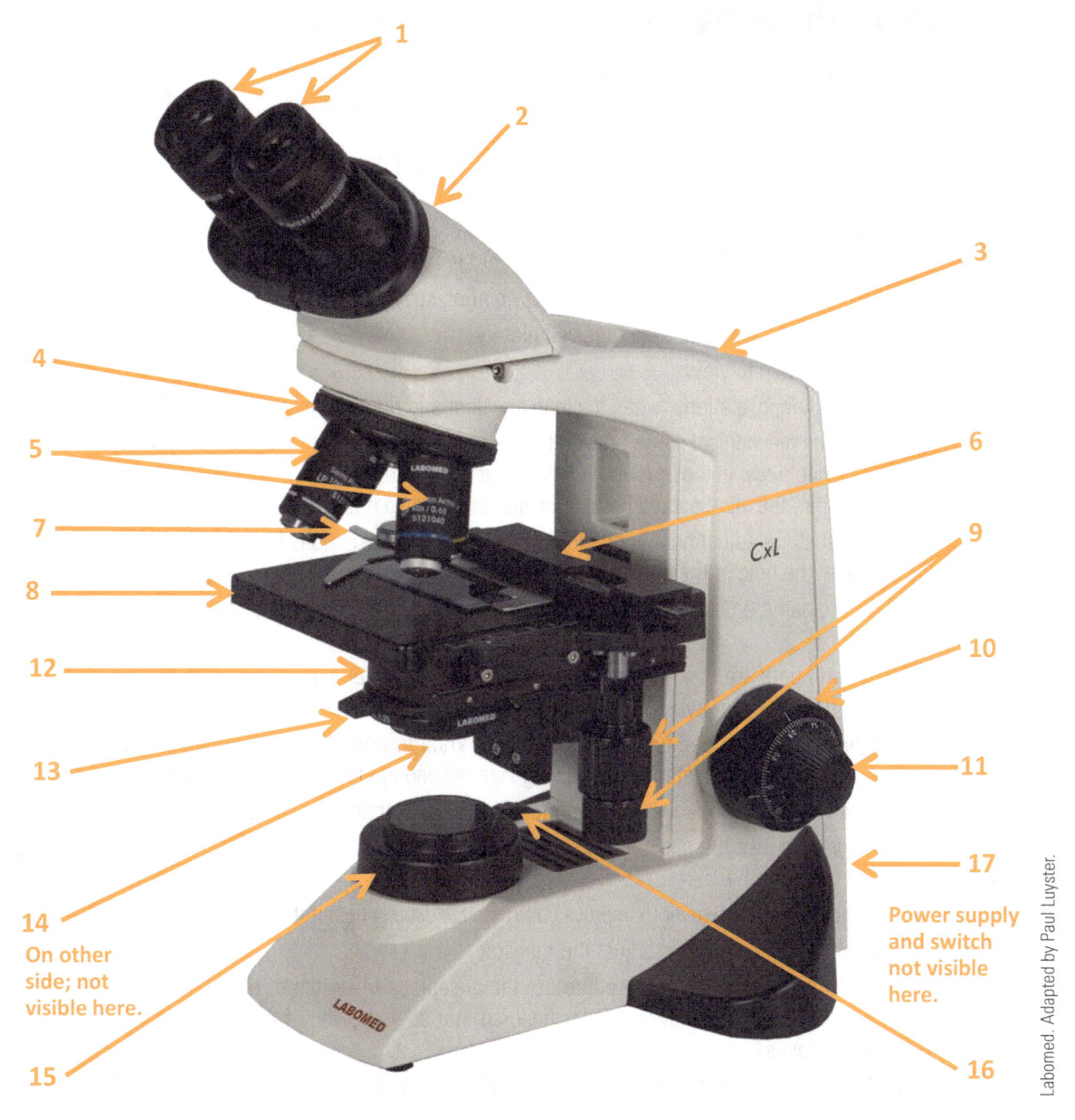

Laboratory Exercise 5 Microscopy: Observing Life at the Cellular Level **119**

Figure 5.6 A compound microscope with parts numbered. Some parts such as #14 are not visible from this angle, and you will need to study an actual microscope in lab to find them.

Lab Skill 5.1 Properly Using a Light Microscope

The following steps should always be followed when using a microscope, in this or any laboratory. Hopefully these steps will become routine and habitual as you use the microscope throughout the semester and in future classes.

1. Always store the microscope so the arm faces out. When lifting the microscope, always use the arm and base to move it. Never grab a microscope by the ocular lenses, head, or stage.
2. Ensure the microscope is on a firm, flat surface. Plug it in and turn it on. The lamp in the base should come on. If it does not, turn the rheostat and check that the power cord is fully inserted into the base.
3. Rotate the nosepiece until the scanning objective is pointing straight down.
4. Use the course focus knob to lower the stage completely.
5. Obtain the slides you wish to view.
6. If using a prepared slide, use a lens wipes to clean any dust and fingerprints from the slide.
7. Place your slide in the mechanical stage bracket by opening the mechanical stage clip. DO NOT force the slide under the clip. The slide should rest in the bracket and move freely when the mechanical stage adjustment knobs are turned. If the slide has a label, the label should be face up and oriented so it is readable.
8. Use the mechanical stage adjustment knobs to move the slide into position. Try to ensure that the specimen on the slide is located over the hole in the stage.
9. Use lens wipes to clean the ocular lenses and objective lenses. (Until this point you should not be looking through the microscope.)
10. Turn the oculars until both are in all the way and level with each other. Place your eyes to the ocular lenses. Adjust the oculars to the width of your eyes. You should use both eyes and see one circular image. (This takes practice.)
11. While looking through the oculars, turn the coarse focus knob to bring the specimen into view. Turning the coarse focus knob raises and lowers the stage quickly, so go slowly enough to watch for the specimen to come into focus.
12. Once the specimen is roughly in focus, use the fine focus knob to sharpen the image.
13. Once the slide is visible, use the mechanical stage adjustment knobs to center the specimen or desired part of the specimen in your field of view.
14. Adjust the condenser diaphragm lever to improve contrast. It is usually best if the condenser is raised close to the stage.
15. If the image is still too bright or dark, adjust the rheostat.
16. Once the specimen is focused and centered, rotate the nosepiece to switch to the next higher objective if desired. Do not lower the stage. The objectives are designed so that as you switch from one to the next the image should already be nearly in focus.
17. Repeat steps 12 through 16 as needed. Do not use the oil immersion objective without oil. (See Lab Skill 5.4 for how to use the oil immersion objective.)

Troubleshooting

- If you have trouble finding the specimen, go back to the scanning or low-powered objective and start over.
- If you can't see details (or you have poor resolution), adjust the condenser diaphragm lever and the condenser focus knob.
- If you can't see clearly using both eyes, adjust the width of the oculars.
- If one eye can see clearly but the other eye is blurry, one of the ocular lenses is out of adjustment. Determine which ocular has the diopter adjustment markings. Close the eye using that ocular lens and use the fine focus to focus the image using the other eye. Then close that eye and look through the ocular with the adjustment markings. Turn the adjustable ocular lens until that eye can see clearly. Then check to see if both eyes can focus clearly on the image. (This takes some practice, but it helps avoid headaches.)
- If the specimen appears partially in focus and partially out of focus, it is because the depth of focus is too great at that magnification. Use a lower powered objective.

- If the image is blurry no matter how much you try to focus, the image is either too far out of range for the microscope, which can occur if the slide is upside-down, or the slide is too thick (two slides may be stuck together). If on the oil immersion objective, it is because there is not enough (or no) oil between the objective and the slide.
- If the image is very dark but the light is on, slide the condenser diaphragm lever to open the diaphragm. If it is still dark, check the rotating nosepiece. If the nosepiece is not fully clicked into place, the view will be obstructed.

Lab Skill 5.2 Calculating Magnification and Estimating Specimen Size

It is often necessary to estimate the size of an object viewed under the microscope. Some slides actually contain microscopic markings that can serve as rulers. However, if you know the diameter of the field of view, you can estimate the size of an object being viewed.

In Figure 5.7, Volvox is being viewed at low power. The small blue-green circles are colonies of single-celled protists, and the tiny darker circles inside the larger circles are new "daughter" colonies forming. The size of one adult colony can be estimated by imagining colonies lined up across the field of view and determining how many fit. In this case, it appears about six would fit across the field of view. (Figure 5.8).

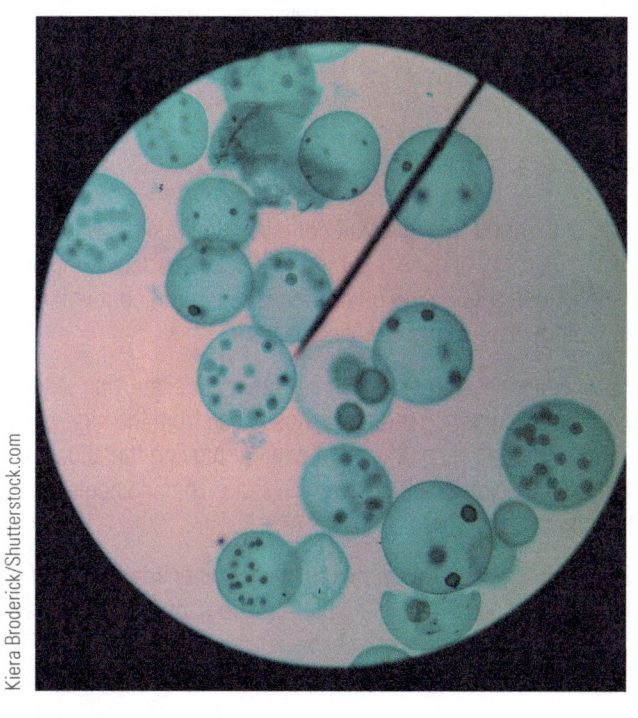

Figure 5.7 Volvox colonies (at low power, 100X).

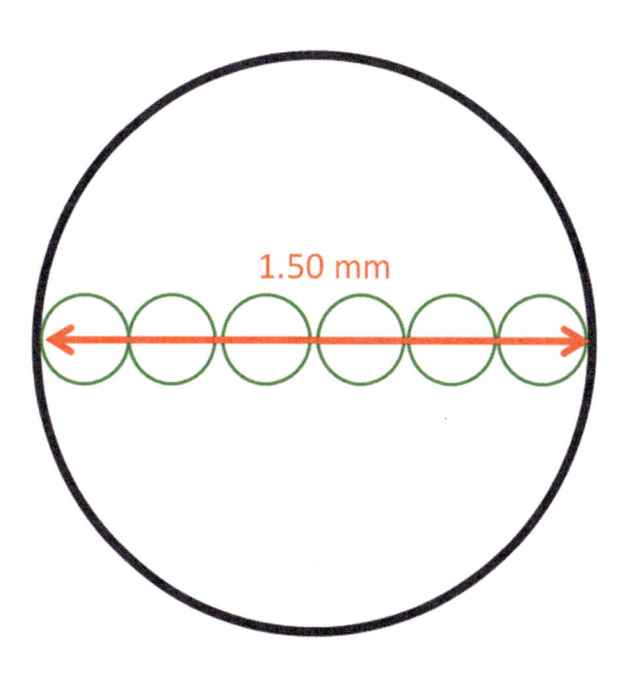

Figure 5.8 Imagining that about six specimens could fit across the field of view.

Recall from Table 5.3 that each objective has a specific field of view. Divide the field of view for the objective being used by the number of specimens that fit across. In this case, the field of view for the low-power objective is 1.50 mm (from Table 5.3). About six colonies fit across the field of view. The size of one colony is approximately 1/6th of the field of view. Therefore, the size of one colony is approximately 0.25 mm (or 250 μm).

Lab Skill 5.3 Using the Oil Immersion Lens

At high magnification, the resolution of the image seen through the microscope drops. This is due to the wave properties of light and how it is diffracted (spread) as it passes through the glass, the specimen, and the air. Placing a drop of oil between the objective and the slide reduces the scattering and improves resolution. See Figure 5.9. Oil should only be used with the oil immersion lens, and only when looking at extremely tiny specimens such as bacteria. It is extremely hard to see bacteria with the scanning, low-, or high-power objectives. Bacteria may be visible but very blurry on the oil immersion objective *without* oil.

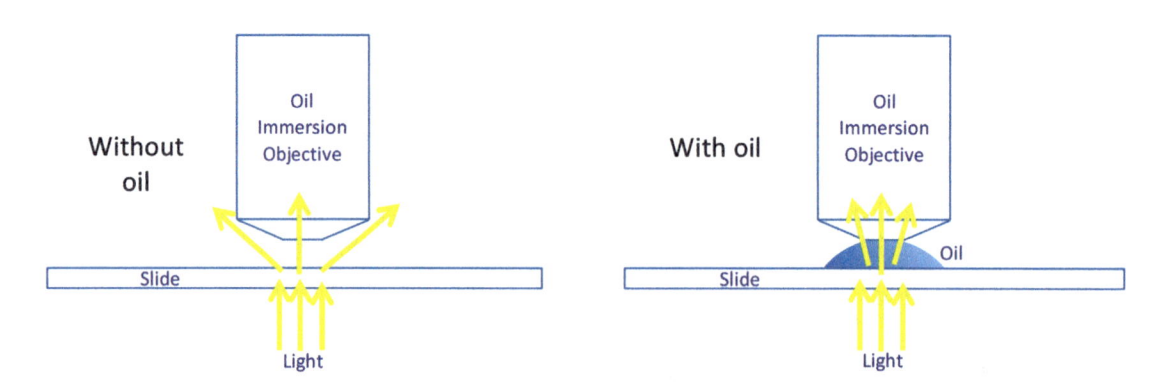

Figure 5.9 The benefit of using the oil immersion lens. The oil reduces the scattering of light rays and improves the resolution of the image.

The oil immersion objective lens can magnify 1000X. However, without oil this lens will not give good results. But oil can only be used with this lens. Using the oil immersion objective can be tricky and messy. It typically takes patience and practice. In addition, the slide and objective must be carefully wiped clean after using oil or the oil can permanently damage the objective lens. This is why oil immersion is rarely used in biology labs unless studying bacteria.

To use the oil immersion lens, you must first follow the same procedures used for viewing any other slide, starting with the scanning objective and working up to the high-power objective. However, as you rotate the nosepiece from the high-power objective to the oil immersion objective, you must carefully place a drop of oil on the slide without moving the slide. Then you must continue rotating the oil immersion objective so that it moves into position over the slide. The oil should form a bridge between the surface of the slide and the bottom of the objective.

If you accidentally move the slide or stage, you may need to go back to a lower objective to find the specimen again. If so, you must clean the oil from the slide. Any oil on the slide will blur the image when using the scanning, low-, or high-power objectives.

Lab Skill 5.4 Making a Wet Mount Slide

A wet mount slide is made from a fresh specimen. The specimen may be suspended in liquid and then placed on the slide, or the specimen may be placed on the slide first, and then a few drops of water are added afterwards. The specimen is then covered with a thin plastic coverslip, and the water helps hold the coverslip in place. See Figure 5.10. Note that the coverslip is placed so that one edge is touching the slide, and then the coverslip is dropped in place. This reduces the number of air bubbles that can get trapped under the coverslip.

The procedure for making a wet mount varies with the type of specimen being viewed. In some cases, you will need to stain the specimen. Most animal cells are usually small and clear. With the exception of red blood cells, most are difficult to see without staining them. Plant cells may be easier to see due to the green pigments found in most plant cells.

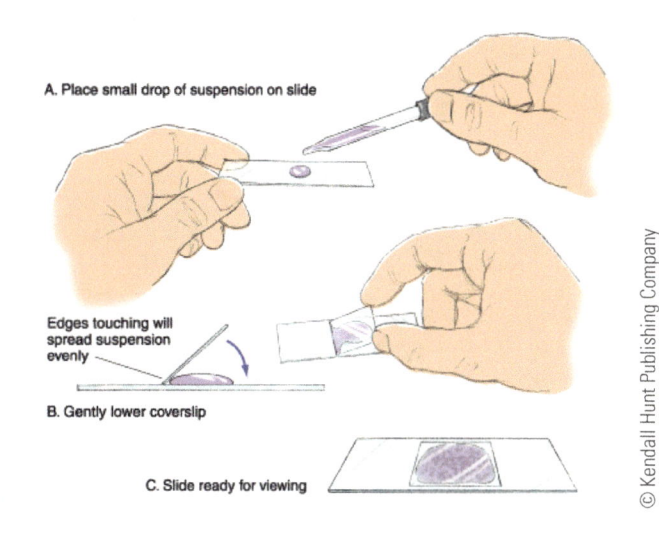

Figure 5.10 Preparing a wet mount slide.

In some cases, the cells are stained *after* the cell suspension is covered with a coverslip. This is to create a gradient of stain so that at least some of the cells are stained enough to be visible, but there is not so much stain that the cells are difficult to see.

To do this, you need to prepare the slide as shown in Figure 5.10, but with a little extra water under the coverslip so that the coverslip is resting on a layer of water. Then a tiny amount of stain, such as methylene blue, must be added to one edge of the coverslip as shown in Figure 5.11. A small piece of paper towel should be placed on the opposite edge of the coverslip to draw the stain under the coverslip. Then excess stain and water must be removed by gently dabbing away excess liquid. The final result should be a slide with a darkly stained area on one side of the coverslip, and a lightly stained area on the other, as shown in Figure 5.11. The stained cells should be clearly visible somewhere in between the extremes.

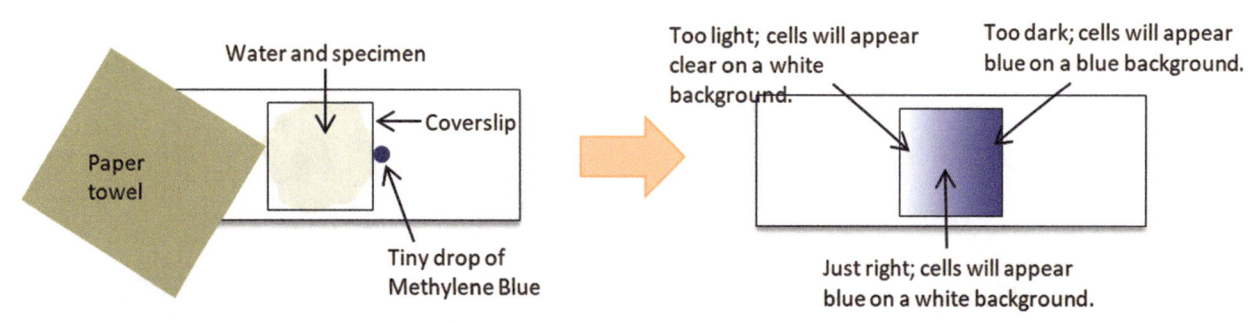

Figure 5.11 Making a wet mount and applying a gradient of stain.

Name: Bethany Williams   Date: 9/14/2021

PRE-LAB QUIZ 5: Are you ready to proceed?

1. What are *domains* and *kingdoms* in biology?
   Domains are a large grouping of organisms that share fundamental characteristics. Kingdoms are the largest subdivisions of a domain.

2. Complete the table by checking which types of cells could have the features listed. Some characteristics apply to more than one cell type.

Feature	Prokaryotic — Bacterial cell	Eukaryotic — Animal cell	Eukaryotic — Plant cell
Membrane-bound nucleus		X	
Ribosomes	X	X	X
Large central vacuole			
DNA	X	X	
Mitochondria		X	X
Chloroplasts			X
Golgi apparatus		X	X
Pili	X		
Cell wall	X	X	X
Flagella	X		
Plasma membrane	X		X
Plasmid	X		
Plasmodesmata			X
Microtubules		X	X

3. Clearly explain the difference between *magnification, resolution*, and *contrast* as they relate to a microscope.

4. What are two ways to adjust the amount of light being viewed through the microscope? Use the correct names of the parts.
   The condenser diaphragm lever controls amount of light passing through the stage to the specimen.
   The Rheostat controls the brightness of the lamp.

5. What are at least three things you should do *before* you start looking through the microscope at a slide?

6. What is the purpose of the condenser, and what are the two ways it can be adjusted?

7. How do you calculate the magnification of any ocular-objective combination?

8. You observe this blue specimen using the *high-power* objective. The circle represents the field of view. Estimate its size in millimeters. Show how you obtained your answer.

9. When is it appropriate to use the *oil immersion* objective? *When* specifically should you add the oil? Why? *How* should you add it? Why?

10. If you need to stain suspended cells so they can be viewed through the microscope, why can't you just put a drop of stain on the cells before covering the sample with a coverslip?

Problem to Solve 5.1 Can I Visually Identify Types of Cells?

The purpose of this exercise is threefold:

- Practice with the microscope to view prepared slides, including using the oil immersion lens to view very tiny specimens
- Prepare and view wet mount slides, both unstained and stained.
- Study the visual difference between cells of the different biological domains and kingdoms. (However, examples of Archea will not be viewed.)

Obtain a microscope and prepared slides as instructed. Use the Lab Skills sections to help you set up and properly use the microscope to view prepared slides (Lab Skill 5.1). Follow the instructions for using the oil immersion lens (Lab Skill 5.4) to view the bacteria slide. Record your work on the following pages, paying special attention to sketching and labeling the cells you observe.

Prepare an unstained wet mount slide of a tiny leaf from the aquatic plant, *elodea* (Lab Skill 5.3). Ensure that the leaf is unfolded and flat on the slide. View the plant cells and identify the specific structures associated with plant cells.

Prepare a wet mount slide of your own human cheek cells. Use a toothpick to scrape the inside of your cheek and prepare a stained wet mount.

Note that prepared slides must go back into the proper trays or slots from which they came. Do not leave them on the lab bench, microscope, or counters. They must be put back properly to ensure they will be available for other students.

Wet mount slides are disposable and should be discarded in the glass box after use. (Do not discard the cheek cell slide until after completing Problem 5.2.)

Laboratory Exercise 5 Microscopy: Observing Life at the Cellular Level

Title

Practicing with the microscope and identifying types of cells by domain and kingdom

Purpose

Practice with prepared slides, oil immersion objective, unstained wet mount slides, and stained wet mount slides. Observe cell types from the different biological domains and kingdoms, and learn to identify the differences between them.

Protocol

Materials:
Microscope
Prepared slides:
  Mixed bacteria
  Mixed protozoa (protists)
  Rhizopus (fungus)
Immersion oil
Clean blank slides (2)
Coverslips (2)
Elodea leaf
Toothpick
Water
Methylene Blue
Paper Towel

Protocol:
View prepared slides.

1. Obtain a microscope, prepare it, and adjust oculars as directed in Lab Skill 5.1.

2. Obtain the three prepared slides.

3. Start with the Rhizopus and Mixed Protozoa slides. View with the scanning objective and work up to high power (Refer to Lab Skill 5.1 if necessary.)

4. Observe the characteristics of the cells of these specimens. Sketch and label your observations in the Data section.

5. Obtain the Mixed Bacteria slide.

6. View on scanning and then work up to high power. (Bacteria are hard to see, so be patient and find them.)

Continued from previous page.

Protocol

7. Using the immersion oil, use the oil immersion objective to observe them. (Refer to Lab Skill 5.3 if necessary.)

8. Sketch and label what you see in the Data section.

Prepare an unstained wet mount of plant cells. (Refer to Lab Skill 5.4 as needed.)

9. Place a small drop of water on a clean, blank slide.

10. Pluck one small leaf from the elodea and place it in the drop of water. Ensure it is flat and unwrinkled.

11. Place a coverslip over the leaf and water. Place one edge of the coverslip on the slide before dropping it to reduce bubbles.

12. Use a paper towel to dab away excess water so the cover slip does not move.

13. View through the microscope with the scanning objective. Work up to high power to see individual cells. Observe specific plant cell characteristics including cytoplasmic streaming and organelles.

14. Sketch observations in the Data section.

Prepare a stained wet mount of animal cells.

15. Place a clean, blank slide on the bench top.

16. Place several drops of water on the slide to make a small circle of water about the size of a dime.

Purpose

Continued from previous page.

Protocol

17. Use a toothpick to scrape the inside of your cheek. Be careful not to injure yourself.

18. Dab the toothpick in the water on the slide to release the cheek cells.

19. Throw away the toothpick. (Do not put it back in the container, leave it on the lab bench, or leave it on the counter!)

20. Place a coverslip over the water and cells. The coverslip should move freely on the water.

21. Add a very small drop of Methylene Blue adjacent to the edge of the coverslip. (See Figure 5.11.)

22. Use a paper towel to blot the water from the *opposite* side of the coverslip. (It also helps to slide the coverslip a little toward the Methylene Blue as you blot the water.)

23. Continue to remove excess water and dye until the coverslip no longer moves and the slide has a gradient of dark and light blue areas. See Figure 5.11. (This may take more than one try.)

24. Observe the cells with the scanning objective and work up to high power.

25. Sketch observations in the Data section.

26. Write a summary and your conclusions.

Continued from previous page.

Data:

Slide: **Rhizopus**

Domain:

Kingdom:

Visible characteristics:

Sketch of cells is at magnification:

Slide: **Mixed Protozoa**

Domain:

Kingdom:

Visible characteristics:

Sketch of cells is at magnification:

Laboratory Exercise 5 Microscopy: Observing Life at the Cellular Level

Title

Purpose

Continued from previous page.

Protocol

Slide: **Mixed Bacteria**

Domain:

Kingdom:

Visible characteristics (shapes):

Sketch of cells is at magnification:

Slide: **Elodea**

Domain:

Kingdom:

Visible characteristics:

Sketch of cells is at magnification:

Continued from previous page.

Slide: **Human cheek cells**

Domain:

Kingdom:

Visible characteristics (shapes):

Sketch of cells is at magnification:

(Do not dispose of cheek cell slide until after completing Problem to Solve 5.2)

Summary and Conclusions:

Problem to Solve 5.2 Can I Estimate the Size of One Human Cell?

This activity requires that you have already completed Problem 5.1. You must have a slide of your stained cheek cells prepared, and you should have found some good examples of properly stained cheek cells on your slide (Figure 5.12).

Locate a good representative of one cheek cell using the *low-power* objective.

- Ensure the cell you selected is not clumped or overlapping other cells.
- Ensure it is not wrinkled or folded.
- It should have distinct blue coloration and clearly visible borders.
- The nucleus should be darkly stained, and other organelles should appear as dark blue specks within the cell.
- The background should be white or almost white, and the cell should stand out clearly.

Estimate the size of the cheek cell as viewed with the low-power objective. Refer to Lab Skills 5.2 if necessary. Record your observation and calculations in the following Data section.

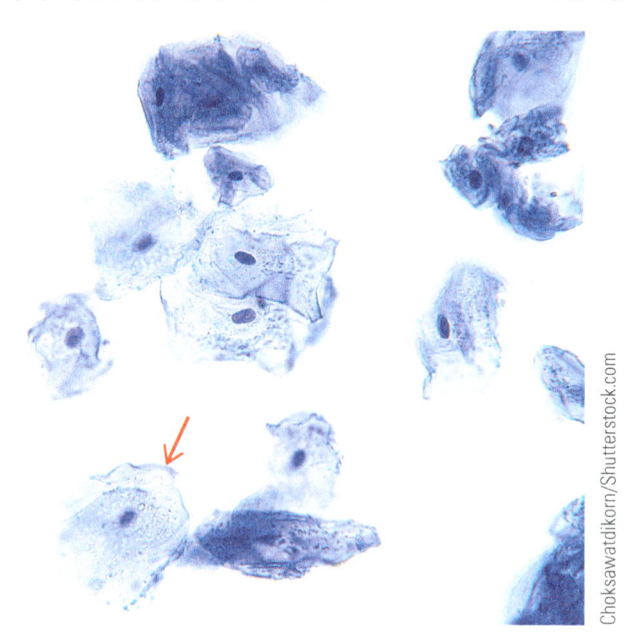

Figure 5.12 Human check cells stained with Methylene Blue. The indicated cell is fairly flat, isolated, and easily seen. It would be a good choice for estimating cell size. Many of the other cells are wrinkled or overlapping each other.

Repeat this process with two more cells. Record your observations and calculations in the Data section.

Switch to the high-power objective and repeat the process. Find three good cheek cells and estimate their size. Record the data in the data table. In the conclusion section, respond to the questions provided.

It is important to note that this activity estimates the size of human cheek epithelial cells. The size of these cells is typical for the types of cells that make up internal and external *surfaces* of the human body. Many other types of cells, such as neurons, muscle cells, and blood cells, may be much larger or smaller than the ones observed in this activity.

Title

Estimating the size of a human cheek cell.

Purpose

Learn how to estimate the size of specimens viewed through a microscope.

Protocol

Materials:
Microscope and cheek cell slide from previous activity, Problem 5.1.

Protocol:

View three different cheek cells with the low-power objective and independently estimate the size of each cell (Refer to Lab Skill 5.2 if necessary).

View three different cheek cells with the high-power objective and estimate their sizes.

Average the sizes of the cells viewed at each magnification.

Data:

Low-Power Objective	Field of View in mm (See Table 5.3)	Approximate number of cells that could fit across field of view	Estimated size (mm) = $\dfrac{\text{field of view}}{\text{number of cells}}$
Cell 1			
Cell 2			
Cell 3			
		Average size of cells	

High-Power Objective	Field of View in mm (See Table 5.3)	Approximate number of cells that could fit across field of view	Estimated size in mm: = $\dfrac{\text{field of view}}{\text{number of cells}}$
Cell 1			
Cell 2			
Cell 3			
		Average size of cells	

Date | Laboratory Exercise 5 Microscopy: Observing Life at the Cellular Level

Title

Purpose

Continued from previous page.

Protocol

Conclusions:

Based on your observations, about how big are human cheek cells? (Include units.)

Are your cells about the same size as the cheek cells of other students in the class?

Was there an *apparent difference* in size when the cells were viewed with the low- and high-powered objectives?

When cell size was calculated, was there a difference in the *actual* size of the cells when viewed with the low- and high-powered objectives?

Which objective provided a more *accurate* and *precise* estimation of size? Explain.

Name: _____   Date: _____

POST-LAB QUIZ 5: What have you learned?

1. What is the field of view and total magnification of the microscope on low power?

2. What is the field of view and total magnification of the microscope on oil immersion? Why is the field of view smaller at higher magnification?

3. What are at least three distinct visible characteristics that should allow you to easily identify bacteria versus other type of cells when viewing specimens with a microscope?

4. Complete the diagram:

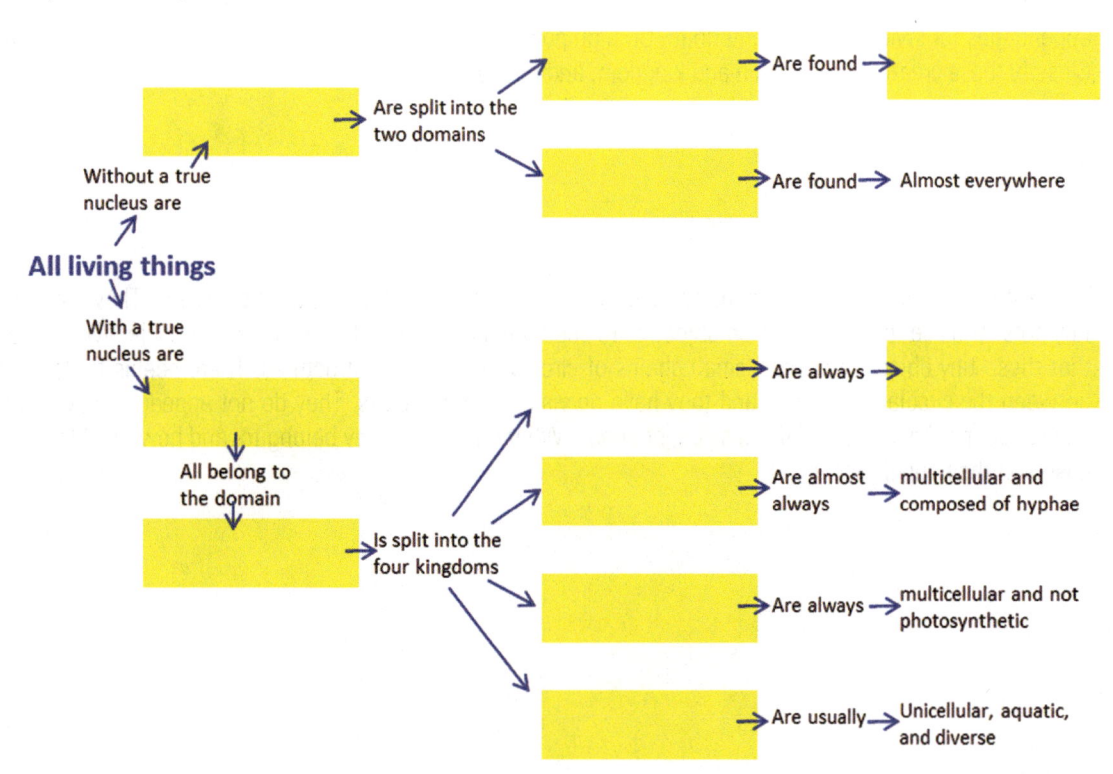

5. How can you distinguish plant cells from other types of cells?

6. Name at least two cellular structures found only in plant cells.

7. Which groups of organisms typically have cell walls?

8. Does it matter which objective you use when calculating the size of a specimen under the microscope? Explain your answer using your results from Problem 5.2.

9. You notice tiny particles moving around in an aquarium with the fish. You make a stained wet mount of the particles and look through the microscope. The particles are easily visible on scanning and low power. They are organisms that appear to have multiple moving parts, They move in and out of your field of view. They appear to have simple eyes, appendages for swimming, and antennae. On high power, the individual cells are visible and appear to have nuclei. Classify these creatures by domain and kingdom, and justify your answer.

10. While studying the creatures described above, you notice even smaller particles in the water. They are only visible as tiny dots. You use the oil immersion objective to see them better. With the oil immersion objective you can observe that these tiny objects are really small chains of circular or spherical structures. There seems to be no difference between the circular structures, and they have no visible internal parts. They do not appear to move on their own. They appear to be connected like a string of beads. What domain do they belong to, and how would you name them based on their appearance?

Laboratory Exercise 6

Cell Biology I: Cellular Membranes and Structures

Concepts to Understand

After completing this lab, the student should be able to:

1. Describe the molecular components of the plasma membrane.
2. Describe the conditions for cellular diffusion and osmosis.
3. Explain how temperature, particle size, charge, and polarity affect diffusion.
4. Describe the process of osmosis and how it occurs at the cellular level.
5. Differentiate between active and passive cell transport.
6. Apply the concept of the *fluid mosaic model* to other membranous organelles.

Skills to Learn

After completing this lab, the student should be able to:

1. Infer the size of molecules with dialysis tubing.
2. Demonstrate selective permeability of a membrane using dialysis tubing.
3. Demonstrate the effects of osmosis on living cells.
4. Design a cell with components needed to do a specific task.

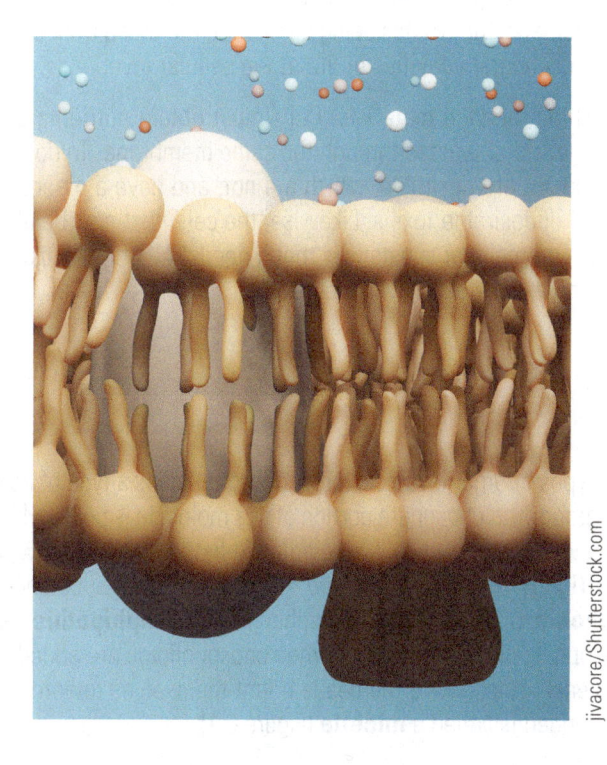

SEE ALSO

Campbell Biology - Chapter 6, section 6.7; Chapter 7, sections 7.1-7.5.

Others:

Important Terms to Know

- Active transport
- Amphipathic
- Aquaporin
- Attachment protein
- Barium chloride (BaCl$_2$) (indicator)
- Bulk transport
- Concentration gradient
- Crenate
- Cytosol
- Dialysis
- Dialysis tubing
- Endocytosis
- Exocytosis
- Extracellular matrix
- Facilitated diffusion
- Flaccid
- Fluid mosaic model
- Glycocalyx
- Glycolipids
- Glycoproteins
- Hydrophilic
- Hydrophobic
- Hypertonic
- Hypotonic
- Integral protein
- Isotonic
- Liposome
- Lyse
- Membranous organelles
- Micelle
- Osmosis
- Passive transport
- Peripheral/surface protein
- Phagocytosis
- Pinocytosis
- Plasma membrane
- Plasmolysis
- Precipitate
- Receptor protein
- Receptor-mediated endocytosis
- Semi/selectively permeable
- Silver nitrate (AgNO$_3$) (indicator)
- Simple diffusion
- Transport protein
- Turgid

Pre-lab Preparation and Background Information

Humans need oxygen, nutrients, and a way to remove wastes from the body. These are a few of the fundamental requirements of all living things. But *why* does a human breathe? *Why* does a human eat? Where do the wastes that are released from the body actually come from?

The reality is that each and every one of the trillions of *cells* in a multicellular animal need nutrients and oxygen and to get rid of wastes. The reason humans and animals eat is to get food to *every cell* of the body. The reason humans and animals take in oxygen through their lungs, gills or skin is to provide oxygen to *every cell* of the organism. When wastes are eliminated by the excretory system (primarily the kidneys in humans) the wastes being eliminated are those from *every cell* of the body. Every cell is alive, and as such, conducts the fundamental activities of life at the cellular level.

Yet cells do not have lungs, mouths, or complex systems to obtain needed materials. Cells must obtain nutrients and oxygen through their outermost membrane, and they must also get rid of wastes through the same membrane. To do this cells, must be able to distinguish what materials in the environment are desirable, which are not, and have a means of getting the right materials into the cell while preventing entry of undesirable toxins. Likewise, the cell must also be able to selectively get rid of waste materials without losing important intercellular substances that it needs. All of this must be done through the plasma membrane, making it a critical and complex part of all living cells.

The Structure of the Plasma Membrane

Properties of Phospholipids

As discussed in Lab Exercise 3, lipids are one of the four major types of organic molecules. A phospholipid is a type of lipid composed of a glycerol, two fatty acid chains, and a phosphate group. The phosphate portion of this molecule is polar, while the fatty acid chains are non-polar. Polar substances easily mix with water, which is also polar. Substances that mix with water, such as polar covalent molecules and ions, are **hydrophilic**. (*Hydro* = water; *phileo* = love.) Non-polar substances do not easily mix with water. Non-polar substances are **hydrophobic**. (*Phobia* = fear.) Phospholipids are **amphipathic**—a single molecule that is both polar and non-polar, and both hydrophilic and hydrophobic. When phospholipids are added to water, the tails of the phospholipids are naturally attracted to each other, while the heads orient themselves toward the surrounding water. A cluster of amphipathic molecules thus arranged is called a **micelle** (Figure 6.1).

LIPIDS IN WATER

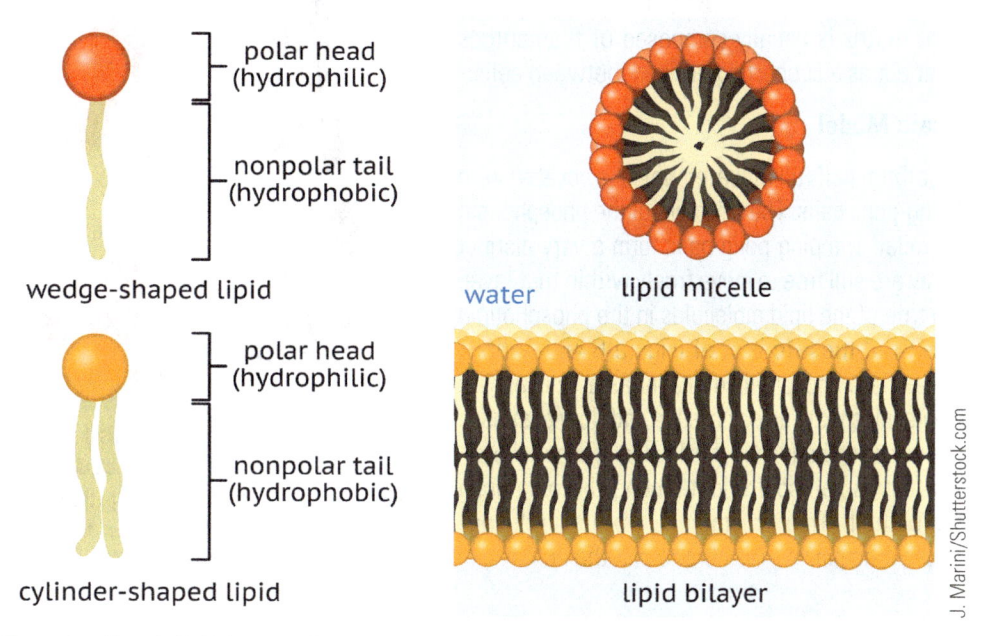

Figure 6.1 Phospholipids in water. These molecules naturally orient themselves so that non-polar (hydrophobic) tails are together, and polar (hydrophilic) heads face water.

If water is injected into the center of a micelle, the structure of the phospholipid layer would change. Some of the phospholipids would turn around with their hydrophilic heads toward the water in the center. This would create a double-layered phospholipid sphere, or **liposome** (Figure 6.2). The double layer of lipids is appropriately called the *phospholipid bilayer*.

All cells, from bacteria to human cells, have a *plasma membrane* (or cell membrane) composed of two layers of phospholipid molecules. The inside of a cell contains **cytosol** (water and dissolved molecules), and cells are surrounded by extracellular fluid (also water and dissolved molecules). At a minimum, all cells are composed of a *plasma membrane*, *cytoplasm*, and *nucleic acids*. The cytoplasm contains organelles, which vary considerably with cell type and function. Some organelles are also made of phospholipid bilayers, such as the *endoplasmic reticulum*, *mitochondria*, *vesicles*, *lysosomes*, and *Golgi apparatus*. These are referred to as **membranous organelles** because they are made of phospholipid membranes much like the plasma membrane.

A blob of lipids in water is as fragile as drops of oil in dishwater. In fact, many membranous organelles can merge and split apart much like blobs of oil, or like the blobs of wax in a lava lamp (Figure 6.3). The inner workings of a cell are dynamic. Vesicles, like the blobs in a lava lamp, shuttle materials throughout the cell.

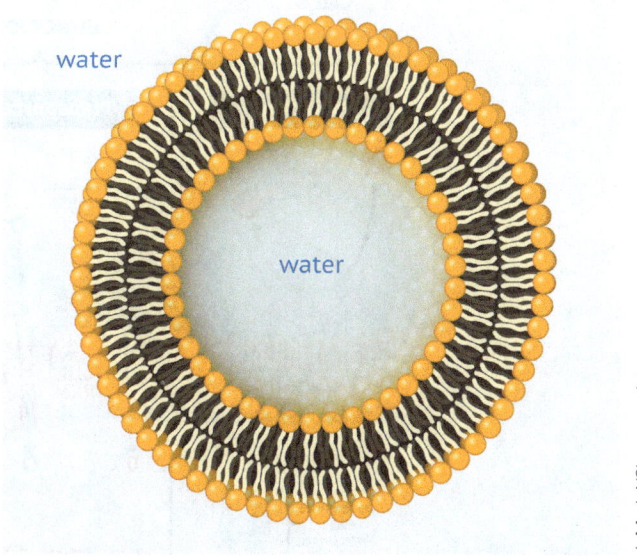

Figure 6.2 Liposome. Like a micelle, lipids orient themselves naturally in this way due to the polar nature of water and the amphipathic nature of phospholipids. A liposome is a plasma membrane without the rest of the cell.

Some membranous organelles and the plasma membrane are not as fragile as oil droplets in water. Proteins embedded in the plasma membrane help hold it together.

These proteins vary in structure and function depending on the cell type, but in multicellular organisms they help attach cells to each other and to the **extracellular matrix.** The extracellular matrix is usually composed of filamentous proteins that act as a supporting network between cells.

Fluid Mosaic Model

Imagine a bathtub half full of water which is covered with a layer of ping-pong balls. The structure of the phospholipid bilayer is similar. The ping-pong balls form a very distinct layer but they are still free to move freely within that layer. This is also true of the lipid molecules in the phospholipid bilayer. The **fluid mosaic model** (Figure 6.4) describes the structure of the phospholipid bilayer as a "mosaic" of separate molecules that can easily slide and flow past each other. Proteins are embedded in this layer of phospholipids, much like a tennis ball could float among the ping-pong balls in the bathtub. These membrane proteins influence the behavior of the plasma membrane and the cell.

Figure 6.3 Lava lamps. Popular lamps like these contain non-polar wax and water. As the wax warms at the base, it becomes less dense and floats to the top. As it cools at the top, the wax becomes denser and sinks. As the blobs move up and down, they often merge and split apart as they bump into each other.

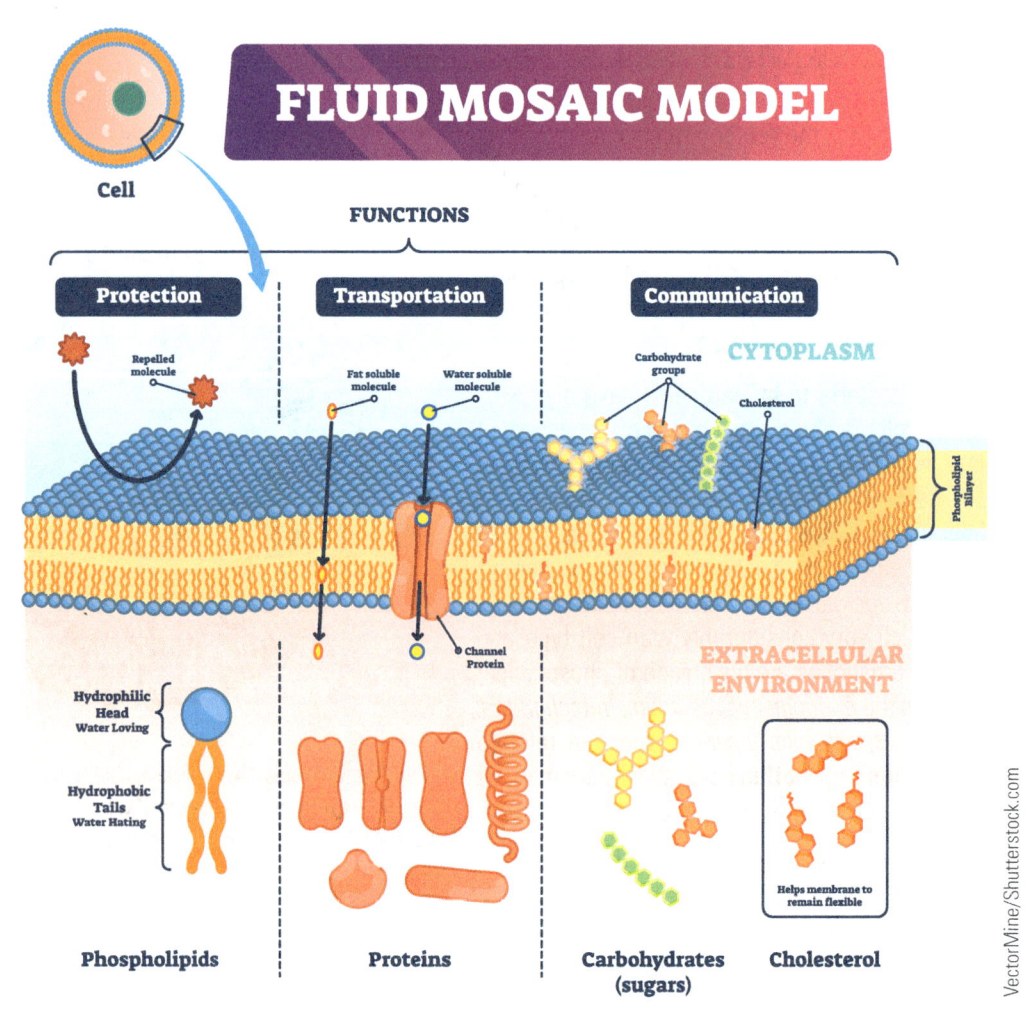

Figure 6.4 Fluid Mosaic Model. Proteins and other molecules associated with the phospholipid bilayer perform many functions that affect the overall behavior of a cell.

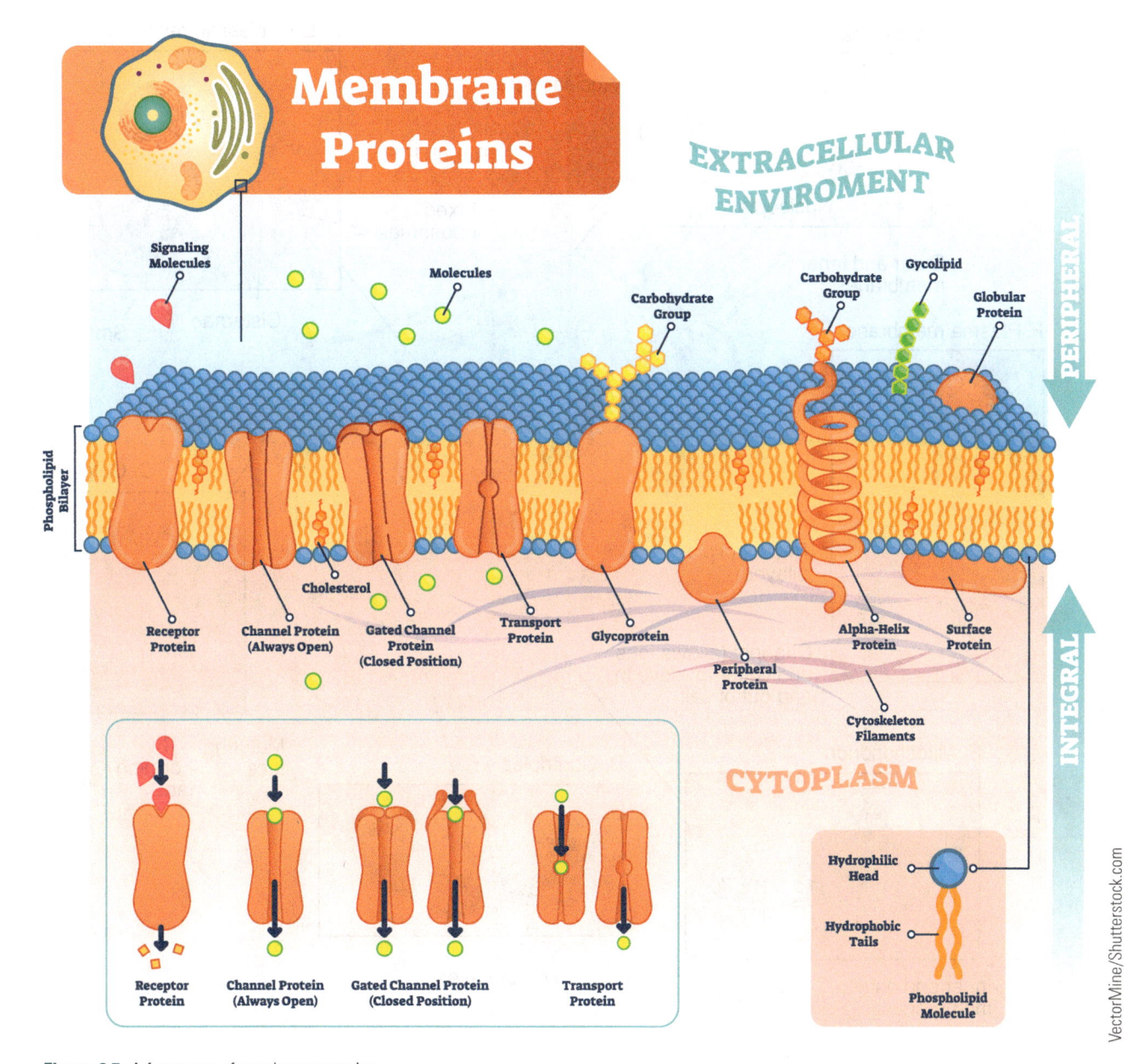

Figure 6.5 A few types of membrane proteins.

Figure 6.5 illustrates some of the structural components of the plasma membrane. **Integral** proteins pass all the way through the phospholipid bilayer, while **peripheral** and **surface** proteins are only on the inner or outer surface of the cell. The outside of a cell is studded with carbohydrate chains that serve as "name tags" for the cell. The **glycocalyx** is the collective term for all of the carbohydrate chains. **Glycoproteins** and **glycolipids** are carbohydrate–protein and carbohydrate–lipid complexes that help make up the glycocalyx.

Functions of Membrane Proteins

Proteins perform many cellular functions, including activities occurring at the plasma membrane. There are three broad *functional* categories of proteins associated with the plasma membrane. **Attachment proteins** attach the plasma membrane to other cells or the extracellular matrix. **Receptor proteins** initiate cellular activities when activated by a signaling molecule. And **transport proteins** allow materials to move through the plasma membrane.

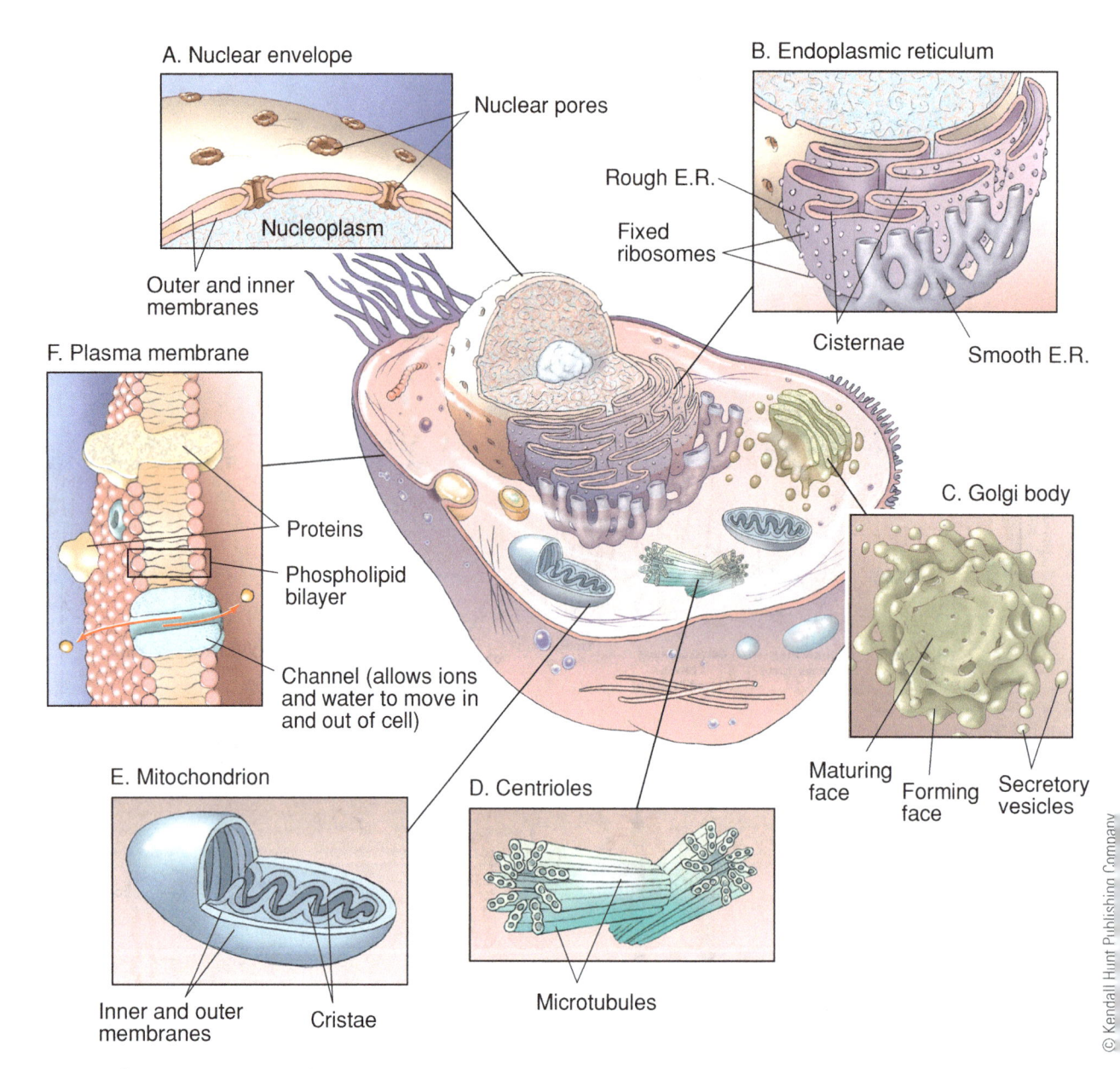

Figure 6.6 Some specific structures of a simple eukaryotic cell.

Membrane proteins are also found in the membranes of organelles. The combination of proteins embedded in the plasma membrane and in the organelle membranes directs much of the cellular metabolism. Figure 6.6 illustrates some of the internal membranes of a generic eukaryotic cell.

Movement Through a Phospholipid Bilayer

Simple Diffusion

Recall from Lab Exercise 2 that all molecules and atoms move all of the time. When *solutes* are added to a *solvent*, the solute particles will gradually spread out within the solvent until the concentration of solute is equal throughout the solution. **Simple diffusion** always occurs along a **concentration gradient**. Solute particles move from an area of higher solute concentration toward an area of low solute concentration until the concentration reaches equilibrium. *Brownian movement* of the molecules in the solution provides the kinetic energy for diffusion to occur. Recall from Exercise 1 that the *temperature* of a material is actually a measurement of the average kinetic energy of the molecules, or Brownian movement. The greater the kinetic energy, the faster diffusion will occur.

Simple diffusion occurs when sugar is added to iced tea. At first, the sugar may sit on the bottom of the glass, but over time the sugar will diffuse (or dissolve) into the water, causing the tea to become sweetened. There are three factors that can speed the process of simple diffusion: agitation, heat, and the size of the solutes. Obviously stirring (agitation) speeds diffusion, like stirring tea after adding sugar. The sugar would diffuse eventually, but stirring speeds up that process. Likewise, sugar diffuses faster in hot tea than iced tea. Heat and stirring increase the kinetic energy of the solute particles. And powdered or granulated sugar will dissolve faster than an undisturbed sugar cube. Smaller particles move more freely than large particles or particles that are stuck together.

Diffusion Through a Membrane

Cells obtain some of what they need by diffusion. However, this diffusion must occur through the plasma membrane. Some substances can easily pass through the plasma membrane because they are very small and can pass between the phospholipid molecules. A phospholipid bilayer is *permeable* to small, non-polar molecules. But large particles will not diffuse through a phospholipid bilayer. Charged ions and polar molecules are repelled by the non-polar middle region of the bilayer, which is packed full of hydrophobic fatty acid chains. A phospholipid bilayer is therefore *impermeable* to large and highly charged molecules. Some substances like water or other small, polar molecules can sometimes slip through the phospholipids. Thus, a phospholipid bilayer is slightly permeable to small, non-polar molecules. The phospholipid bilayer is generally described as **semi** (or **selectively permeable** to reflect the fact that some materials easily pass through it, and some do not (Figure 6.7).

If the plasma membrane is impermeable to large molecules, how can cells obtain needed fuels and raw materials like carbohydrates and proteins? Cells have a number of methods for moving materials through the plasma membrane. Proteins embedded in the plasma membrane can transport some materials through the phospholipid bilayer by providing a tube-like opening all the way through the membrane. These transport proteins can allow for diffusion of materials through the plasma membrane if the solute particles can fit through the protein's opening. One of the key features that distinguishes one cell type from another is the presence or absence of certain proteins in the plasma membrane. **Facilitated diffusion** is diffusion of normally impermeable solutes through a plasma membrane through specific transport proteins. Therefore, the permeability of any cell's plasma membrane depends on the types of proteins present in it at any one time.

SELECTIVE PERMEABILITY OF LIPID BILAYER

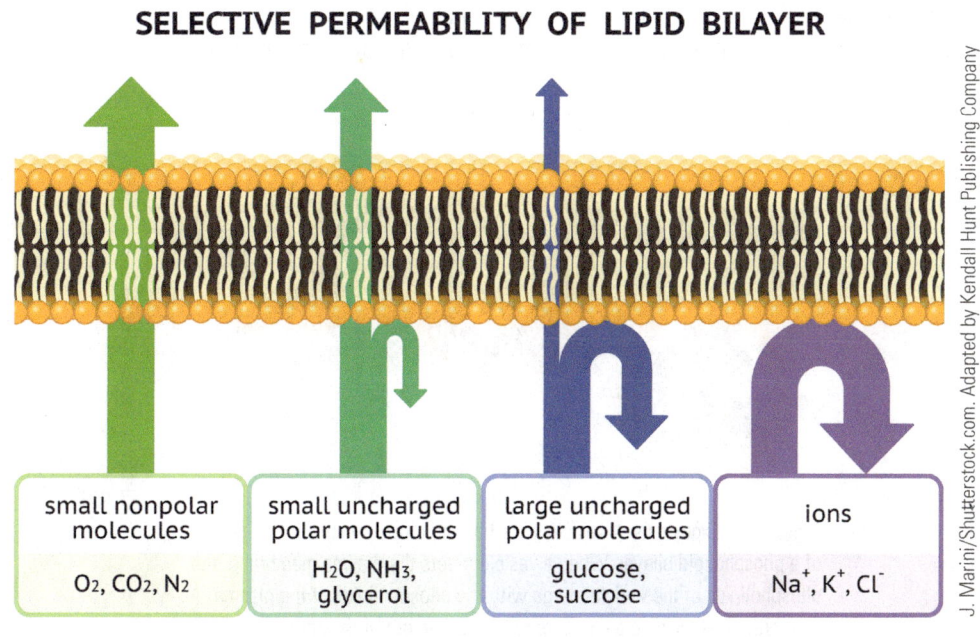

small nonpolar molecules	small uncharged polar molecules	large uncharged polar molecules	ions
O_2, CO_2, N_2	H_2O, NH_3, glycerol	glucose, sucrose	Na^+, K^+, Cl^-

Figure 6.7 Diffusion of some materials through a phospholipid bilayer.

THREE MAJOR TYPES OF ENDOCYTOSIS
Phagocytosis   Pinocytosis   Receptor-mediated endocytosis

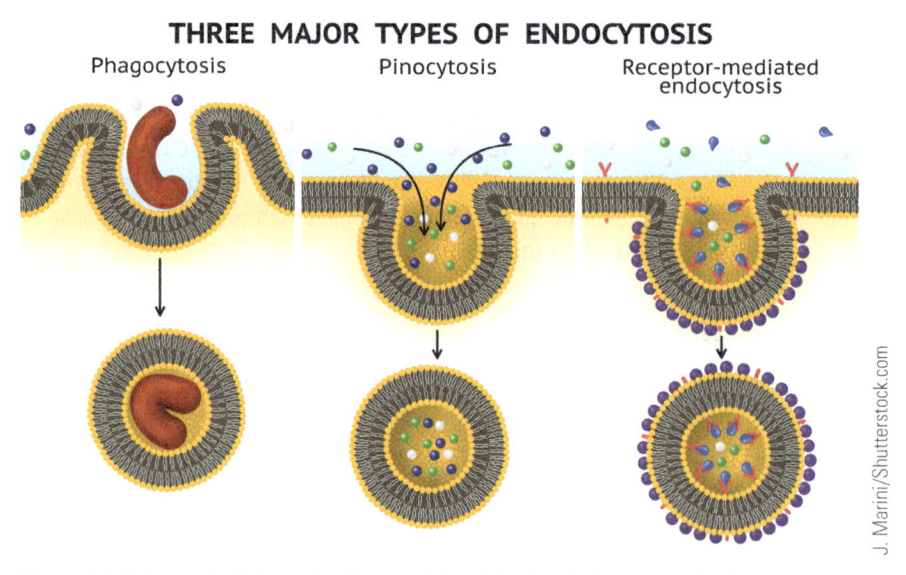

Figure 6.8 Endocytosis. Endocytosis of large solid particles is called phagocytosis. Endocytosis of liquids and dissolved materials is called pinocytosis. Endocytosis triggered by receptor protein activation is called receptor-mediated endocytosis.

Bulk Transport

Cells often need to take in large molecules or other substances that can't fit through a transport protein. This is called **bulk transport**. **Endocytosis** (*endo* = into; *cyto* = cell) occurs when cells form a vesicle around extracellular material. Endocytosis of solids is referred to as **phagocytosis**, endocytosis of liquids as **pinocytosis**, and endocytosis triggered by activating specific receptor proteins as **receptor-mediated endocytosis** (Figure 6.8).

When cells want to expel materials, the same strategies apply. Small non-polar materials diffuse directly through the plasma membrane. When a cell needs to release many particles that may be difficult to diffuse (such as hormones or ions), the cell undergoes a reverse version of endocytosis called **exocytosis** (Figure 6.9).

Simple and facilitated diffusion are **passive transport** processes, meaning that no metabolic energy is needed for movement to occur. Brownian movement from the heat of the organism or environment provides the necessary energy. But sometimes cells need to move materials against their concentration gradient. For example, kidney tubule cells must selectively remove and excrete wastes from the bloodstream while keeping nutrients and electrolytes in the bloodstream. The movement of

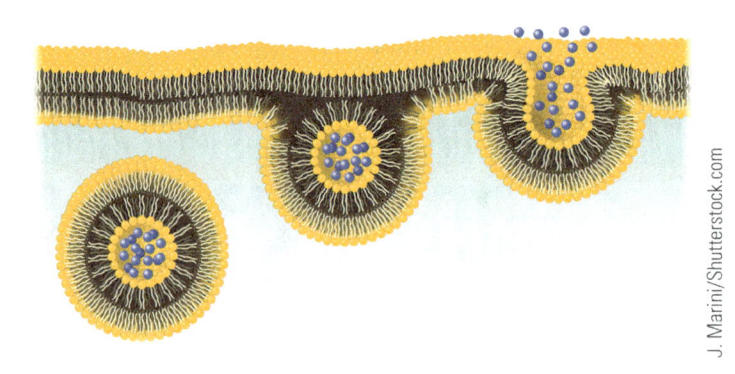

Figure 6.9 Exocytosis. A vesicle (here holding blue particles) is composed of a phospholipid bilayer. When a vesicle meets the plasma membrane, the phospholipids of the vesicle merge with the phospholipids of the plasma membrane, releasing the contents of the vesicle out of the cell.

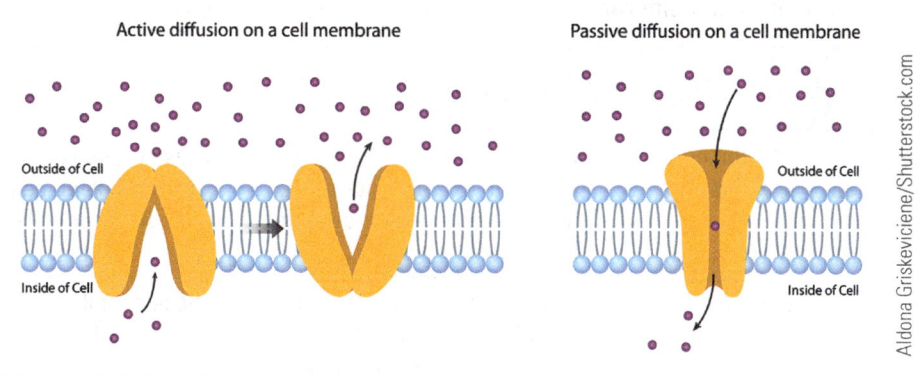

Figure 6.10 Active and passive transport. Active transport occurs when a carrier protein uses ATP to move substances through the phospholipid bilayer against the concentration gradient. Passive diffusion, shown here "facilitated" by a channel protein, occurs when materials pass through the membrane along the concentration gradient. No ATP is needed.

materials against the concentration gradient requires additional energy. **Active transport** processes require metabolic energy in the form of ATP (Figure 6.10).

It is often easy to assume that the plasma membrane of a cell is simply the "skin" of the cell, and that it doesn't significantly contribute to a cell's specific functions. However, this is far from true. Table 6.1 summarizes some of the many ways in which the plasma membrane contributes to cellular activities.

Osmosis

The plasma membrane is semi-permeable. However, water molecules, though polar, are small enough to pass through the phospholipid bilayer. Cells are surrounded by extracellular fluid and filled with cytosol, so water is on both sides of the plasma membrane. When there is a concentration gradient across the membrane, but the solute particles cannot go through the membrane, the system is not at equilibrium. Since the particles cannot diffuse, water, which can pass through a plasma membrane, goes through the membrane instead of the solute in order to achieve an equal concentration of solute on both sides of the membrane. Cells that fill with water become plump and **turgid**, like a water balloon. If this causes the cell to gain too much water, it may **lyse** (burst). If a cell loses too much water, it cell may dehydrate and become **flaccid** (limp), or eventually **crenate** (shrivel up) to the point it can no longer function. In plants, the cell wall is rigid. The plasma membrane pulls away from the cell wall, causing the cells to look like mostly empty boxes containing small areas of cytoplasm. In plants, the effect of dehydration is called **plasmolysis**, and even though the cell walls are rigid, the plant will wilt.

Osmosis is often defined as the "diffusion of water" through a semi-permeable membrane. While this is technically correct, it is an awkward definition since we typically think of diffusion as the movement of *solutes*, not *solvents* like water. **Osmosis** is more easily thought of as the movement of water through a semi-permeable membrane from an area of low solute concentration to an area of high solute concentration. This occurs when the solutes cannot diffuse through the membrane, and water molecules move to attempt to reestablish an equal concentration of solutes on either side of the membrane (Figure 6.11). This may seem counterintuitive since water appears to move "uphill" against gravity. Yet this occurs due to osmotic pressure created by the difference in solute concentrations.

The side of a semi-permeable membrane with a higher solute concentration is described as **hypertonic** (*hyper* = more; *tonicity* = concentration), while the side with the lower solute concentration is described as **hypotonic** (*hypo* = below/lower). If the concentration of solutes is equal on either side of the membrane, they are **isotonic** to each other and water can pass freely through the membrane in either direction by Brownian movement. We can understand that osmosis is just a special case of the *diffusion of water* if we consider that solutes occupy space in a solution. The greater the concentration of solutes, the lower the concentration of water molecules. Water diffuses from an area of higher concentration of water

Table 6.1 A summary of plasma membrane activities, structures, and functions

Major membrane function	General structures involved	Specific molecules involved	Functions
How cells connect to each other or to an extracellular matrix	Attachment proteins	Integral proteins can be attached to extracellular matrix fibers or intracellular cytoskeleton fibers.	These proteins provide critical support to keep cells from individually or collectively falling apart.
	Desmosomes	These protein–carbohydrate complexes link the plasma membranes of adjacent cells.	These complexes provide loose attachments between cells, keeping cells connected while allowing fluid to pass between them.
	Gap junctions	Channel proteins on one cell line up with channel proteins on another cell, creating tubes that connect their cytoplasm.	These tubular connections between cells allow small molecules like ions to pass quickly from one cell to the next.
	Cell Adhesion Molecules (CAMs)	Integral proteins of one cell can directly connect to integral proteins of another cell, acting as cellular "glue."	These proteins tightly link cells. Fluids and other materials are unable to pass between cells.
How cells communicate with or respond to each other	Receptor Proteins	Receptor proteins have an active site that can bind to specific extracellular signal molecule such as hormones or neurotransmitters.	A chemical messenger outside the cell can trigger a receptor protein to activate cellular enzymes, open channels, initiate mitosis, or synthesize proteins.
	G-Protein Linked Receptors	A receptor protein, when triggered by an appropriate signal molecule, releases a *G-protein* inside the cell.	G-protein receptors relay a chemical message from outside the cell to inside the cell, activating other cellular processes.
	Cell Recognition Proteins	These proteins on one cell can fit the shape of another cell's *glycocalyx*.	These proteins allow cells to identify "self" cells versus invading "non-self" cells.
How cells can move materials into the cell	Direct Diffusion	Small, non-polar molecules and water move directly through the membrane down their concentration gradients.	Molecules like O_2 and CO_2 diffuse directly into cells down their concentration gradients.
	Transport Proteins	*Channel Proteins* have a hole through them, allowing any molecules that fit through the holes diffuse through the plasma membrane.	Proteins "facilitate" the diffusion of particles into the cell. Some "gated" channel proteins can be triggered to open or close.
		Carrier proteins change shape when triggered, which transports any materials that fit the carrier through the membrane.	Carrier proteins may move particles down their concentration gradient (facilitated diffusion), or they may use ATP to pump particles against their concentration gradient (active transport).

	Endocytosis	*Pinocytosis* occurs when proteins on the inner surface of the membrane form a vesicle to engulf liquids.	A cell forms a vesicle, capturing liquids and dissolved molecules from outside the cell.
		Phagocytosis occurs when proteins on the inner surface of the membrane form a vesicle to engulf solid particles.	A cell extends *pseudopodia* to engulf a large solid, such bacteria or cell fragment, and then forms a vesicle around it.
		Receptor-mediated endocytosis occurs when receptors trigger inner surface proteins to form a vesicle.	A cell forms a vesicle, as in phagocytosis or pinocytosis, but selectively—only when a receptor protein is activated.
How cells can move materials out of the cell	Direct Diffusion	Small, non-polar molecules and water move directly through the membrane down their concentration gradients.	Molecules like O_2 and CO_2 diffuse directly out of cells down their concentration gradients.
	Transport Proteins	Channel or carrier proteins move materials out of the cell.	Allows for "facilitated" diffusion of particles out of the cell. Some "gated" channel proteins can be triggered to open or close.
	Exocytosis	A vesicle from within the cell merges with the plasma membrane, causing the material in the vesicle to be released from the cell.	Much like a bubble popping, the vesicle reaches the surface of the cell and pops open, releasing hormones, neurotransmitters, or other compounds generated by the cell.

OSMOSIS

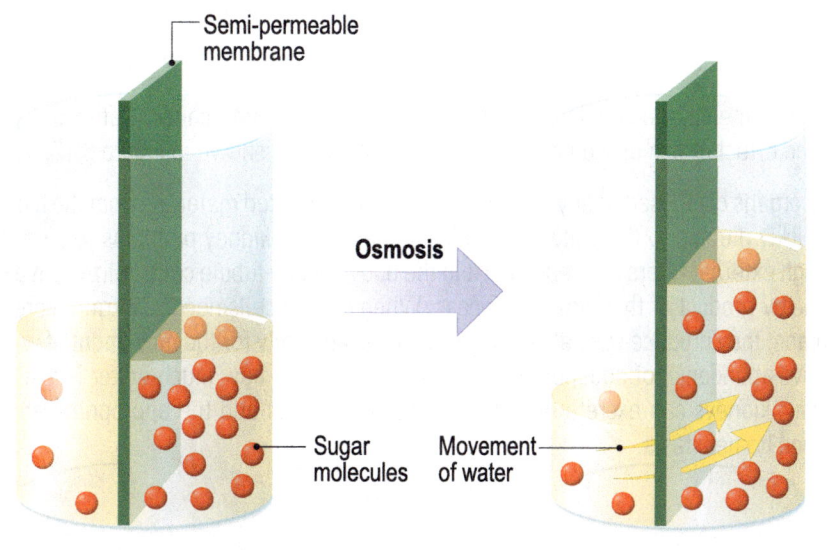

Figure 6.11 Osmosis. Water moves through a semipermeable membrane from the side with more water and less solute to the side with more solute and less water until the two sides of the membrane have the same solute concentration.

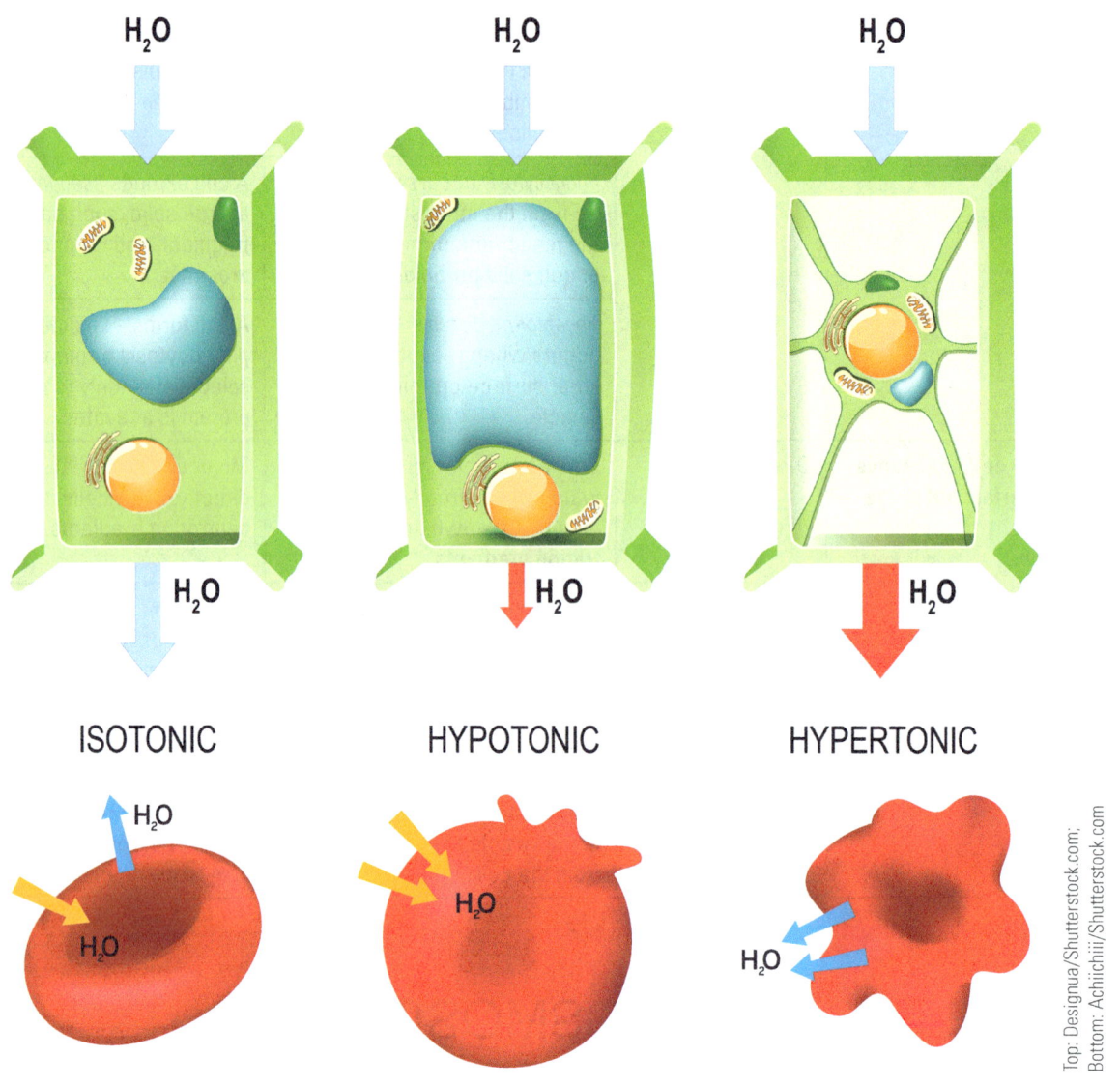

Figure 6.12 Plant cells (top) and animal blood cells (bottom) in isotonic, hypotonic, and hypertonic solutions.

(the hypotonic side of the membrane with less solute) toward an area of lower concentration of water (the hypertonic side with more solute). The effect of osmosis on cells in different solutions is shown in Figure 6.12.

Kidneys are complex organs composed of tiny tubules that remove unwanted materials from the blood. If they stop functioning, the body can quickly die due to the buildup of toxins. Patients with kidney problems require *dialysis*, which involves pumping blood through external filters and replacing it in the body. Kidney tubule cells utilize active transport, osmosis, and diffusion. Urine is the by-product of this filtering process. When certain cells need to permit more water to move through their plasma membrane, they produce special channel proteins specifically for the movement of water called **aquaporins**. In kidneys, the relative abundance of aquaporins in tubule cells controls how much water is lost or retained during urine formation. In this way, kidneys can excrete unwanted dissolved solutes and the appropriate amount of water from the body to maintain blood homeostasis.

Lab Skill 6.1 Using Dialysis Tubing

Dialysis refers to selective filtration. Kidneys perform dialysis as they selectively filter wastes. Dialysis machines function as artificial kidneys for people experiencing kidney failure. **Dialysis tubing** is a synthetic material commonly used in clinical settings and cell biology labs to selectively filter tiny molecules out of a solution.

Dialysis tubing is made of cellulose acetate. It is typically manufactured as a tube that can be tied or clamped at one end to form a bag (Figure 6.13). Cellulose acetate is porous. Molecules with a mass less than 14,000 AMU can typically fit through the holes. Molecules larger than 14,000 AMU are too large to fit through the holes and remain on their side of the membrane. Dialysis tubing therefore is a selectively permeable membrane. Water molecules, with a molecular mass of 18 AMU, are able to pass through the holes in the membrane. (However, since water molecules are attracted to each other by hydrogen bonds, if a bag formed from dialysis tubing is filled with water and the bag is not submerged, water tends to stay in the bag rather than leak out.)

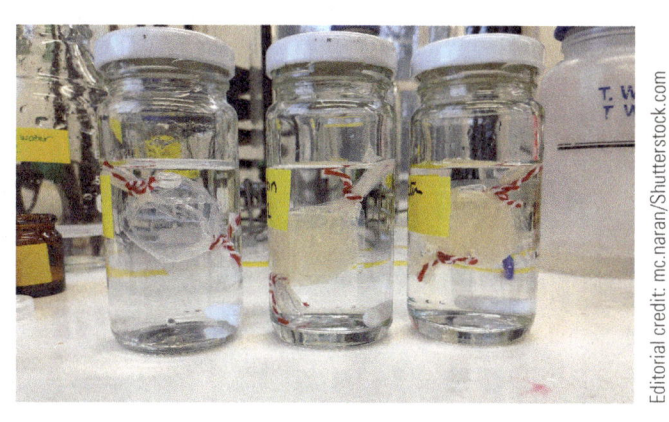

Figure 6.13 Three dialysis tubing bags containing different solutions in three jars.

Dialysis tubing typically comes as a long roll. It tends to look like tape, but is actually a flat tube. In order to use it, it must be soaked for a few minutes in water. Then the tube can be opened by gently rubbing it between your fingers. One end of the dialysis tubing can be tied or clamped to make a bag. The tubing can be filled with a solution and tied off at the opposite end to simulate the semipermeable plasma membrane of a cell.

Dialysis tubing can be used as a simple model of the plasma membrane. However, like any model, it is important to note its limitations. Since dialysis tubing is not composed of phospholipids, it does not have a non-polar inner region like the phospholipid bilayer. It does not selectively filter molecules based on their charge or polarity. But it does selectively filter particles based on molecular weight, blocking the movement of large particles and allowing water and other small particles to pass through.

Lab Skill 6.2 Using (More) Chemical Indicators

Recall that indicators are chemicals that indicate the presence of a certain compound by producing a visible color change. There are many types of chemical indicators. Review Lab Skill 3.1. You have already used *iodine*, *Benedict's solution*, *Albustix test strips*, and *blue litmus*. In this lab exercise, you will use some of these again and learn two more, **silver nitrate** and **barium chloride**. These indicators produce a white precipitate. A **precipitate** is a solid substance that separates and settles out of a solution. In this case, if the indicator gives a positive result, the solution will appear to be "snowing."

Indicator/reagent	Normal reagent color	Tests for the presence of	When chemical is present, changes color to	Special considerations
Silver Nitrate ($AgNo^3$)	Clear and colorless	Chloride ions (Cl^-)	White precipitate	Works instantly
Barium Chloride ($BaCl^2$)	Clear and colorless	Sulfate ions ($SO_4)^{-2}$	White precipitate	Works instantly

Problem to Solve 6.1 Can I Predict When Diffusion or Osmosis Will Occur?

Your goal is to work as a group to design an experiment that shows which of five types of compounds can diffuse through dialysis tubing. You will need to develop a hypothesis and design an experiment to test it. You will need to determine if the following compounds can move through the dialysis tubing membrane.

- Water
- Starch (a polysaccharide)
- Na_2SO_4 (a salt)
- NaCl (a salt)
- Albumin (a protein)

These compounds have been mixed so that the Na_2SO_4 and starch are mixed together and the NaCl and albumin are mixed together. The concentrations are undetermined. You will have to use them as they are provided.

Unlike other activities, this one requires you to think independently, develop a hypothesis, design your experiment, and write out your own protocol.

The goal and materials have been provided, but you will need to work with your group to quickly develop a *specific* hypothesis. You must be more specific than "some things will move through the dialysis tubing." Hypothesize which ones! (Hint: review Lab Skill 6.1.)

Make specific predictions based on your hypotheses. A prediction is an "if…then…" statement.

Design a controlled experiment, and write out the protocol. Determine your variables and the data you will collect. A data table has been started for you. You will need to complete it or design your own.

Hints:

Consider how you will test for the presence of compounds on the inside or outside of the dialysis tubing membrane (Review Lab Skills 5.1 and 6.2). Also consider how you will determine if water moves through the membrane. There are multiple ways to do this.

Ensure your experiment has a control.

Take precautions to prevent cross-contaminating your solutions. Ensure all glassware and pipettes are clean, and rinse all equipment between uses.

Wear safety glasses.

Title: Demonstrating Osmosis and Diffusion

Purpose

Goal: Determine which of five compounds diffuse through a "fake" cell membrane made of dialysis tubing.

Hypothesis: _____

Predictions: _____

Materials:

(Per group)
- 2 beakers (100-150 ml)
- 2 pieces of dialysis tubing (approximately 8 cm each)
- String or clamps to close off tubing (DO NOT throw away clamps after the experiment!)
- Small graduated cylinder
- 2 Disposable transfer pipettes (3 ml)
- Distilled water bottle
- 8 Test tubes
- Test tube rack
- Wax pencil or tape for labeling tubes
- Mass scale and weigh boat

(Per class)
- Albumin/NaCl solution (concentration undetermined)
- Na_2SO_4/starch solution (concentration undetermined)
- Dropper bottle of $BaCl_2$ solution
- Dropper bottle of $AgNO_3$ solution
- Dropper bottle of iodine solution
- Albustix test strips

Protocol:

Continued from previous page.

Date

Title

Purpose

Continued from previous page.

Protocol

Data:
You may need to modify this table depending on how you designed your experiment.

BEFORE

Tube #	Sample from	Indicator	Pos./neg. result	Implication
1				
2				
3				
4				
5				
6				
7				
8				

AFTER

Tube #	Sample from	Indicator	Pos./neg. result	Implication
1				
2				
3				
4				
5				
6				
7				
8				

	Mass of _____
Before	
After	
Implication:	

Researcher signature | Date | Supervisor/Instructor signature | Date

Continued from previous page.

Conclusions:

Which materials diffused through the dialysis tubing? How do you know?

Which materials did not diffuse through the dialysis tubing? How do you know?

Did osmosis occur? How do you know?

What can you infer about the molecular weight of each of the five compounds you tested?

Problem to Solve 6.2 Can I Demonstrate Differences in Osmotic Pressure?

In this problem, your goal is to work with your group to show evidence of osmotic pressure and demonstrate that osmotic pressure varies with both the concentration of the solute and the temperature of the system.

You will have to make starch solutions at two specific concentrations. Review Lab Skill 2.2 if necessary.

You will need an *osmometer*, or a device that measures osmosis. You will create a simple version of an osmometer using a disposable pipette and dialysis tubing as shown below.

Refer to the figure below and follow these steps to build an osmometer:

1. Cut pipette bulb at the very end of the bulb.
2. Slide a moistened 12cm piece of dialysis tubing over the pipette bulb.
3. Tie a string tightly around the dialysis tubing and pipette at the base of the bulb. Do not crush pipette shaft but ensure liquid in the dialysis tubing will not leak around shaft.
4. With tip pointed down and firmly plugged with a finger, add 8 ml syrup/food coloring solution to dialysis tubing bag.
5. Fold over the open end of dialysis tubing and seal with a dialysis tubing clamp.
6. Turn osmometer so it is pointing up. Mark the location of the syrup solution on the pipette shaft with a permanent marker.

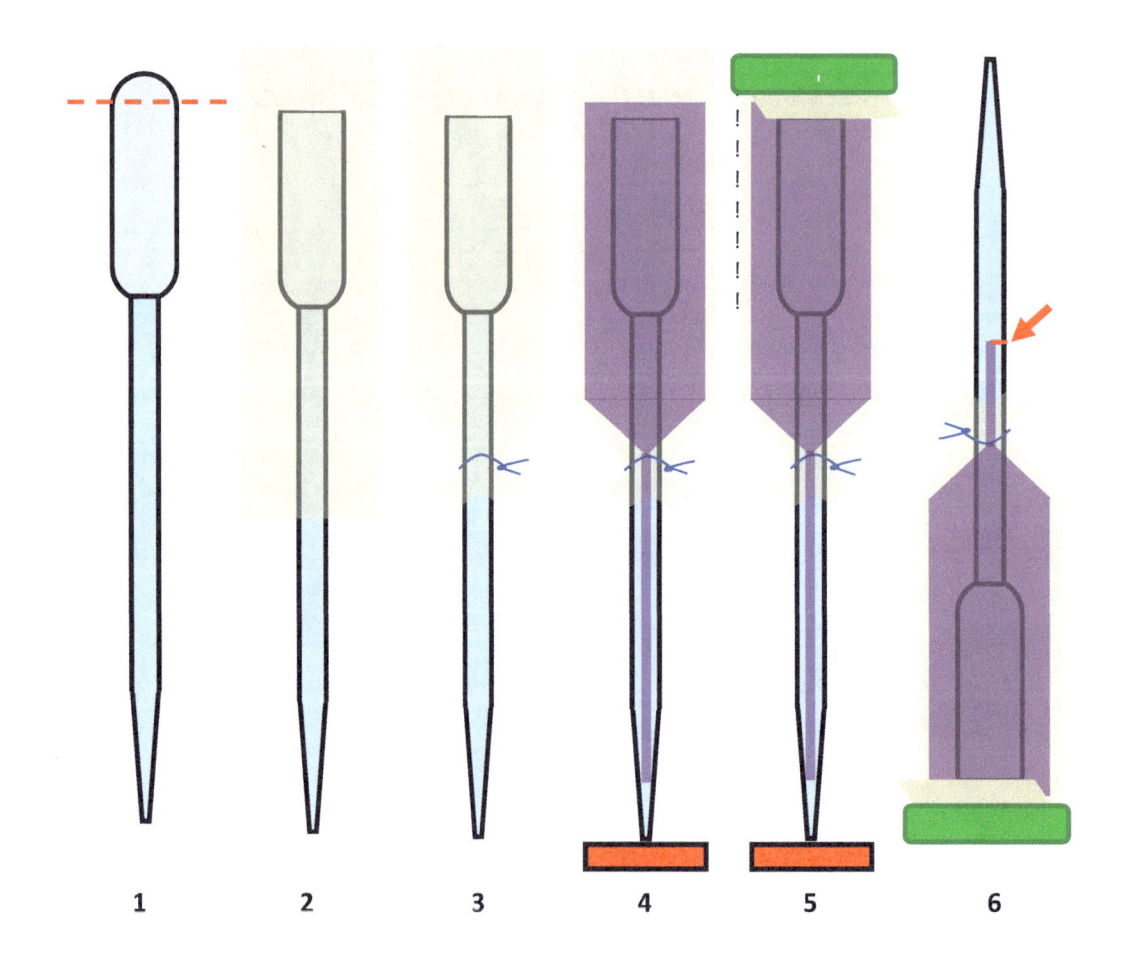

Laboratory Exercise 6 Cell Biology I: Cellular Membranes and Structures

Title: Demonstrating osmotic pressure at two temperatures

Goal: Demonstrate that osmosis creates osmotic pressure, and that the pressure varies with the solute concentration and the temperature of the system.

Hypothesis: _____

Prediction: _____

Materials:
(Per group)
- 6 Disposable transfer pipettes (3 ml)
- 5 Beakers (2x 100 ml, 3x 600 ml)
- 4 Pieces of dialysis tubing (12 cm each)
- 4 Tubing clamps (Do NOT throw clamps away after the experiment!)
- 4 pieces of string (approximately 15 cm each)
- Scissors
- Small graduated cylinder
- Small centimeter ruler
- Distilled water bottle
- 2 Glass stir rods.
- Tape and marker to label beakers

(Per class)
- Large flask containing 1 liter of 10% Karo syrup solution
- Large flask containing 1 liter of 30% Karo syrup solution
- Food coloring dropper bottles (multiple colors)
- Hot plates

Protocol:
1. Fill large beaker half full of distilled water and place on hot plate to begin heating. Set hotplate to high. While waiting for water to heat, move on to the next steps. But once water begins to boil turn down to 90°C.

2. Label the two 100 ml beakers "10% syrup" and "30% syrup." Obtain approximately 50 ml of each syrup solution in each beaker.

3. Add 3 drops of food coloring (any color) to the 10% syrup solution. Stir well with a glass stir rod.

4. Record the color in the data table.

5. Add 3 drops of a different color to the 30% syrup solution. Stir well with a glass stir rod.

Title

Purpose

Continued from previous page.

Protocol

6. Record the color in the data table.

7. Pour 300 ml distilled water into a clean 600 ml beaker. Label beaker with group number and 90° C.

8. Pour 300 ml distilled water into a clean 600 ml beaker. Label the beaker with group number and 27° C.

9. Prepare two osmometers with 8 ml of the 10% syrup solution in each.

10. Prepare two osmometers with 8 ml of the 30% syrup solution in each.

11. While keeping the osmometers pointed straight up, mark the level of the syrup solution in each osmometer by marking the shaft of the pipette with a permanent marker

12. Use the third large beaker and squirt bottle to gently rinse the outside of the osmometers with distilled water.

13. Place one osmometer with 10% syrup and one osmometer with 30% syrup in the beaker labeled 90° C that is on the hot plate. Record the start time.

14. Place the remaining osmometer with 10% syrup and the remaining osmometer with 30% syrup in the beaker labeled 27° C. Leave it at room temperature. Record the start time.

15. **Wait about 50 minutes**, checking to make sure the syrup solution does not come out of the tip of the pipette. If the syrup solution gets close to the tip before 50 minutes, go on to the next step.

16. Carefully remove each of the osmometers and immediately mark the level of the syrup solution on the shaft of the pipette with a permanent marker. Record the stop time in the data table.

17. Measure the distance between the marks on the pipette shaft (in mm). Record this in the data table. Then divide the distance by the number of minutes you ran the experiment to estimate a relative rate of osmosis (mm/minute).

Continued from previous page.

Data:

Osmometer Concentration	10%	30%	10%	30%
Temperature	27° C	27° C	90° C	90° C
Color				
Start time (T_i)				
End time (T_f)				
Distance (D) solution moved up the shaft of the pipette (mm)				
Difference in time ($\Delta T = T_f - T_i$)				
Relative rate of osmosis ($R = D/\Delta T$)				

Title

Purpose

Continued from previous page.

Protocol

Conclusions:

How did concentration of the syrup solution and the heat of the system relate to the distance the solution traveled up the pipette shaft? Explain clearly.

Was the 10% and/or 30% solution hypotonic, hypertonic, or isotonic to the liquid in the beaker? How do you know?

Is the 10% solution hypertonic or hypotonic to the 30% solution? How can you tell based on your experimental results?

In the four osmometers, what was the "osmotic pressure" pushing against?

Which of the four osmometers demonstrated the greatest rate of osmosis (or highest osmotic pressure per unit of time)? Why?

Which osmometer demonstrated the lowest rate of osmosis? Why?

Could this experiment demonstrate how trees might get water to the tips of their highest branches? If so, what would this experiment imply about tree roots?

Problem to Solve 6.3 Can I Recognize the Effects of Hypotonic, Hypertonic, and Isotonic Solutions on Cells?

If someone was adrift on a life raft out in the ocean, you might think that the last thing they should worry about is dying of thirst. However, this is not true.

A human body can only go a few days without replenishing the water regularly lost in breathing, sweating, and urinating. Dehydration causes a cascade of physiologic effects that can quickly lead to death. For example, loss of too much water causes loss of blood volume and lower blood pressure, which then increases blood viscosity and chance of stroke. The heart has to work harder to try to pump less blood, resulting in excessive heart rate. This can lead to hypovolemic shock, unconsciousness, and death.

If stranded on a life raft in the ocean, water is plentiful. But it is saltwater. The concentration of salt in seawater is about 3.5%. Drinking sea water leads to further dehydration and an overabundance of salt, called *hypernatremia*. Excessive salt in the body can disrupt electrolytes, causing increased muscle and nerve activity, spasms, and convulsions. Eventually, hypernatremia will lead to confusion, lethargy, and coma (Figure 6.14).

Recall from Lab Exercise 2 that the concentration of salt in body fluids and cells is 0.9% NaCl. Saline is an *isotonic* salt solution. It matches the salt content of body fluids.

What would happen to a person's cells if they ingested sea water? What would happen to a person's blood if they were accidentally injected with distilled water? Create a hypothesis of what you expect to see and then observe the effects of hypertonic, isotonic, and hypotonic solutions on elodea, dog blood, and human cheek cells.

Figure 6.14 Don't drink the ocean! Drinking salt water speeds the effects of dehydration and can cause electrolyte imbalances that rapidly lead to death.

Laboratory Exercise 6 Cell Biology I: Cellular Membranes and Structures

Title

Title: Observing the effects of hypertonic, isotonic, and hypotonic solutions on cells

Purpose

Goal: To observe living cells subjected to different solution concentrations.

Hypothesis: (If living cells are placed in a hypertonic/isotonic/hypotonic solution...?)

Protocol

Materials:

(Per student)
- 3 Glass slides and 9 coverslips
- 3 Elodea leaves
- Drops of dog blood (from a vet clinic)—obtain from instructor when needed
- 3 Toothpicks
- Wax pencil
- microscope

(Per table)
- Distilled water bottle
- 10% NaCl solution
- 0.9% NaCl solution
- Methylene blue dropper bottle

Protocol:

(Review Lab Skill 5.4 for how to make a wet mount slide.)

1. Label three slides 1, 2, and 3 with wax pencil.
2. Place one elodea leaf on each slide. Ensure it is flat. Do not add water.
3. Add one drop of each salt solution:
4. To slide 1, add one drop of 10% NaCl.
5. To slide 2, add one drop of 0.9% NaCl.
6. To slide 3, add one drop of distilled water.
7. Add a coverslip to each slide.
8. Observe the slides with the microscope. Work up to the high power objective.
9. Sketch the cells. Label the sketches appropriately.
10. Remove the coverslip and elodea leaves and discard in the trash

Continued from previous page.

11. Rinse and dry the slides. If necessary, re-label the slides 1, 2, and 3.
12. Use a 1 ml transfer pipette to place a very tiny drop of dog blood on each slide.
13. Repeat steps 3 through 6 with the blood slides.
14. Clean the slides and re-label if necessary.
15. Use a toothpick to scrape the inside of your cheek. Dab the toothpick on slide 1. Dispose of the toothpick in the trash. Repeat with other toothpicks and slides.
16. Repeat step 3 and 4. Then stain the cells as described in Lab Skill 5.4.
17. Repeat steps 5 and 6.
18. Discard the slides in the glass disposal box.

Laboratory Exercise 6   Cell Biology I: Cellular Membranes and Structures

Title

Purpose

Continued from previous page.

Protocol

Data and Observations:

Elodea in 10% NaCl   Elodea in 0.9% NaCl

Elodea in distilled water

Continued from previous page.

Blood cells in 10% NaCl

Blood cells in 0.9% NaCl

Blood cells in distilled water

Date **Researcher name**

Title

Purpose

Continued from previous page.

Protocol

Cheek cells in 10% NaCl

Cheek cells in 0.9% NaCl

Cheek cells in distilled water

Researcher signature **Date** **Supervisor/Instructor signature** **Date**

Problem to Solve 6.4 Can I Design a Cell?

Goal: Demonstrate a clear understanding of how the presence of specific cellular structures determines a cell's functions. You and your group must **design a cell** (or tissue) that is well equipped to perform a specific job.

Your cell must
- a. **maintain homeostasis,**
- b. **demonstrate the basic characteristics of living things,**
- c. **perform the specific function assigned.**

Your *base model* cell comes with a plasma membrane, glycocalyx, nuclear membrane, chromatin, a few ribosomes, and a few mitochondria. It has no specialized membrane proteins and no other internal organelles. You need to purchase everything else. If necessary, use Lab Exercise 5 and your textbook to review the functions of different organelles.

You have a supply of parts from which to choose. Each part costs $1. You only have a budget of $10. Choose wisely. You can't just give your cell everything—that is *way too expensive*. If you need more parts than your budget allows, prioritize your needs and defend your reasons. Your project manager may grant you the extra funds if you have good rationales. Don't leave anything out that the cell needs to stay alive and perform its job.

Complete the Cell Specifications Sheet. Use colored markers or pencils to sketch a "blueprint" of your specific cell. Label all structures you add. You are encouraged to look up different types of cells to get ideas.

Cell Parts:

Active transport proteins (pumps)
Cell recognition proteins
Chloroplasts
Cilia
Cytoskeletal attachment proteins
Cytoskeletal proteins (all three kinds)
Desmosomes
Extra mitochondria
Extra ribosomes
Extracellular matrix
Flagella
Gap junctions
Golgi apparatus

Intracellular joining proteins/CAMs
Lysosomes
Membrane enzymes
Microvilli
Motor proteins
Transport proteins (channels/carriers)
Peroxisomes
Receptor proteins
Rough ER
Signal transduction proteins
Smooth ER
Vacuoles

(You may "special order" cellular structures mentioned in your biology textbook if you need a cell structure that is not listed above.)

Each team will be assigned one of the following:

Goals for Cell Design Teams: Your group must design...	Group Number
A) Cells that form a tough, impermeable, outer membrane of an organism.	
B) Cells that secrete peptide hormones into the blood when needed.	
C) Cells that can find, recognize, and engulf invading bacteria.	
D) Cells that can check for and destroy any toxins and poisons in the blood.	
E) Cells that can absorb and pass on selected nutrients from ingested food.	
F) Cells that can carry genetic information to another organism for reproduction.	
G) Cells that can make organic food molecules from sunlight and inorganic carbon.	
H) Cells that can work in unison to cause movements at the organism level.	

Name: _____   Date: _____

CELL SPECIFICATIONS SHEET

Specific Function of Cell:	
Types of organisms and organs where this kind of cell might be found:	
Standard Features (already included):	Plasma membrane (no proteins), glycocalyx, nuclear membrane, chromatin, a few mitochondria, a few ribosomes.

PRIORITY	CELL PART	REASON
1.		
2.		
3.		
4.		
5.		
6.		
7.		
8.		
9.		
10.		
11.		
12.		
13.		
14.		
15.		

Every item costs $1. Allowed budget is $10. Total cost of requested parts: _____

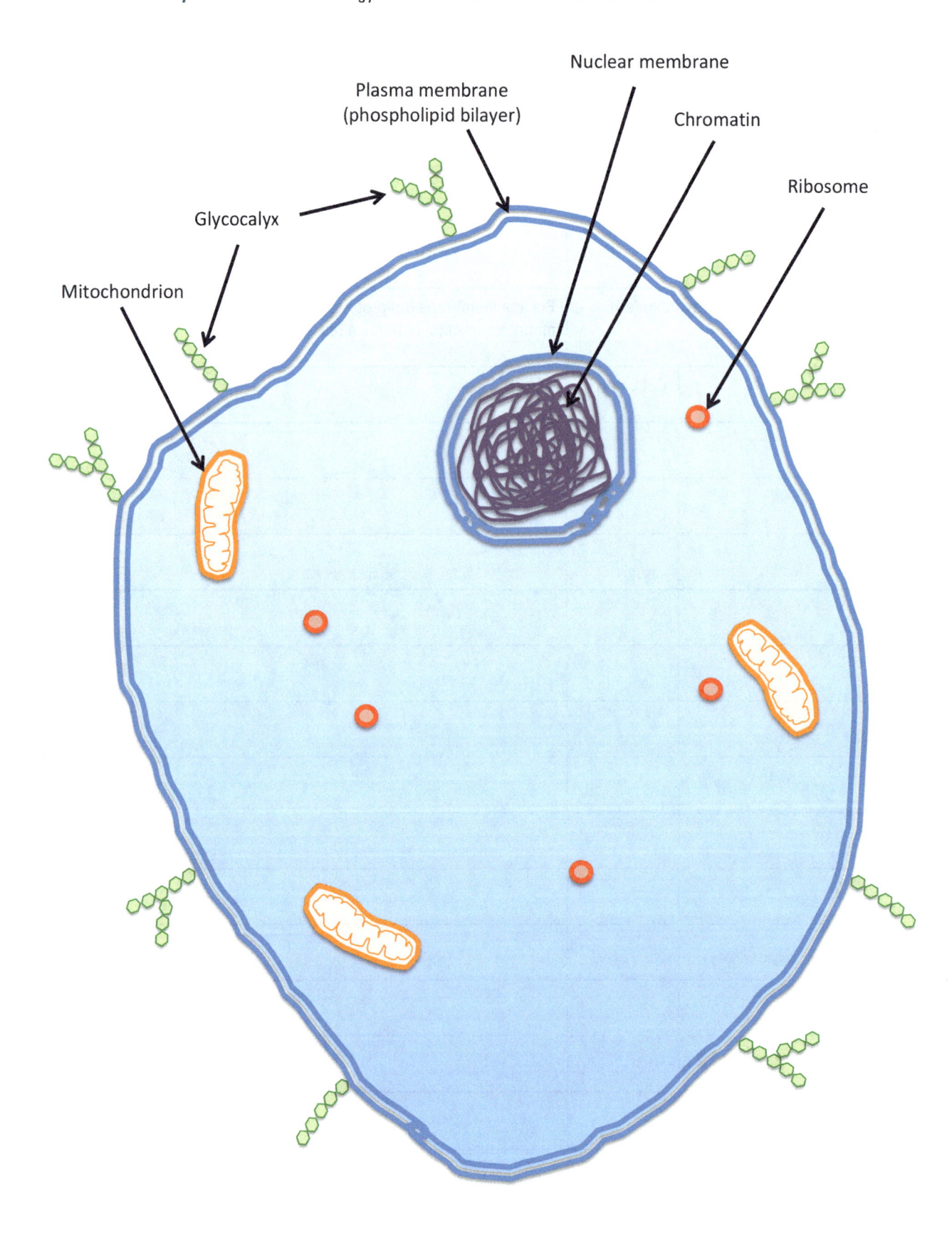

Laboratory Exercise 6   Cell Biology I: Cellular Membranes and Structures

- Mitochondrion
- Glycocalyx
- Plasma membrane (phospholipid bilayer)
- Nuclear membrane
- Chromatin
- Ribosome

Name: _____   Date: _____

POST-LAB QUIZ 6: What have you learned?

1. List at least eight specific molecular structures that may be present in the plasma membrane of an animal cell.

2. What does *not* diffuse through phospholipid bilayers? Give at least three ways cells get around this problem.

3. Imagine a section of dialysis tubing was clamped at one end, filled with 20% NaCl, and then clamped at the other end. It was then placed in a beaker of distilled water.

    a. Which would occur—osmosis only, diffusion only, both osmosis and diffusion simultaneously, or neither? Why?

    b. Would the dialysis tubing gain mass, lose mass, or stay the same? Why?

    c. After letting this dialysis tubing sit for 30 minutes, how could you test for the diffusion of NaCl in the beaker water?

4. An unconscious construction worker was brought in to the emergency room by his coworkers. They were all working outside in over 100° F weather, and it appears the man became dehydrated and overheated. Based on your understanding of the material covered in this lab, explain the likely consequences of taking each of the following actions.

    a. Hooking the man up to an IV of distilled water:

    b. Injecting the man with 0.9% NaCl solution:

    c. Pouring a sports drink in his mouth and hoping he will swallow it:

    d. Hooking the man up to an IV of a 10% NaCl solution:

5. Define osmosis using the terms hypertonic and hypotonic.

6. Some spas advertise that you can lose weight by tightly wrapping yourself in fabric strips soaked in Epsom salts (or other kinds of salts). If someone lost two pounds this way, when will they gain those two pounds back? Why?

7. Grocery stores frequently spray their fresh fruits and vegetables with water. Which would make the most sense—to spray vegetables with distilled water, tap water, or salt water? Why?

8. A powdered aspirin product claims to relieve headaches much faster than aspirin taken in tablet form. Would this be a reasonable claim? Why or why not?

9. Based on your answer to number 8, why should parents avoid giving a young child any time-release medication in tablet form?

10. Which membranous organelles of a cell often merge with each other and break apart into vesicles or smaller membranous structures? What purpose does this intracellular activity serve?

Laboratory Exercise 7

Molecular Biology II: Manipulating DNA

Concepts to Understand

After completing this lab, the student should be able to:

1. Describe restriction enzymes and their origins.
2. Explain the concept of gel electrophoresis.
3. Describe other new advances in DNA manipulation.
4. Explain the role of electrophoresis in analyzing DNA.
5. Interpret a DNA "fingerprint."

Skills to Learn

After completing this lab, the student should be able to:

1. Create an agarose electrophoresis gel.
2. Load a horizontal "submarine" electrophoresis gel.
3. Set up and operate an electrophoresis chamber.
4. Interpret DNA electrophoresis results.

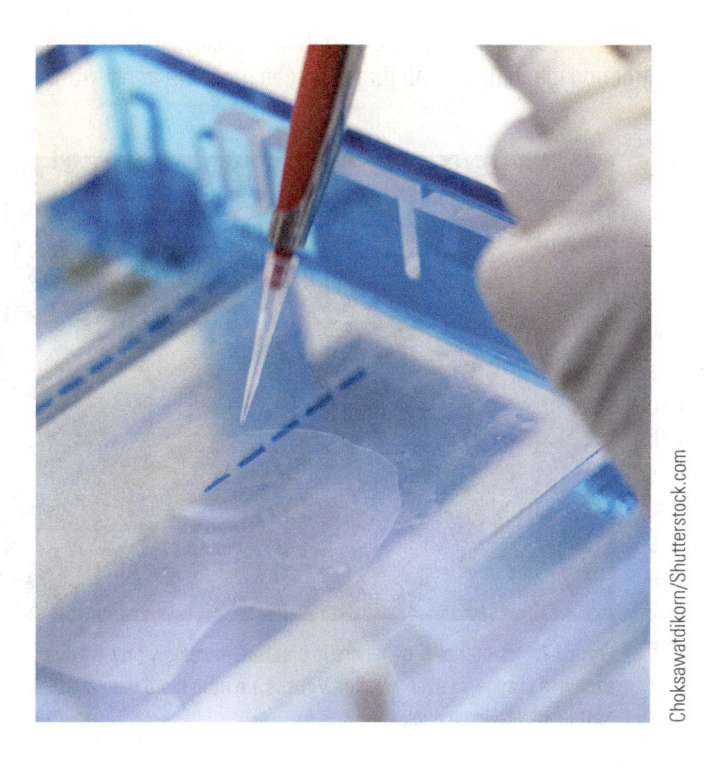

SEE ALSO

Campbell Biology—
   Chapter 19, section 19.2;
   Chapter 20, section 20.1.

JoVE (video)—DNA Gel
   Electrophoresis

JoVE (video)—Restriction Enzyme
   Digests

YouTube—Forensic Files season 1 Ep
   4 The footpath murders

Others:

Important terms to know

Agarose	DNA fingerprint	Plasmid
Anode	DNA ladder/ruler	Polymerase Chain Reaction
Bacteriophage	Electrophoresis	Recognition site
Bands (on a gel)	Electrophoresis buffer	Restriction enzyme/nuclease
Barcoding	Forensics	Restriction Fragment Length Polymorphism (RFLP)
Base pairs	Kilobase	
Blunt ends	Lambda	Sex pilus
Catenanes	Lanes (on a gel)	Sticky ends
Cathode	Loading/tracking dye	Submarine loading
Comb (in electrophoresis)	NEBcutter	Supercoiled
Conjugation	Nicked	Thermocycler
CRISPR-Cas9	Palindrome	Well (in a gel)

Pre-lab Preparation and Background Information

Bacteria are really amazing little bits of life. They are extremely tiny prokaryotes, yet have all of the characteristics of life. Compared to eukaryotic cells, they are very simple. They are unicellular, so there is no cellular differentiation or specialization. Yet, this little bag of DNA, proteins, and ribosomes maintains homeostasis, reproduces, eats, excretes, and even defends itself against other bacteria and viruses.

We normally don't think of bacteria as having enemies. But bacteria do continually deal with a threat from *bacteriophages*, or *phages* for short. **Bacteriophages** are viruses that infect bacteria (Figure 7.1). Viruses will be covered in greater detail later, but for now we need to know that bacteriophage viruses infect bacteria by inserting their nucleic acid into the bacterial cell. This changes the bacterial cell's own genetic "instruction manual." Recall that DNA can replicate itself, transcribe RNA, and provide instructions for making proteins. Viruses themselves are nothing more than nucleic acid (DNA or RNA) in a protein container. And the DNA or RNA in a virus is generally nothing more than a set of instructions to reprogram a bacteria to make more *viral* nucleic acids and proteins.

Bacteriophages do not infect eukaryotic cells, though many other kinds of viruses do. Bacteriophages are ubiquitous (found everywhere), and they are a constant threat to bacteria. When the bacteriophage comes in contact with a bacteria, it attempts to bind the cell. If it can, the virus then injects its viral nucleic acid into the host bacterial cell. The viral nucleic acid becomes integrated with the bacterial DNA, and the bacteria starts to synthesize viral proteins and nucleic acids. Eventually, the bacteria will fill up with viral components and burst open, killing the bacteria and releasing many more viruses to infect other nearby bacteria (Figure 7.2).

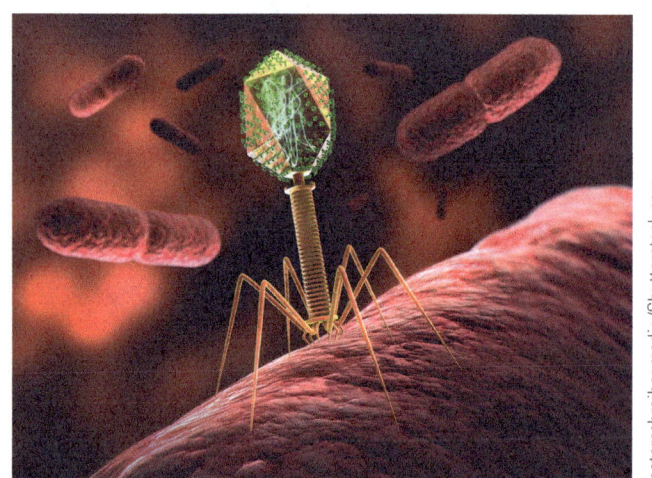

Figure 7.1 Bacteria, shown here in pink, are extremely tiny. A virus, here shown in orange, is even smaller. Viruses are little more than a protein container carrying a nucleic acid.

Restriction Enzymes

Bacteria have defenses against bacteriophages. Many different kinds of bacteria contain special **restriction enzymes** (also called **endonucleases**) that bind to and cut DNA. These enzymes recognize specific sites on DNA called **restriction sites**. Restriction sites on bacterial DNA are blocked by methyl groups. This prevents the restriction enzymes from destroying bacteria's own DNA. Viral nucleic acids are not protected by methyl groups. Restriction enzymes bind to restriction sites and deactivate the viral DNA.

Geneticists have isolated these enzymes and can use them to cut DNA like molecular scissors. Restriction enzymes are named after the bacteria from which it was isolated. The first three letters of the restriction enzyme name refer to the bacteria genus and species; if there is a fourth letter, it refers to a strain of that bacteria, and the Roman numeral at the end refers to the order in which it was discovered in that bacteria. For example, EcoRI came from *E. coli*, strain RY, and was the first restriction enzyme isolated from it. Each restriction enzyme has a particular restriction site at which it cuts. This is useful, because the ends of the cut fragments are specific. Figure 7.3 illustrates how EcoRI

LIFE CYCLE OF BACTERIOPHAGE

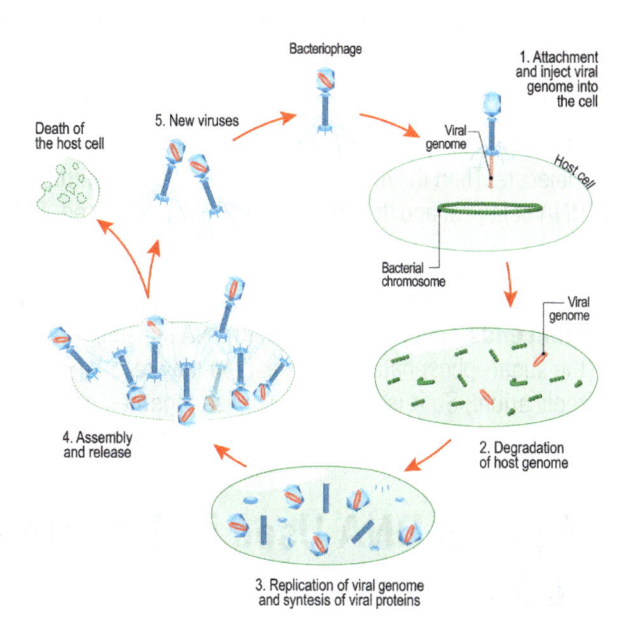

Figure 7.2 Bacteriophages attack bacteria by binding to the bacterial cell and injecting viral nucleic acid into the bacteria. The bacteria then produces viral parts until it bursts open, releasing the new viruses.

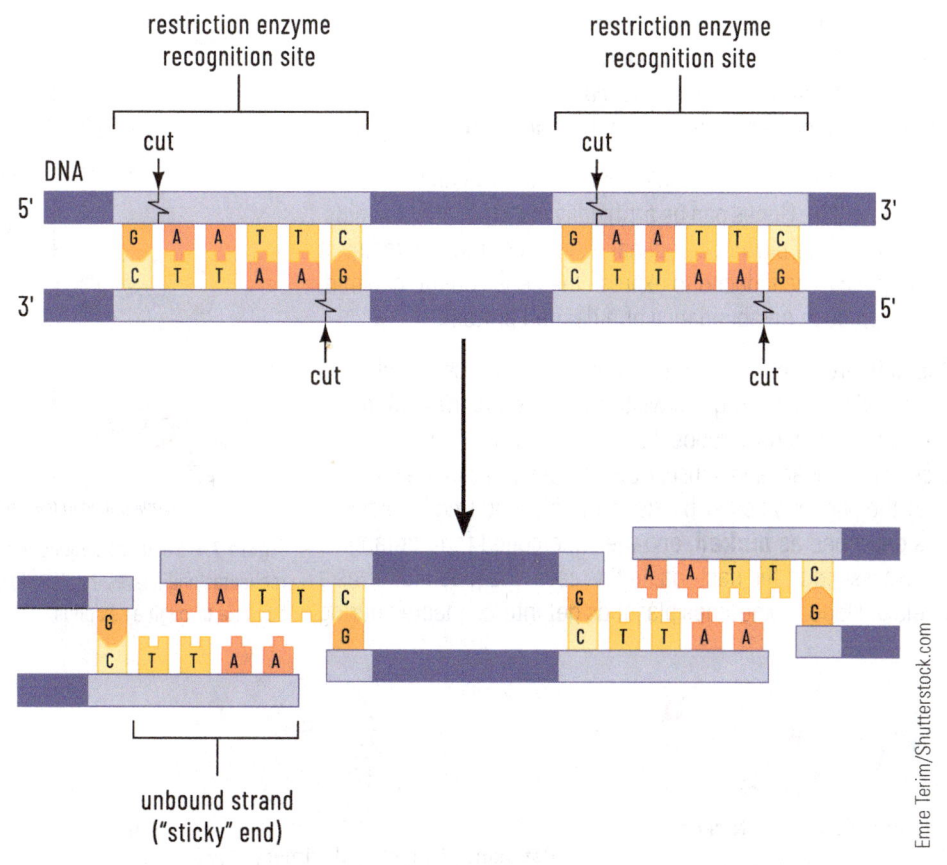

Figure 7.3 Recognition site for EcoR1. It cuts between the G and A in a sequence GAATTC. Note that the order of bases is the same going from 5' to 3' on either strand.

would cut a strand of DNA at two restriction sites. The DNA between the cuts may contain a gene of interest, and could then be isolated from the rest of the DNA fragments.

Restriction sites are typically four to eight base-pairs long and **palindromes**. A palindrome is a set of letters that read the same forwards or backwards, like "RACECAR" or "A TOYOTA" or, a favorite, "DAMMIT I'M MAD." Note that the DNA strand in Figure 7.2 has the same order of nitrogen bases when reading in the 5' to 3' direction on either strand.

Restriction enzymes break the phosphodiester bonds between the nucleotides in the sugar–phosphate backbone of the molecule. Then the hydrogen bonds between the nitrogen bases are easily separated. Restriction sites vary depending on the DNA being cut and the restriction enzyme being used. The larger the DNA, the more likely there will be more restriction sites.

When the ends of the cut DNA strand are staggered, they are called **sticky ends**. Any other DNA fragments cut with the same restriction enzyme, regardless of origin, will have the same sticky ends. If there are no overhangs, the ends are called **blunt ends**. Other enzymes, called DNA ligases, can connect DNA fragments, re-establishing the phosphodiester bonds in the sugar–phosphate backbone. (You may recall in Lab Exercise 4 that DNA ligases connected Okazaki fragments in DNA replication.) So it is possible to splice together, or "recombine," DNA from multiple sources into new strands. The ability to cut and recombine DNA is incredibly powerful, and much of the ongoing research in genetics utilizes these techniques.

Types of DNA Used in Research

Plasmids

Recall from Lab Exercise 5 that **plasmids** are a small, double-stranded, loops of DNA found in bacteria. A plasmid can be anywhere from 1,000 to 200,000 *base pairs* long, or 1 to 200 *kilobase pairs*. A **base pair** refers to a nitrogen base pair, such as adenine–thymine or cytosine–guanine. One **kilobase** is 1,000 base pairs. Plasmid DNA is not the genomic DNA of the bacteria. It is a small, non-essential circle of DNA that may provide extra benefits to the bacteria, but is not part of its *genome*. In bacterial **conjugation**, some plasmids (called *F-plasmids*, or *fertility plasmids*) can be transferred from one bacterium to another by a **sex pilus** (Figure 7.4).

Plasmids are important in molecular biology because they can be used to carry genetic information. Genes can be artificially inserted into plasmids. Bacteria will replicate the plasmids and make copies of the recombinant DNA. If the bacteria can be induced to express the genes on the plasmid, the bacteria may produce an abundance of a desired protein.

Because plasmids are circular, they can take on a number of shapes (Figure 7.5). Typically, a plasmid gets twisted up, or **supercoiled**, much like what can happen if you roll a rubber band between your fingers. In this case, the plasmid is compact and dense. However, enzymes in the bacteria may cut one of the phosphodiester bonds of the plasmid ring. A plasmid in this state is described as **nicked**, and the supercoiled form untangles and becomes accessible to the bacteria. In this case, the plasmid forms a large, open, relaxed loop. Sometimes plasmids get interconnected during

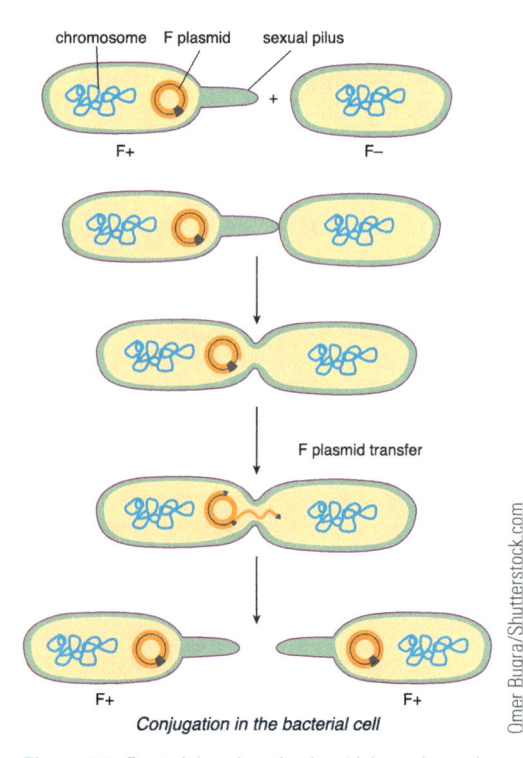

Figure 7.4 Bacterial conjugation in which one bacteria replicates and shares its plasmid DNA with another bacteria through a sex pilus.

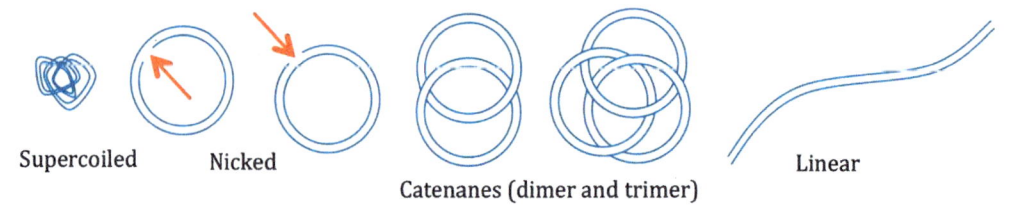

Figure 7.5 Forms of plasmids.

replication, like links in a chain. These forms of a plasmid are called **catenanes**. A catenane of two plasmids is a *dimer*, and a catenane of three plasmids is a *trimer*. If an enzyme completely cuts the plasmid, it becomes a linear piece of DNA, just like cutting a rubber band makes it a linear piece of rubber.

Lambda DNA

Lambda is a specific type of virus. Viruses can contain single-stranded or double-stranded DNA or RNA. Lambda contains double-stranded DNA. It is a bacteriophage that infects *E. coli* bacteria (Figure 7.6). Lambda DNA is about 49 kilobase pairs long. Lambda DNA is well characterized and understood, and it is used as a standard type of DNA in many areas of molecular biology. It can be used as a marker or standard for comparison, it can be cut with restriction enzymes and used as a **DNA ladder** (set of DNA fragments of known size used as a standard for comparison), and it can be attached to unstable fragments of DNA to make them more stable.

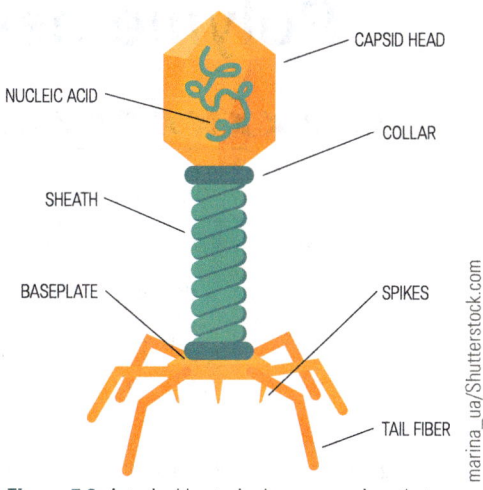

Figure 7.6 A typical bacteriophage, or a virus that infects bacteria.

Using Bacterial Enzymes in Genetics

You have already learned in Lab Exercise 4 that DNA can be used in many ways. Future biology courses will introduce you to many of those techniques. However, we will briefly discuss three of them here.

Polymerase Chain Reaction (PCR) is a process that uses DNA polymerase to make many exact copies of a DNA sample. A device called a **thermocycler** (Figure 7.7) uses heat cycles to facilitate enzyme action. DNA samples, along with a mix of free nucleotides and enzymes, are loaded into the thermocycler. The thermocycler is then programmed appropriately and allowed to run for a few hours. The enzymes open the DNA strand, replicate it, and then repeat the cycle. At the end of the run, the samples of DNA will have been replicated millions of times (Figure 7.7). This is how a single hair follicle or drop of blood can provide enough DNA evidence to place a suspect at the crime scene.

Restriction Fragment Length Polymorphism (RFLP) literally means "DNA fragments that are of different sizes, cut by restriction enzymes." Species of organisms have portions of their DNA that are highly conserved (unchanging) among the members of their species. It is possible with this technology to compare DNA samples to identify a species and to compare how close their DNA sequences are to determine their level of evolutionary relatedness. In biology, this is called **DNA barcoding**, where the banding patterns produced by RFLP serve as the bar code. This technique also allows researchers to determine relatedness in humans such as in ancestry, ethnic origins, and paternity testing. Much of human DNA is very similar from person to person. However, there are certain regions that are more variable among humans. If this DNA is isolated, and cut them with restriction enzymes, the pattern produced by the fragments is as good as a fingerprint.

RFLP is also used to make **DNA fingerprints**. A person's DNA fingerprint has nothing to do with their fingers or with touching objects. It is the RFLP pattern produced by subjecting a person's DNA to electrophoresis. The banding pattern produced is as unique to them as their fingerprint. DNA evidence is now used **forensically** (in criminal investigations). It is important to realize that the presence of one's DNA at a crime scene *does not prove their guilt*. It does indicate that they (or at least some of their body fluids or cells) were recently *present* at that location. The only people who theoretically share the same DNA fingerprint are identical siblings (though this has become more complicated with stem cell therapy and tissue/organ transplants).

CRISPR-Cas9 (*clustered regularly interspaced short palindromic repeats* and crisper-associated protein 9) is a fairly new technique for editing DNA. Bacteria that survive bacteriophage infection store fragments of the viral DNA in their cells. This "library" of viral DNA allows the bacteria to immediately recognize and destroy viral DNA if it is re-infected, much like our immune response. Since the Cas9 enzymes that cut viral DNA are "programmed" by short sequences of stored DNA and RNA, researchers can adapt these proteins to cut any DNA. Now, Cas9 can be "programmed" with a guide RNA of the researcher's choosing, and the Cas9 protein will find that sequence in a host cell and cut the DNA at that location. Cellular DNA ligases will attempt to repair the cut DNA, at which point new DNA fragments may be inserted (Figure 7.8).

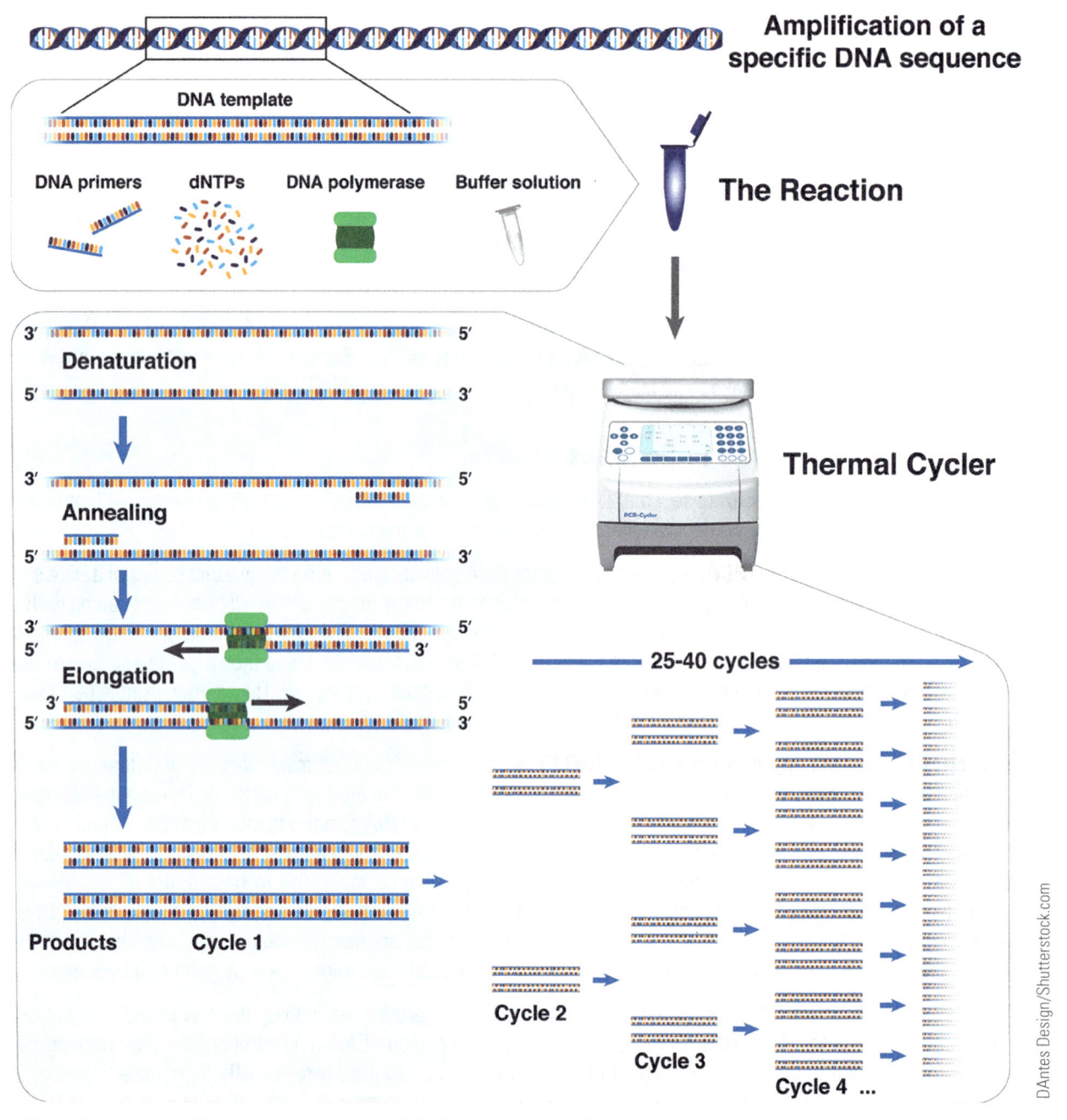

Figure 7.7 The steps of Polymerase Chain Reaction (PCR). This is a way of making many copies of a very small sample of DNA.

New discoveries in molecular biology and genetics are made daily. This is an exciting time to work with DNA. You will likely have opportunities to learn about and use some of these technologies as you pursue your degree. Many people worry that the technology is progressing faster than the legal and ethical regulations needed to keep this technology from being abused. This is also an important and highly active area of bioethics.

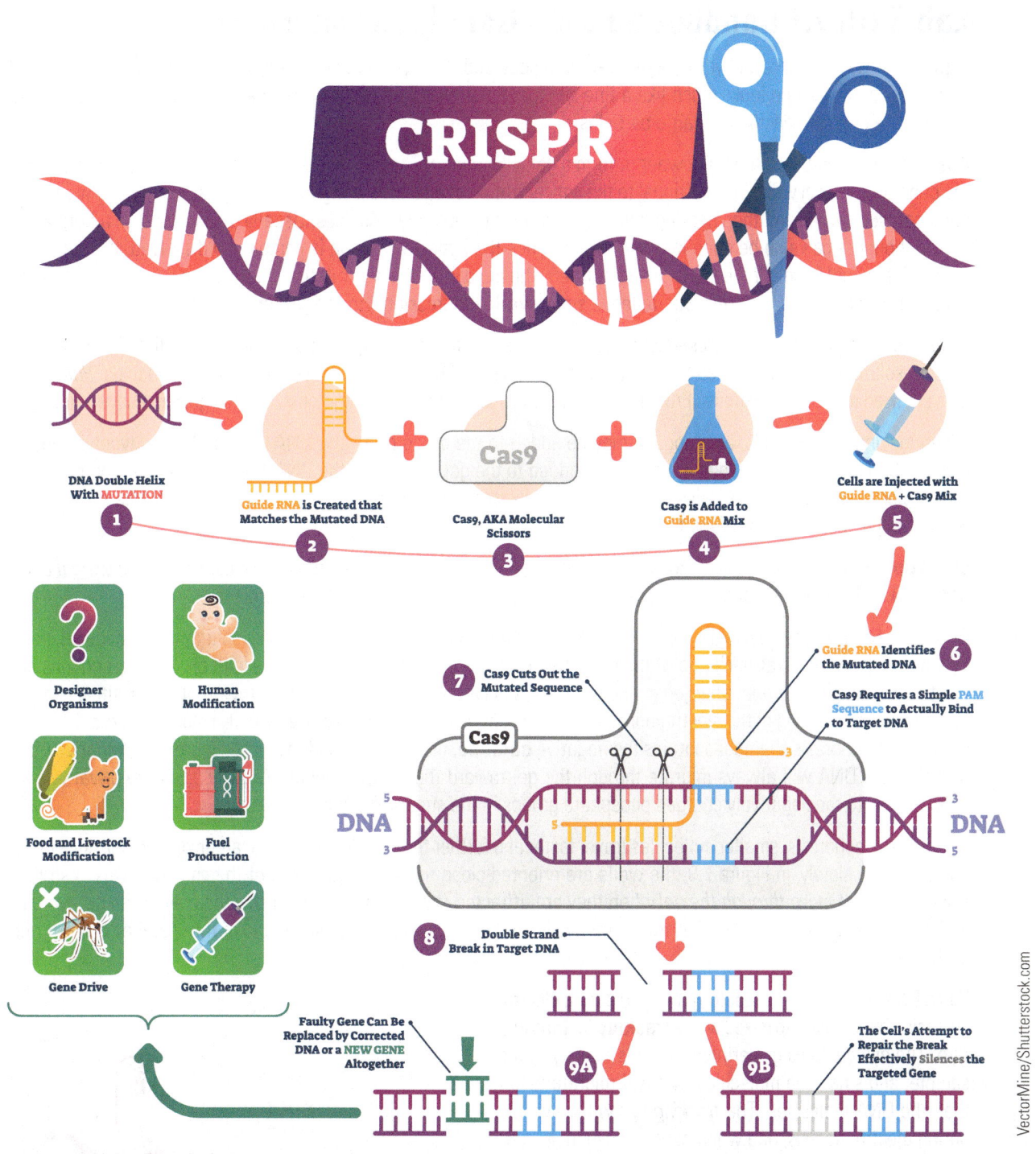

Laboratory Exercise 7 Molecular Biology II: Manipulating DNA 183

Figure 7.8 CRISPR-Cas9. CRISPR is an extension of the use of restriction enzymes, allowing researchers to not just cut at restriction sites, but at any specific target site they choose.

Lab Skill 7.1 Conducting DNA Gel Electrophoresis

Agarose is a complex carbohydrate extracted from seaweed. An agarose gel is a porous mesh of polysaccharide chains that under a microscope looks much like a fine sponge. Small particles can pass between the polysaccharide chains and through the gel more easily than large particles.

Agarose can be prepared from a powder to form a gel, like gelatin. The concentration of the agarose is important because the density of the gel will be proportional to the concentration of agarose. An agarose gel is made by mixing water, a buffer, and agarose powder, and then heating it until the agarose completely dissolves. Usually, this is done on a hotplate with a stirrer, though it can also be done in a microwave. Once the agarose solution is completely dissolved, the liquid must be poured into a mold and allowed to cool undisturbed. The mold includes a **comb** that sticks down into the gel. After the gel solidifies, the comb must be removed to leave small **wells** (depressions) in the gel for loading samples.

A 0.7-1% agarose gel is usually used to separate DNA fragments. You may wish to review Lab Skill 2.2 for how to make a percent solution, as you may need to do that later. Other types of gels are used to separate other molecules. *Polyacrylamide* gels are used to separate proteins. Protein fragments are smaller than DNA fragments, and the gel must have a finer porosity.

When making agarose gels, a compound may be added to the gel to make the DNA visible. Several types of fluorescent markers and stains are available. If a marker is added to the gel, it binds to the DNA fragments as the fragments migrate during electrophoresis. Some stains are added after electrophoresis is finished. Fluorescent DNA markers are only visible when the gel is exposed to UV light.

Electrophoresis is separation of molecules based on three possible characteristics: the mass (*molecular weight or number of base pairs*) of the fragments, the *shape* of the molecules, and the *charge* of the molecules.

An electrophoresis chamber is a plastic tub containing **electrophoresis buffer**, an ionic solution that conducts electricity. When turned on, wires at each end of the chamber create an electrical field. A buffer solution is ionic, and ions conduct electricity through the liquid. Charged molecules in the gel will migrate through the gel toward the oppositely charged end of the chamber. One end of the electrophoresis chamber contains the positive **anode**, usually color-coded red. The other end of the electrophoresis chamber contains the negative **cathode**, usually color-coded black. (Figure 7.9). Negatively charged particles like DNA will always migrate through the gel toward the positive anode. *Nucleic acids are always negatively charged* due to the negatively charged phosphate groups in the sugar–phosphate backbone.

During electrophoresis, smaller molecules travel faster and further through the gel, while larger, or more massive molecules migrate more slowly. In Figure 7.9, the wells are oriented close to the negative end of the chamber. This is so the DNA fragments *must move through the gel* when they are attracted to the positive end of the chamber. If the gel were placed in the chamber backwards, the DNA fragments would still travel toward the positive end of the chamber, but exit right out of the gel.

Since DNA molecules are usually not visible during electrophoresis, the samples are mixed with a **tracking** (or **loading**) **dye**. The dye contains glycerin which adds density to the sample, and small dye molecules that will migrate faster than the DNA molecules. The *tracking dye* will be visible during electrophoresis, and when bands of dye molecules approach the opposite end of the gel, the chamber should be turned off. This indicates that the DNA fragments must be between the tracking dye and the wells.

In Figure 7.9, the gel is shown as a block submerged in a buffer solution. The samples are loaded into the well this way. Usually an adjustable micropipette is used to load the samples into the wells while the gel is submerged. The tracking/loading dye helps keep the samples in the wells. If the electrophoresis buffer were added after the gels were

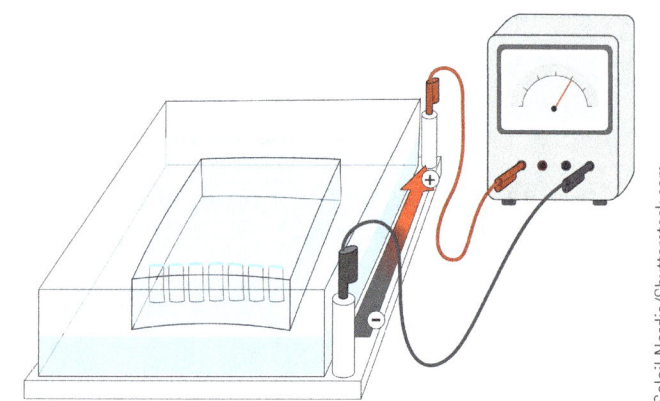

Figure 7.9 Electrophoresis chamber setup. Note that if negatively charged samples, such as DNA, are loaded into the wells, the molecules will migrate through the gel in the direction of the arrow.

loaded, the samples would be washed out of the wells. This method of adding samples to the gel while it is submerged in buffer is referred to as **submarine loading**. It takes a steady hand to get the sample into the wells without missing the well or stabbing the gel (Figure 7.10). It helps to use both hands to steady the hand or the pipette.

After electrophoresis has been conducted, the gel will need to be removed from the chamber and analyzed. This may include staining the DNA, viewing the gel under UV light, and/or using imaging equipment and software to visualize the DNA fragments (Figure 7.11).

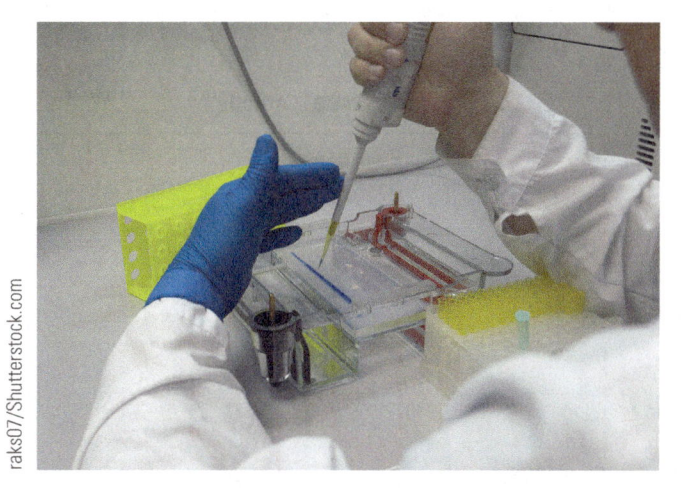

Figure 7.10 Loading an electrophoresis gel.

Figure 7.11 DNA bands fluoresce under UV light when stained with a green fluorescent dye.

Lab Skill 7.2 Interpreting Electrophoresis Data

The end result of gel electrophoresis is a set of data that look something like Figure 7.12. However, the image usually shows a set of wells, usually oriented across the top (but not always) and a series of **bands** and **lanes** (Figure 7.13).

Each lane represents a sample of DNA. The lane extends from the well where the sample was deposited, to the opposite side of the gel. The sample in a well may be from a crime scene or from a particular organism. The DNA in the sample was cut with restriction enzymes before loading the gel. So the gel is acting as a thick filter of sorts as the electrophoresis field moves the DNA through the gel.

Loading/tracking dye was added to the samples to increase their density and to make it easier to see that the samples were loaded properly. It is easy to confuse the tracking dye with the DNA, but the DNA is always somewhere *between* the tracking dye and the well. In many cases, the DNA will not be visible until after the run is complete.

Bands containing larger molecules (or more base pairs) are near the wells. Smaller molecules (or fewer base pairs) migrate further and will be near the bottom.

This assumes all of the particles are negatively charged. The bottom of the gel is positively charged, so negatively charged DNA will migrate in this direction.

If the samples contained both positive and negative molecules, the wells could be formed in the center of the gel. Positively charged molecules would migrate up toward the cathode and negatively charged molecules would migrate down toward the anode.

The bands in Figure 7.13 are purely hypothetical. But this could represent a number of biological or forensic projects. In one case, we could assume that the DNA fragments are from several species. Note that lane 2 and 5 are identical, suggesting that the DNA came from two species are very closely related or from two representatives of the same species.

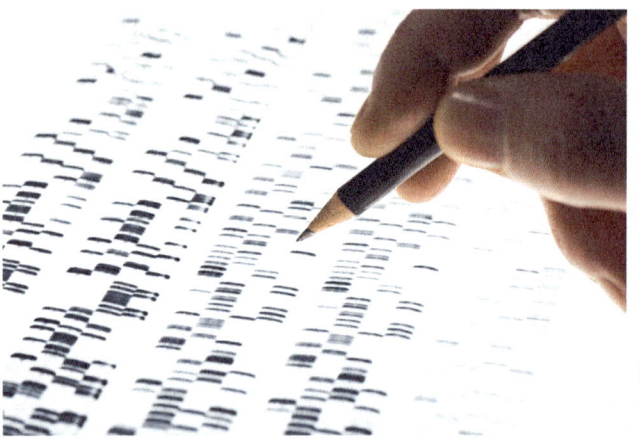

Figure 7.12 Interpreting DNA electrophoresis. Charts like this look intimidating until you understand what the marks actually represent.

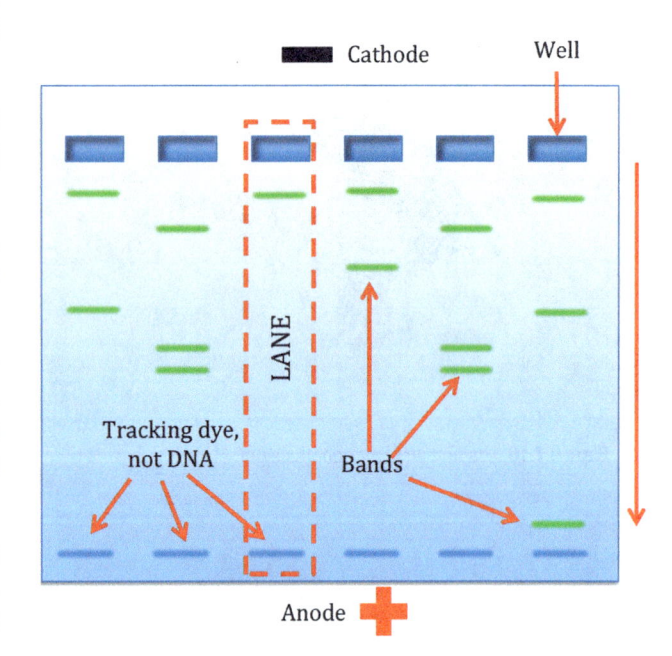

Figure 7.13 Reading an electrophoresis gel.

We might also see something like this in a paternity test. Imagine the first lane contains some of the mother's DNA, the second lane contains a corresponding part of the suspected father's DNA, and the last lane contains the child's DNA. Is the man represented in lane 2 the father of the child in lane 6? If you look at the bands, the child in lane 6 has two bands that match the mother in lane 1. This makes sense. However, the child has none of the bands that the suspected father has, and he has an additional band that he did not get from his mother that does not match the suspected father's DNA. Based on this evidence, it is unlikely that the man represented in lane 2 is the father of the child in lane 6.

Imagine this evidence was presented in a crime investigation. Lane 2 represents the DNA found at the crime scene (say blood from an injury the perpetrator sustained while fleeing the crime). Suspects were brought in, and their blood samples taken. Lanes 4 through 6 represent blood from the suspects. What does that indicate? The DNA pattern in lane 5 matches the DNA from the crime scene, indicating that the suspect was present. DNA doesn't tell us what someone was doing, but it can indicate that they were there.

Problem to Solve 7.1 Can I Predict Restriction Enzyme Recognition Sites?

Before you start working on this problem, do the SETUP steps in Problem to Solve 7.2. You can work on this problem while your gel cools.

The activity requires access to **NEBcutter**, an online database of DNA sequences and locations of restriction sites. New England BioLabs, a biotech company that researches and sells enzymes used in molecular biology research, created NEBcutter to assist molecular biologists. Scientists all over the world use and contribute to these databases.

In this activity, you will look up information about **lambda DNA** and a specific type of plasmid DNA. You will use NEBcutter to predict the results of cutting these strands of DNA with two restriction enzymes, EcoRI and BglI. You will need to use what you learn to create a hypothesis for the following activity.

The plasmid DNA is called pGPS2.1. It was isolated from *E. coli*, and it is used in cloning applications and for inserting specific genes into target cells. It is about 4,500 base pairs in length, which is an average size for a plasmid.

Lambda DNA is from a bacteriophage virus that attacks *E. coli* bacteria. It is over ten times longer than the plasmid DNA, but recall that plasmid DNA is not genomic. The bacterial genomic DNA strand is much longer than both the plasmid and viral DNA strands.

The restriction enzymes of interest are EcoRI and BglI. Follow the protocol and record all of the data in the following data table.

Then, using the information obtained from NEBcutter, predict the number and general location of the bands of DNA produced by EcoRI and BglI when combined with plasmid pGPS2.1 and lambda DNA. Sketch a hypothetical electrophoresis gel.

Laboratory Exercise 7 Molecular Biology II: Manipulating DNA

Title
Title: Using NEBcutter to predict where restriction enzymes will cut specific DNA samples

Purpose
Goal: To learn how to use NEBcutter, and to predict how two specific restriction enzymes will cut Lambda and pGPS2.1 plasmid DNA.

Protocol

Materials:
Access to NEBcutter at NEB.com

Protocol:
Access the New England BioLabs website at NEB.com
Hover over "Tools and Resources" and select NEBcutter.

Plasmid pGPS2.1 and BglI
- Click the **# Plasmid vectors** pull down menu (top right corner) and scroll down to **pGPS2.1**.
- On the left side of the screen choose **circular**.
- Click **Submit**.
- This image shows the circular plasmid DNA and all of the restriction sites on it.
- Under "Main options" choose **Custom digest**.
- Select the box next to BglI and click **Digest** at the bottom of the screen.
- Sketch the location of the restriction sites and label them with the enzyme name.
- Complete the data table for Plasmid (pGPS2.1) with BglI.

Plasmid pGPS2.1 and EcoRI
- Navigate back to the Nebcutter page.
- Ensure the **# Plasmid vectors** is set to **pGPS2.1**.
- Ensure **circular** is selected.
- Click **Submit**.
- Choose **Custom digest**.
- Scroll down and select **EcoRI** and click **Digest**.
- Sketch the location of the restriction sites on the EcoRI template.
- Complete the data table for Plasmid (pGPS2.1) with EcoRI.

Continued from previous page.

Lambda and BglI or EcoRI
Use the same process for Lambda DNA.
Return to the main NEBcutter page.
Unselect the plasmid (return it to **#Plasmid vector**).
Click **#Viral + phage** and select Lambda.
Choose linear instead of circular.
Click **Submit** and use **Custom digest** to predict the effect of EcoRI
Record the information in the data table and on the template.
Repeat the process for Lambda and BglI

Approximate restriction sites for plasmid pGPS2.1

BglI   EcoRI

_____ bp   _____ bp

Approximate restriction sites for lambda and EcoRI
_____
                                                                                    _____ bp

Approximate restriction sites for lambda and BglI
_____
                                                                                    _____ bp

Title

Purpose

Continued from Previous page.

Protocol

Data:

	Plasmid DNA		Viral DNA	
Specific name of DNA				
Origin of DNA (organism)				
Exact length of DNA (bp)				
Restriction enzymes	BglI	EcoRI	EcoRI	BglI
Cuts at base pair sequence				
# Restriction sites on DNA				
# Fragments produced				
Lengths of fragments (bp) (list by decreasing size)				(for lambda and BglI just give range of sizes)
Total of all fragment lengths (bp)				

Date | Researcher Name | **Laboratory Exercise 7** Molecular Biology II: Manipulating DNA | 193

Title

Purpose

Continued from previous page.

Protocol

Analysis:

Sketch a hypothetical gel based on the NEBcutter information. The gel lanes are labeled. Add the appropriate number of bands about where you expect them to be. (If you return to the "custom digest" screen, you can click on "View Gel" to see how the bands should appear.)

A	B	C	D	E	F
pGPS2.1	pGPS2.1	pGPS2.1	Lambda	Lambda	Lambda
Uncut	+BglI	+EcoRI	uncut	+EcoRI	+BglI

Researcher signature | Date | Supervisor/Instructor signature | Date

Problem to Solve 7.2 Can I Verify Restriction Enzyme Recognition Sites by DNA Electrophoresis?

Do the SETUP steps in this protocol FIRST! After you prepare the gels you will have to wait for them to cool. Use that time to work on Problem to Solve 7.1. Then Prepare your electrophoresis chambers and run them. While the electrophoresis chamber is running, watch the 20-minute video over the Footpath Murders and work on Problem to Solve 7.3.

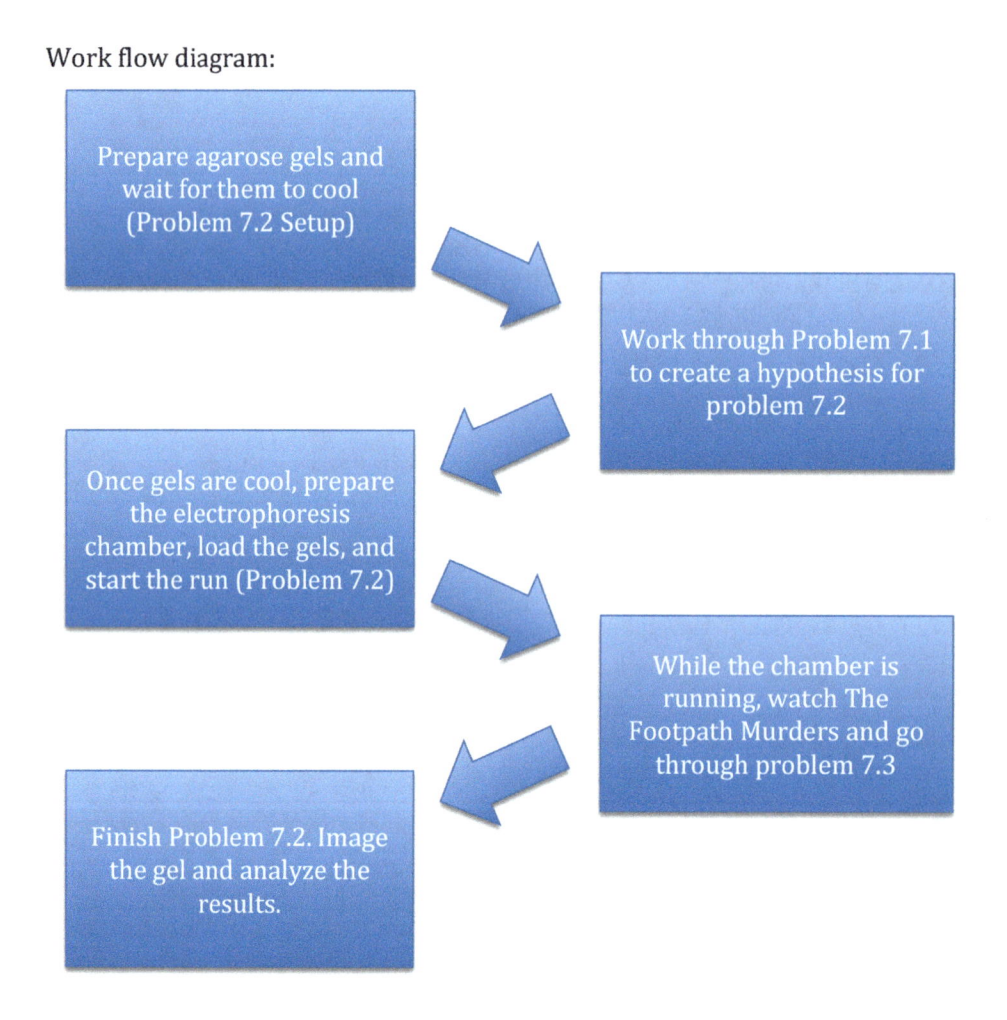

Work flow diagram:

- Prepare agarose gels and wait for them to cool (Problem 7.2 Setup)
- Work through Problem 7.1 to create a hypothesis for problem 7.2
- Once gels are cool, prepare the electrophoresis chamber, load the gels, and start the run (Problem 7.2)
- While the chamber is running, watch The Footpath Murders and go through problem 7.3
- Finish Problem 7.2. Image the gel and analyze the results.

Use Lab Skill 7.2 to help you work through the following protocol. You will not be isolating DNA or adding restriction enzymes to DNA in this activity. The DNA and restriction enzymes have been prepared in advance. Your goal is to practice doing electrophoresis, and to see if your results match the hypothetical gel you create after completing Problem to Solve 7.1.

Electrophoresis is tricky takes practice. If you have trouble the first time, that is OK. That is the point of this lab exercise. Here are some suggestions to keep in mind to help you succeed:
- Read the directions carefully. It is easy to skip a crucial step.
- Do not disturb the agarose as it is hardening. The gel needs to be uniform, and disturbing it in any way can introduce imperfections in the gel matrix.
- Be very careful when removing the comb from the gel and the ends from the dams. If you break your gel, you have to start over.
- Pay attention to labels. If you accidentally get your samples or lanes mixed up, your gel will be worthless.

- Make sure your gel is oriented correctly in the chamber. If it isn't, your DNA samples will disappear into the solution.
- When loading the gel, rest your elbow on the table and use your other hand to steady the micropipette or your hand.
- When you turn on the chamber, make sure you see bubbles coming up from the wires in the chamber. No bubbles means no current.
- When the electrophoresis chamber is running, do not bump the chamber. Vibrations and disturbances will produce unclear bands.
- Watch the tracking dye. Do not let the chamber run so long that the tracking dye exits the gel. If it does, your DNA sample may have exited the gel as well.

Title: Agarose Gel Electrophoresis

Goal: Learn how to operate an electrophoresis chamber.

Materials:
(Per class)
- Agarose (3 g dry powder)
- 7.5 ml Electrophoresis buffer
- DNA fluorescent dye, such as GelGreen (use as indicated on the package)
- Distilled water
- Hotplate with stirring magnet
- Hot pads or insulated gloves

(Per group of four students)
- Electrophoresis chamber (Groups can share a chamber if chambers can hold multiple gels.)
- DNA/restriction enzyme samples labeled A through F (can be obtained from Edvotek)
- Gel mold and comb
- Spatula
- Graduated cylinder
- Micropipette and tips
- Gloves and safety glasses

Protocol:
Class setup

1. For enough agarose gel for a class of six groups, combine 7.5 ml Electrophoresis buffer, 3 g agarose, and 367.5 ml distilled water in a 1 liter flask.

2. Heat mixture on a hotplate on a medium-high setting until it starts to boil and becomes very clear. Stir with magnetic stirrer on medium speed to ensure even heating.

Continued from previous page.

Protocol

Group setup

3. Prepare the gel molds. Insert combs into the molds at one end of the mold.

4. When the agarose is ready, use the hotpads or gloves to carry the flask of agarose to the mold. Do not carry the mold. Fill the mold to the very top of the black dams. (Thin gels break easily.) Be very careful with the hot agarose.

5. Once poured, leave the mold undisturbed so the gel will cool uniformly.

6. If the agarose in the flask has started to cool, heat it back up and stir it before pouring another mold. Repeat until all groups have filled their molds.

7. While the mold cools, complete Problem to Solve 7.1.

After the gel has solidified

8. Prepare the electrophoresis chambers. Add 1,470 ml distilled water and 30 ml Electrophoresis buffer to the chamber. Ensure the buffer and water are thoroughly mixed.

9. The electrophoresis chamber should have plastic guides for keeping the gels separate, straight, and immobile. If they are not already present, insert them now.

10. Carefully remove the gel from the gel mold. Remove the comb slowly and smoothly without ripping the gel. When removing the gel from the mold, it helps to use the spatula to "break the seal" between the gel and the mold. Be particularly careful not to damage the wells.

11. Use the spatula to place the well into the electrophoresis chamber. Orient the gel so that the wells are closest to the cathode. Ensure liquid has filled the wells and that there are no bubbles under the gel.

12. Tracking dye has already been added to the DNA/restriction enzyme samples. The samples will appear dark blue or purple.

13. Use the adjustable micropipette to transfer 35 μl of each DNA sample to wells A through F. Use a new tip for each transfer.

Continued from previous page.

Protocol

14. Ensure the samples are loaded in the correct order. (See Lab Skill 1.4 if necessary for reminders on correctly using the adjustable micropipette.) A good transfer will produce dark rectangles of sample in the wells. Avoid forming "clouds" around the wells.

15. Once the samples are loaded, put the lid on the chamber.

16. Plug in the power supply, connect the lid to the power supply and turn it on the highest voltage setting.

17. Check to see if bubbles form on the wires at each end of the chamber. If so, wait about 40 minutes for the tracking dye to migrate across the gel.

While the gel is running

18. Watch the 20-minute video titled "The Footpath Murders" that discusses the invention of DNA fingerprinting and how it was used to capture a rapist and murderer.

19. Then go on to Problem to Solve 7.3.

Finishing electrophoresis

20. Once the tracking dye has almost reached the far end of the gel, turn off the power supply, disconnect the lid from the power supply, and remove the lid.

21. Use the spatula to carefully remove the gel from the chamber. Place it in a weigh boat or suitable container.

22. If a fluorescent stain was used, place the gel on the UV illuminator. If a GelDoc Imager is available, insert the gel into the gel doc and use Image Lab to capture the image.

23. Compare the image of the gel with the hypothesis you made in Problem to Solve 7.1.

Problem to Solve 7.3 Can I Interpret DNA Fingerprint Data?

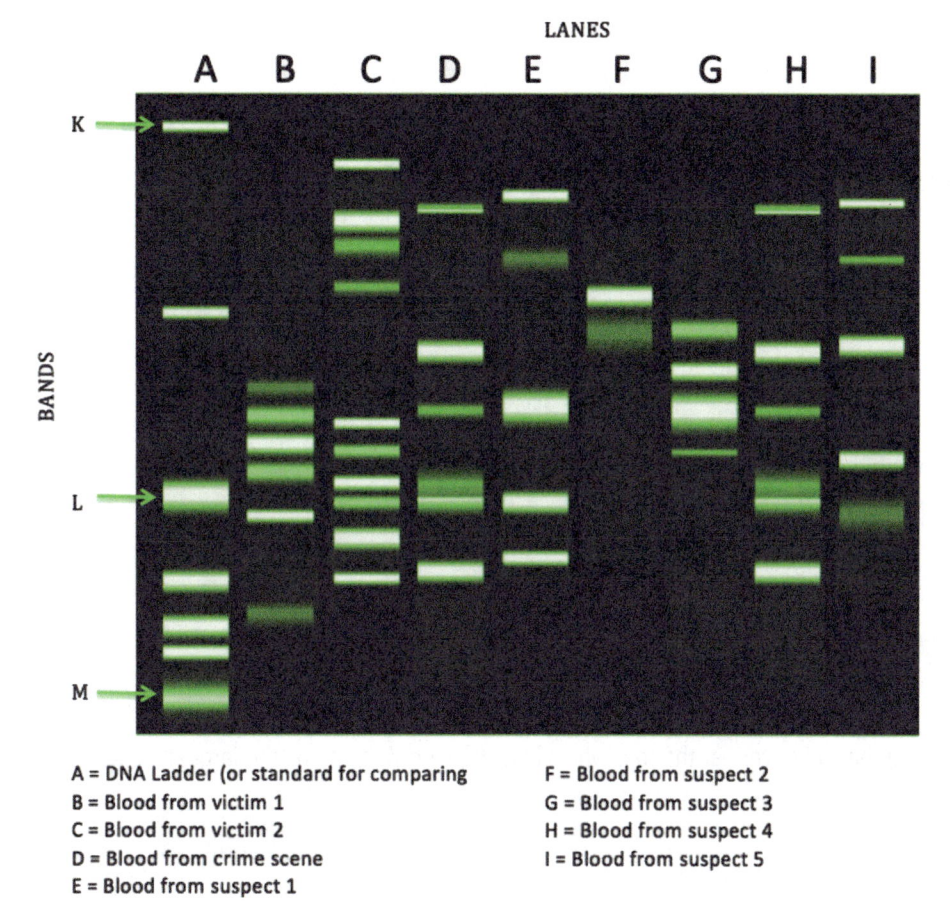

LANES

A = DNA Ladder (or standard for comparing
B = Blood from victim 1
C = Blood from victim 2
D = Blood from crime scene
E = Blood from suspect 1
F = Blood from suspect 2
G = Blood from suspect 3
H = Blood from suspect 4
I = Blood from suspect 5

7.14. *Example of DNA fingerprint.*

To practice working with gel data, study the image in Figure 7.14 and answer the following questions.

1. Which lane contains the DNA Ladder? Why is this used?

2. Which lane contains crime scene DNA?

3. Does the data suggest that one of the suspects is guilty? Explain.

4. If you were a lawyer prosecuting the case, how would you use this DNA fingerprint data?

5. If you were the defense attorney, how would you use it?

6. Assuming the wells were at the top, which band (K, L or M) contains the largest molecular fragments?

7. Which band (K, L or M) contains the smallest molecular fragments?

8. Assuming the wells are at the top, which side of the figure represents the edge of the gel closest to the anode? (Top, bottom, left or right?)

Name: _____   Date: _____

POST-LAB QUIZ 7: What have you learned?

1. What specifically did NEBcutter allow you to do in this lab activity?

2. Who was the scientist who first developed DNA fingerprinting? In what year was it first used to help solve criminal cases?

3. Who was actually the rapist and murderer in Narborough, England?

4. When placing a gel in an electrophoresis chamber (illustrated in the figure), which side (A, B, C, or D) must the wells in the gel be closest to? *Why?*

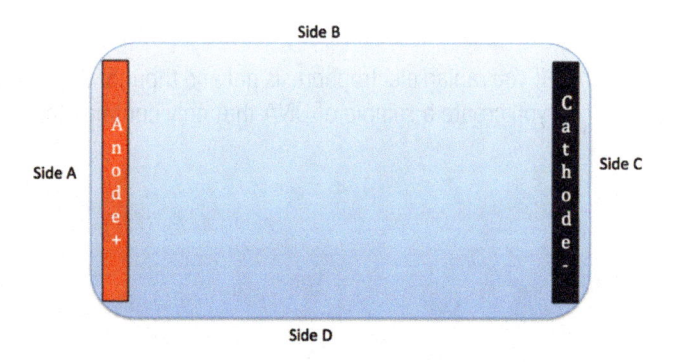

5. Did your actual gel look like your hypothetical gel from Problem 7.1? If not, what was different? *Why* might it have differed?

6. Which laboratory technique would best allow you to make many copies of a desired gene?

7. Which laboratory technique would allow you to compare the DNA of one organism (or person) to another?

8. What are sticky ends? Why are they useful in genetic manipulation?

9. If you wanted to isolate a specific gene from a larger strand of DNA, how could you isolate it? (Be specific.)

10. If you ran an electrophoresis gel and thought that one of the bands contained a DNA fragment you needed, how could you create a sample of DNA that only contains that DNA fragment? (Be specific.)

Laboratory Exercise 8

Cell Biology II: Photosynthesis and Cellular Respiration

Concepts to Understand

After completing this lab, the student should be able to:

1. Describe the reactants, products, and events of cellular respiration.
2. Describe the reactants, products, and events of photosynthesis.
3. Explain the relationship between cellular respiration and photosynthesis.
4. Describe the role of redox reactions in cellular metabolism.
5. Describe the electromagnetic spectrum.

Skills to Learn

After completing this lab, the student should be able to:

1. Extract and identify pigments in plant leaves.
2. Create and interpret an absorbance spectrum.
3. Use a nanodrop spectrophotometer.
4. Infer photosynthesis and cellular respiration from pH changes.

SEE ALSO:

Campbell Biology—Chapter 9, Sections 9.1-9.3; Chapter 10, Sections 10.1-10.3.

JoVE (video)—Introduction to the Spectrophotometer

Others:

Laboratory Exercise 8 Cell Biology II: Photosynthesis and Cellular Respiration

Important terms to know

- Absorbance spectrum
- Acetyl-CoA
- Adenosine
- Adenosine diphosphate (ADP)
- Adenosine triphosphate (ATP)
- ATP synthase
- Autotroph
- Beer-Lambert Law
- Cellular respiration
- Chlorophyll (a and b)
- Chromatogram
- Chromatography
- Citrate
- Citric acid cycle (Kreb's cycle)
- Coenzyme
- Consumer
- Cristae
- Cytochrome complex
- Dark reactions (Calvin cycle)
- Electromagnetic spectrum
- Electron transport chain
- Flavin adenine dinucleotide (FAD^+/$FADH_2$)
- G3P
- Glycolysis
- Grana
- Heterotroph
- Light-dependent reactions
- Mitochondrial matrix
- $NADP^+$ reductase
- Nicotinamide adenine dinucleotide (NAD^+/NADH)
- Nicotinamide adenine dinucleotide phosphate ($NADP^+$/NADPH)
- Oxaloacetate
- Oxidative phosphorylation
- Phosphorylate
- Photosynthesis
- Photosystem I/II
- Pigment
- Producer
- Pyruvate
- Reduction-oxidation (redox) reaction
- R_f factor
- Rubisco
- Spectrophotometer
- Stroma
- Thylakoid

Pre-lab Preparation and Background Information

Photosynthesis and oxidative **cellular respiration** are two of the most important chemical processes in biology. They are complimentary reactions, and the products of each reaction serve as the reactants for the other. Photosynthesis, which is performed in eukaryotic cells by chloroplasts, uses the energy from sunlight to build glucose ($C_6H_{12}O_6$) from carbon dioxide and water. Cellular respiration, which is performed in eukaryotic cells by mitochondria, oxidizes glucose, releasing energy and producing carbon dioxide and water (Figure 8.1). This lab exercise is not intended to cover all of the details of the chemistry involved in these reactions. Instead, it will focus on some of the practical aspects of how these reactions occur in living organisms, and how they can be measured.

The Solar Powered World

Why do plants perform photosynthesis? It might be tempting to assume that they do it to provide us with oxygen and food. But this is not the case. Plants produce sugars for *themselves*. Plants are **autotrophic** (make their own energy-rich food molecules). They photosynthesize in order to make sugars from inorganic carbon (carbon dioxide) and water. But then they "eat" that food. Plant cells constantly need fuel for their metabolism, just like any other organisms. That means plants do BOTH cellular respiration and photosynthesis.

When sunlight is available, plants produce more sugars than they need and store the excess as starch. They use that starch during periods of darkness or during the winter when many plants drop their leaves.

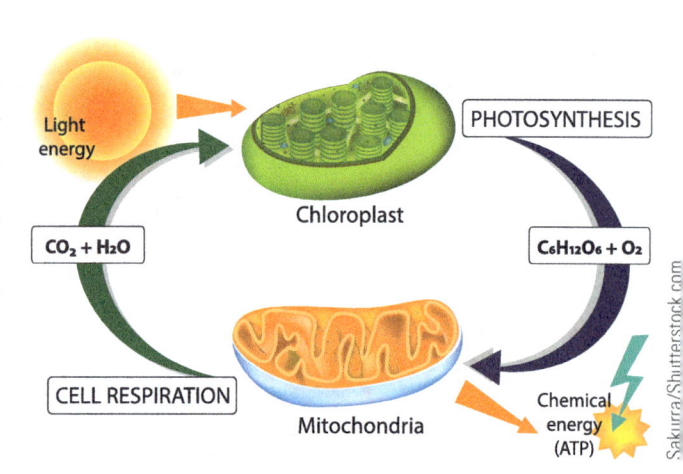

Figure 8.1 The complimentary reactions of photosynthesis and cellular respiration. The products of one reaction serve as the reactants of the other.

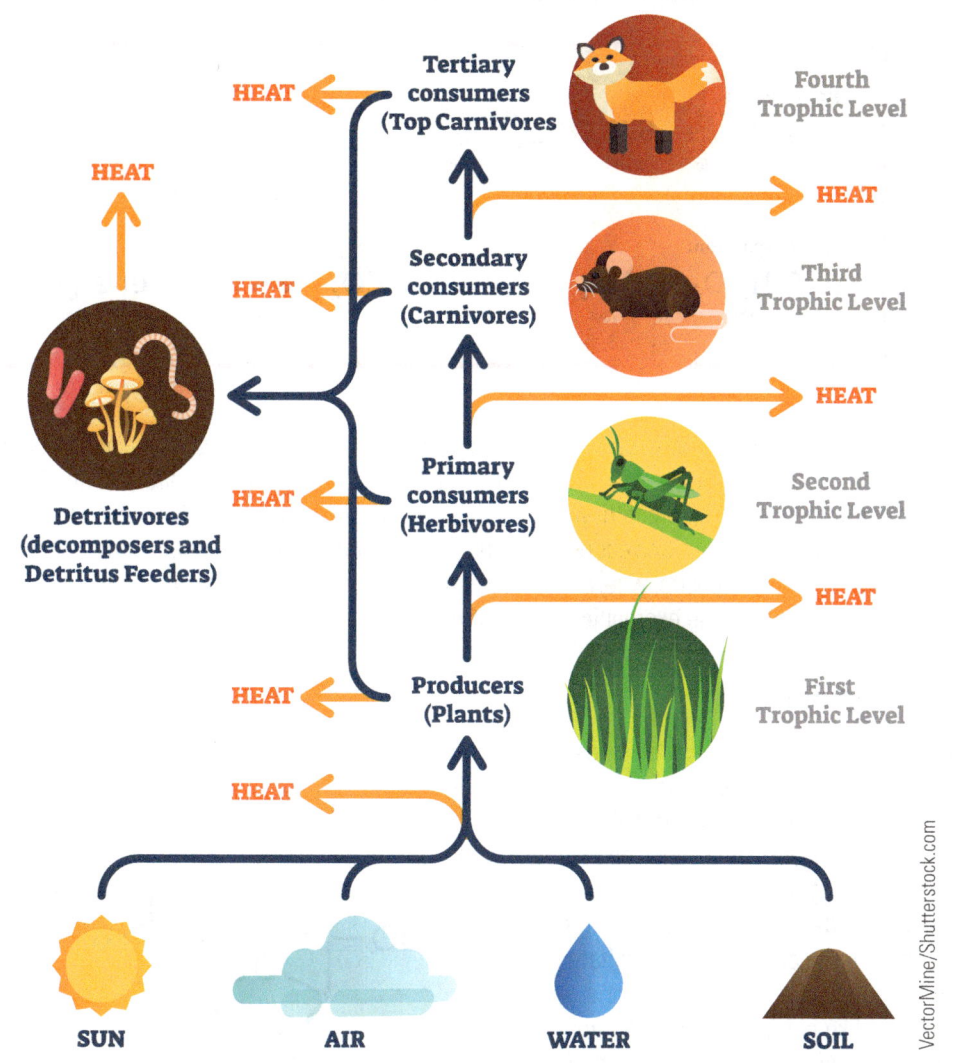

Figure 8.2 Food chain. Almost all food chains are powered by photosynthesis. Even in aquatic ecosystems, photosynthetic plankton provides the base for aquatic food chains.

Humans and other organisms have co-evolved to survive in a world that gets almost all of its energy, directly or indirectly, from the sun. **Heterotrophic** (non-photosynthetic) organisms, including humans, must consume carbon-rich energy molecules through the food chain (Figure 8.2). Plants are described as **producers** because they produce energy-rich molecules. **Consumers** are those heterotrophs that eat the molecules made by producers, or other molecules along the food chain. *Herbivores* are consumers of plants, while *carnivores* consume animals. *Omnivores* consume both plant and animals. Even if a top carnivore eats no plants directly, it is still indirectly dependent on the organisms lower on the food chain that do. Ultimately, the whole food chain depends on photosynthetic producers.

Eating and Breathing

Food is any collection of molecules that can be digested, and in the process of digestion provide energy for cellular activities. You could try to eat a plastic apple or a rock, but they don't count as food since you can't digest them.

Cellular respiration is the fundamental reaction used to break down sugar and other food molecules. It is the reverse reaction of photosynthesis. Instead of using light energy to build energy-rich food molecules, cellular respiration catabolizes those food molecules to release the energy stored in them. Cellular respiration is normally an *oxidative* reaction. The reason we breathe is to provide the oxygen needed for cellular respiration. So if cells dependent on cellular respiration don't get adequate oxygen, oxidative reactions stop. When the reactions stop, cells, and potentially the organism, will die.

The balanced equations for photosynthesis and cellular respiration are shown. It is worthwhile to memorize these.

Photosynthesis:

$$6\ CO_2 + 6\ H_2O \xrightarrow{\text{(light)}} C_6H_{12}O_6 + 6\ O_2$$

carbon dioxide + water with light produce glucose + oxygen

Cellular Respiration:

$$C_6H_{12}O_6 + 6\ O_2 \longrightarrow 6\ CO_2 + 6\ H_2O + \text{energy}$$

glucose + oxygen react to produce carbon dioxide + water + energy for ATP synthesis

Synthesis reactions are *anabolic* and *endergonic*. Decomposition reactions are *catabolic* and *exergonic*. (Review Lab Exercise 2 if necessary.) If you examine the equations for photosynthesis and cellular respiration, all elements in the reactants are accounted for in the products. In photosynthesis, there are twelve reactant molecules and seven product molecules. Photosynthesis is therefore anabolic and endergonic, and light provides the energy needed to fuel the reaction. The reaction for cellular respiration is the opposite, where seven reactant molecules are converted into twelve product molecules. The catabolism of glucose provides the exergonic release of energy that is used to power other reactions in the cell.

Energy Molecules

Cells need glucose or other organic molecules as fuel. But those molecules themselves are not the direct energy suppliers for cellular chemistry. *ATP* powers cellular reactions in all living cells, including microbes, fungi, and plants. Recall from Lab Exercise 3 that nucleic acids are composed of monomers called nucleotides and that a nucleotide is composed of a sugar (deoxyribose or ribose), a phosphate group, and a nitrogen base (adenine, thymine, cytosine, guanine, or uracil). Adenine, like many other organic molecules, has multiple roles beyond storing genetic information in DNA and RNA. **Adenosine** is a partial RNA nucleotide composed of an adenine and ribose (Figure 8.3). Adenosine may have one, two, or three phosphate groups attached to it, and all three forms are biologically important in cells (Figure 8.4). Source: Adapted by Kendall Hunt Publishing Company.

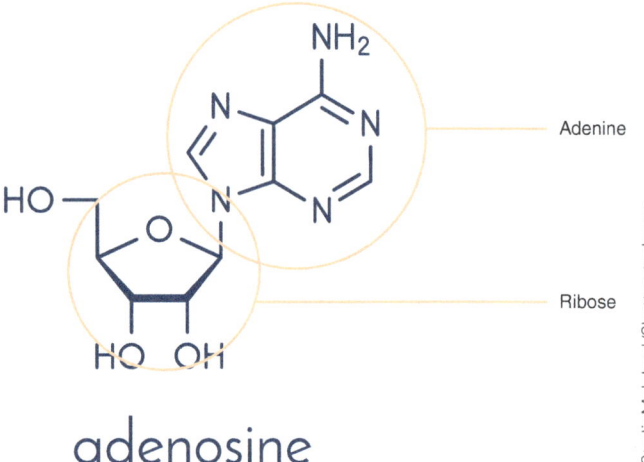

Figure 8.3 Adenosine. Adenosine is made up of a 5-carbon sugar (ribose) and a nitrogen base (adenine).

Adenosine triphosphate (ATP) acts as the major energy "currency" for cellular activities. Free-floating ATP molecules power reactions. When close enough to a suitable binding site, the outermost phosphate group breaks away from ATP and binds to another molecule. This action of binding phosphate to another molecule is called **phosphorylation**, and phosphorylation activates most chemical reactions and enzymatic activities in cells. When ATP loses a phosphate group, it loses energy and becomes **adenosine diphosphate (ADP)**. The major function of a mitochondrion is to use cellular glucose and oxygen to continuously convert ADP molecules back into ATP molecules. Figure 8.5 illustrates the ATP–ADP cycle. Cellular respiration "recharges" low-energy ADP molecules back into high-energy ATP molecules. *Adenosine monophosphate (AMP)* and *cyclic adenosine monophosphate (cAMP)* are other forms of adenosine, but they are not directly involved in cellular respiration or photosynthesis and will not be covered here.

Redox Reactions

You have learned about ionic reactions in which electrons are transferred from one atom to another. A **redox reaction** (short for **oxidation–reduction reaction**) is a reaction in which the *oxidation state* (or hypothetical charge) of an atom

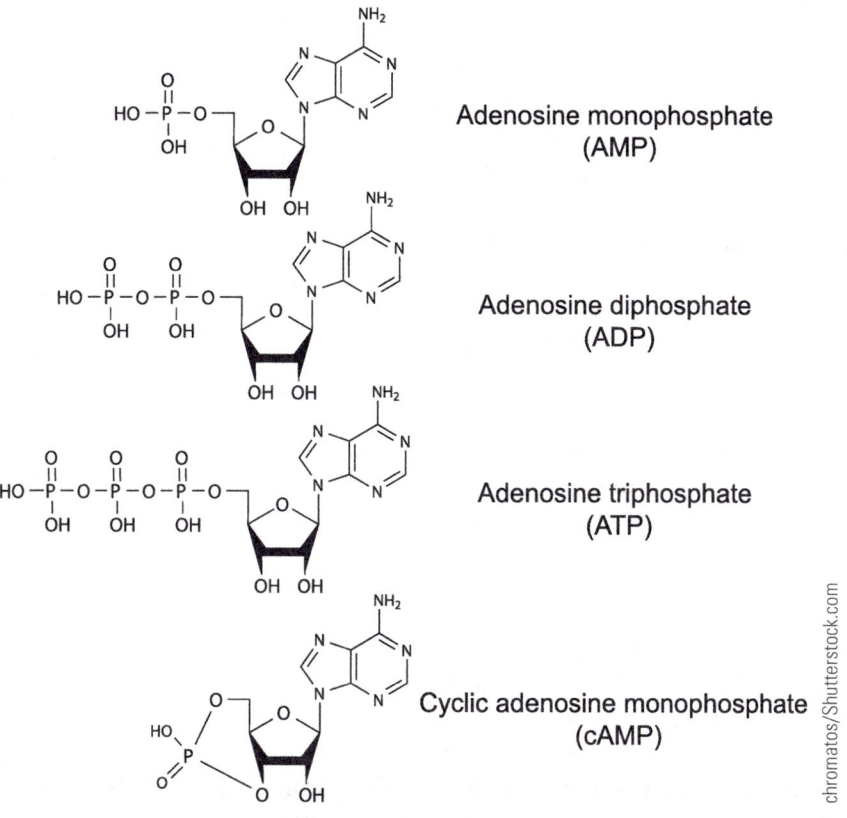

Laboratory Exercise 8 Cell Biology II: Photosynthesis and Cellular Respiration

Figure 8.4 Adenosine phosphates. Adenosine triphosphate (ATP) contains three phosphate groups, Adenosine diphosphate (ADP) contains two phosphate groups, and adenosine monophosphate (AMP) contains one phosphate group. An additional form of adenosine monophosphate, called cAMP, is important in cellular signaling.

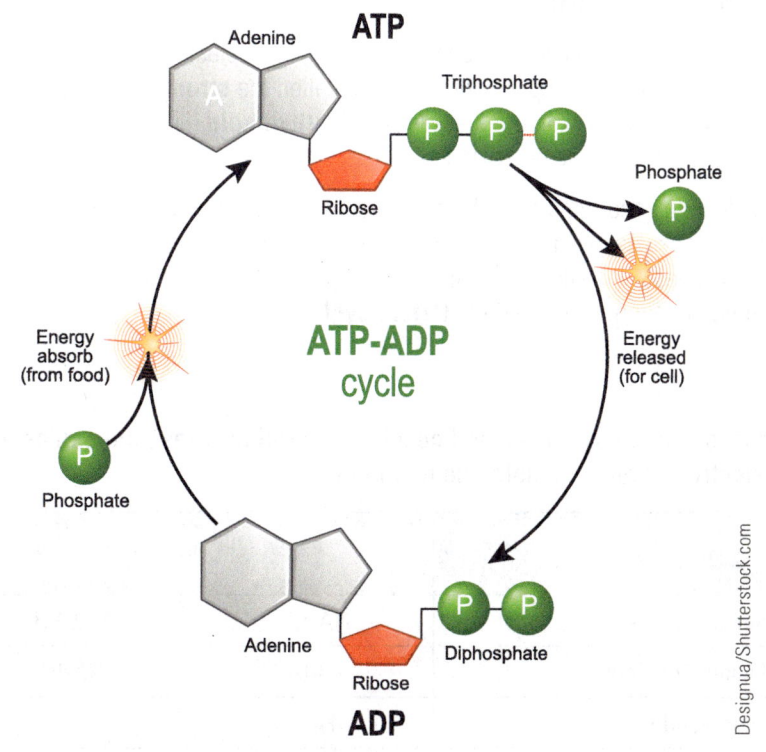

Figure 8.5 ATP-ADP cycle. ATP loses a phosphate, which can bind to other molecules (phosphorylate) to provide energy for cellular reactions. Cellular respiration uses glucose or other food molecules to turn ADP molecules back into ATP molecules.

or molecule is altered by gaining or losing one or more electrons. However, the atom or molecule that gains the electron is *reduced*, and the atom that loses the electron is *oxidized*. The term *reduced* refers to the lowering of the *charge* of the atom or molecule when it gains an electron. The first major element known to "steal" electrons from other elements was highly electronegative oxygen. Thus, the term *oxidation* was used to describe the action of oxygen taking electrons from other atoms or molecules. We now know that other highly electronegative elements can also act as "oxidizing" agents. Any compound that takes an electron is *reduced*, and the compound that lost the electron is *oxidized*. See Figure 8.6. In biological systems, the oxidizing agent is almost always free oxygen (O_2) or a compound containing oxygen. If free oxygen atoms take electrons from other atoms or molecules, available hydrogen ions will bond to the negatively charged oxygen atoms, forming water.

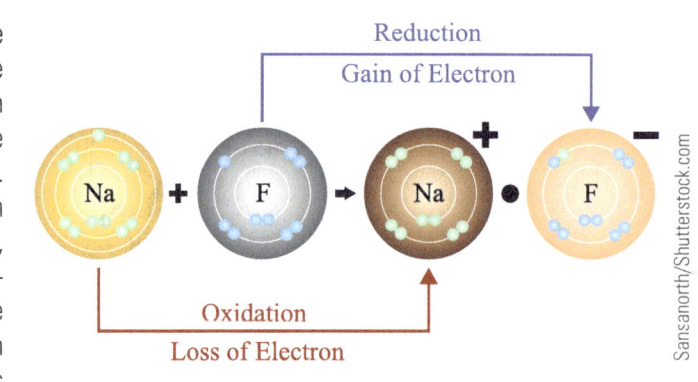

Figure 8.6 Oxidation-reduction (redox) reaction. One atom or compound loses an electron and is *oxidized*, while the other gains an electron and is *reduced*.

Cofactors

Other molecules can also serve as temporary energy carriers. But instead of storing energy in the bonds of phosphate groups, these **coenzymes**, carry high-energy electrons and undergo oxidation–reduction reactions. As electrons move from one molecule to a molecule of higher electronegativity, the electrons lose energy, and the electron acceptor gains that energy. Photosynthesis and cellular respiration use three important electron carriers, **nicotinamide adenine dinucleotide (NAD+/NADH), nicotinamide adenine dinucleotide phosphate (NADP+/NADPH),** and **flavin adenine dinucleotide (FAD/FADH$_2$)** (Table 8.1). Note that like ATP, these are also variations of adenine, ribose, and phosphate nucleotides. These coenzymes are also used in many other cellular reactions in all types of cells in addition to participating directly in photosynthesis and cellular respiration.

Overview of Photosynthesis

Photosynthesis occurs in *chloroplasts* of plant and algae cells. Chloroplasts are double-membrane-bound organelles with their own ribosomes and DNA. A chloroplast contains fluid called the **stroma**, and small stacks of flattened disks. These stacks are called **grana**, and one disk in the stack is called a **thylakoid** (Figure 8.7). The events of photosynthesis occur in the thylakoid membrane and stroma of a chloroplast.

Photosynthesis is a complex and important chemical process in biology, and it should not be oversimplified. However, for the sake of this lab, it has been summarized as a series of steps that should correspond to your textbook description. The process of photosynthesis is split into two major parts, the **light-dependent reactions** and the **dark** (or non-light-dependent) **reactions,** more formally called the **Calvin cycle**.

Table 8.1 Coenzymes in photosynthesis and cellular respiration. These molecules transfer energy within a cell by carrying electrons from one molecule to another

Coenzyme	Oxidized form (Low energy state)	Reduced form (High energy state)	Used in
Nicotinamide adenine dinucleotide phosphate	NADP$^+$	NADPH	Photosynthesis
Nicotinamide adenine dinucleotide	NAD$^+$	NADH	Cellular respiration
Flavin adenine dinucleotide	FAD	FADH$_2$	Cellular respiration

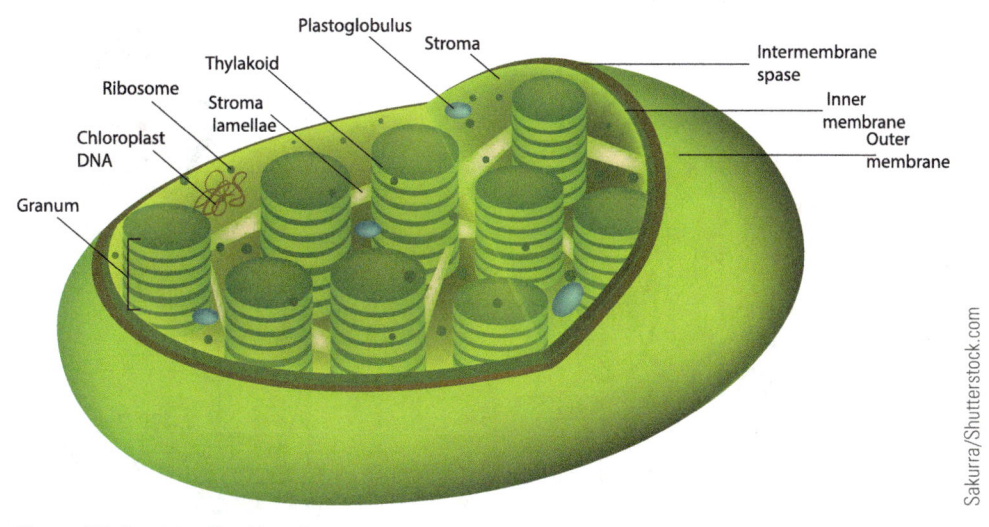

Figure 8.7 Structure of a chloroplast.

Light-dependent reactions

5. Light striking a photosynthetic cell is absorbed by **photosystem II**, a protein complex in the thylakoid membrane of a chloroplast. (The number of a photosystem refers to the order of its discovery.)
6. Photosystems contain **pigments**, molecules that absorb wavelengths light. The most common is **chlorophyll a.**
7. The photosystem splits water into oxygen, two hydrogen ions, and two electrons.
8. The oxygen is released as waste O_2, and hydrogen ions accumulate in the thylakoid.
9. Electrons are shuttled to an **electron transport chain**, a series of proteins in the thylakoid membrane, including a **cytochrome complex**.
10. As electrons move through the cytochrome complex, the complex uses their energy to actively transport more hydrogen ions from the stroma into the thylakoid.
11. The cytochrome complex passes the now low-energy electrons to **photosystem I**.
12. Photosystem I, which also contains chlorophyll a, uses absorbed light energy to re-energize the electrons and pass them to *NADP+ reductase*.
13. **NADP+ reductase**, an enzyme in the thylakoid membrane, uses the electrons and hydrogen ions to reduce and energize NADP$^+$ molecules to NADPH.
14. **ATP synthase**, another enzyme and channel protein in the thylakoid membrane, uses the potential energy of the high concentration of hydrogen ions passing through it to convert ADP to ATP.

Dark (light-independent) reactions of the Calvin Cycle

15. The NADPH and ATP molecules produced in the light reactions power the dark reactions. (Dark reactions do not necessarily occur in the dark; they just don't require direct input of light.)
16. **Rubisco**, an enzyme in the stroma, binds (or "fixes") three carbon dioxide molecules from outside the plant to three 5-carbon molecules called *RuBP*. To do so requires three separate turns of the Calvin cycle, one turn per carbon dioxide molecule. This creates three unstable molecules that immediately split into six 3-carbon molecules called *3-phosphoglycerate*.
17. Six ATP (from the light reactions) phosphorylate each of the molecules into an intermediate compound, and six NADPH convert those molecules into stable molecules called *3-phosphoglyceraldehyde*, or **G3P**. (G3P is half of one glucose)
18. One G3P is released from the cycle. Three ATP convert the rest of the molecules into three 5-carbon RuBP molecules to start the cycle again.
19. A second set of three turns of the Calvin cycle produces another G3P, which combines with the first to form one glucose molecule (Figure 8.8).

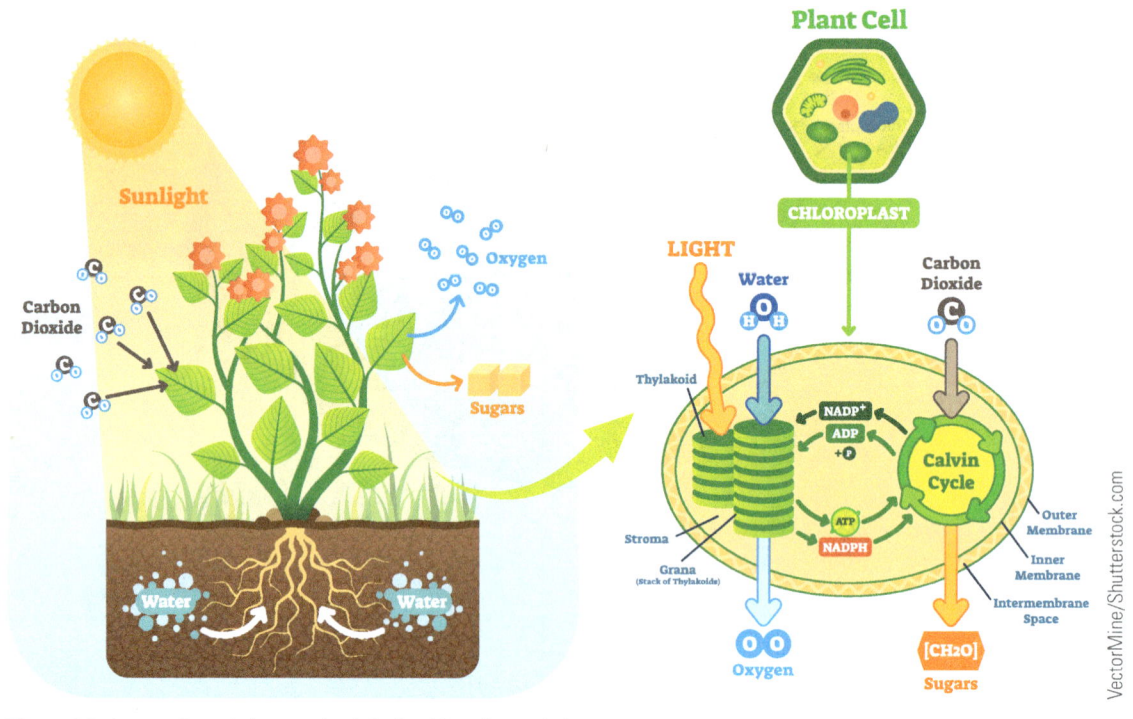

Figure 8.8 An overview of photosynthesis in the chloroplasts of plant cells.

Overview of Cellular Respiration

Cellular respiration occurs in the *mitochondria* of all eukaryotic cells, including plants. Mitochondria, like chloroplasts, have their own ribosomes and DNA. Mitochondria are double-membrane-bound organelles. The inner membrane is folded inward, forming projections called **cristae** that increase surface area for chemical reactions across the inner membrane. The *inner-membrane space* is between the inner and outer membranes. The center of a mitochondrion contains the **mitochondrial matrix**. The processes of cellular respiration occur in the cytoplasm of the cell, across the inner membrane of a mitochondrion, and in the mitochondrial matrix (Figure 8.9).

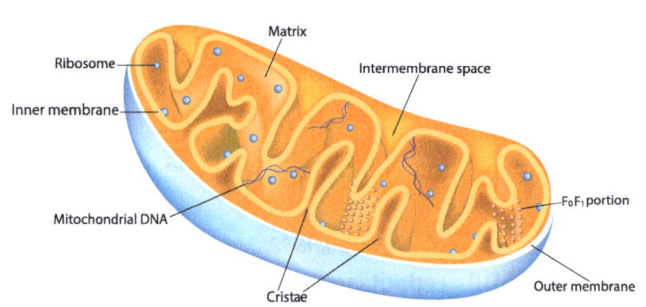

Figure 8.9 Structure of a mitochondrion.

Cellular respiration is typically described as a four-part process. The first is **glycolysis** (*glyco* = sugar; *lyse* = break), the splitting of glucose in the cytoplasm. The second is the oxidation of pyruvate. The third is the **citric acid (Kreb's) cycle**. The fourth is **oxidative phosphorylation**, which uses products of glycolysis and the citric acid cycle to phosphorylate ADPs into ATPs (Figure 8.10 and Table 8.2). The chemistry is more complex than what is described here, but the major events are summarized below.

Glycolysis

1. Glucose from consumed food enters the cytoplasm of the cell.
2. Two ATP provide the activation energy needed to split one glucose into two 3-carbon **pyruvate** (or *pyruvic acid*) molecules.
3. The energy released from splitting sugar is used to convert four ADP into four ATP, and two NAD^+ into two NADH.

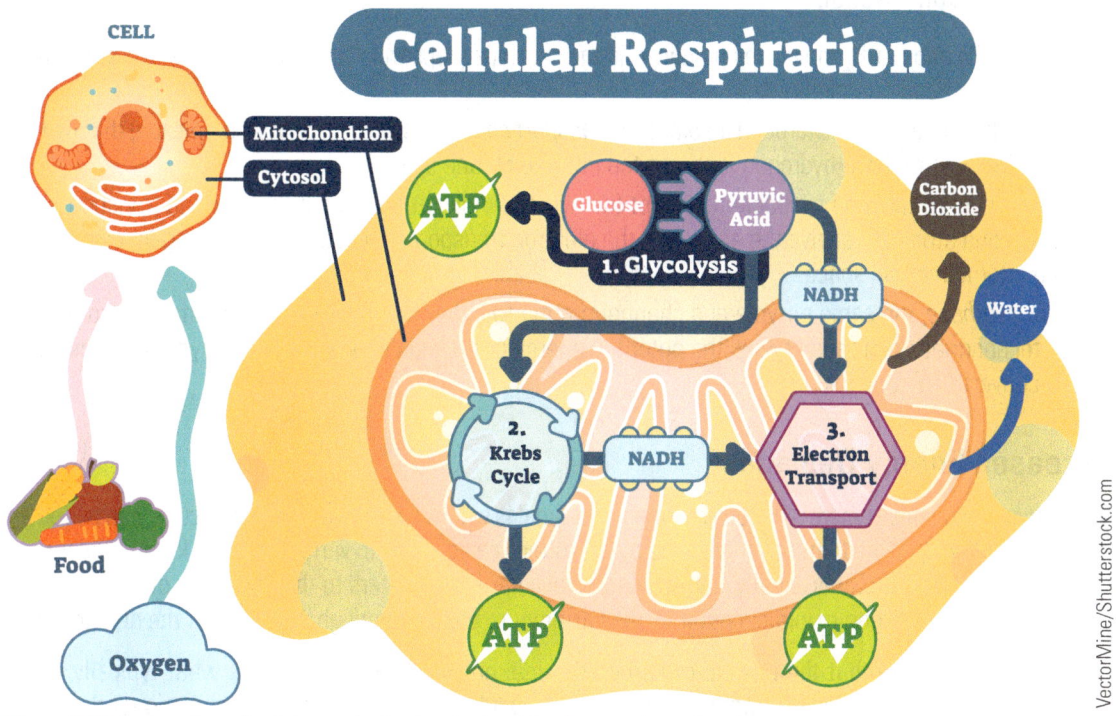

Laboratory Exercise 8 Cell Biology II: Photosynthesis and Cellular Respiration

Figure 8.10 An overview of cellular respiration in mitochondria of most eukaryotic cells.

Table 8.2 Summary of the reactants and products of steps in cellular respiration

Step in cellular respiration	Reactants	Products
Glycolysis of one glucose	1 Glucose, 2 ATP	2 Pyruvate, 4 ATP, 2 NADH
Oxidation of Pyruvate	2 Pyruvate	2 Acetyl CoA, 2 CO_2, 2 NADH
Citric Acid Cycle	2 Acetyl CoA	4 CO_2, 2 ATP, 6 NADH, 2 $FADH_2$
Oxidative Phosphorylation	10 NADH, 2 $FADH_2$	Up to 28 ATP

Oxidation of Pyruvate

4. Pyruvate goes into the mitochondrion, where it is oxidized. One carbon is removed from pyruvate and combined with oxygen to produce carbon dioxide waste.
5. Coenzyme A binds to the remaining 2-carbon molecule, producing **acetyl-CoA**.
6. The energy provided by breaking down pyruvate charges one NAD^+ into NADH.

The Citric Acid Cycle

7. Acetyl-CoA enters the citric acid cycle and binds to 4-carbon **oxaloacetate** to form a 6-carbon **citrate** (*citric acid*).
8. Citrate goes through a series of steps in which two carbons are removed. The carbon atoms are combined with oxygen and released as carbon dioxide.
9. The energy released in the decomposition of citrate is used to charge up one ADP to ATP, one FAD to $FADH_2$, and three NAD^+ to NADH.
10. The end product of the citric acid cycle is an oxaloacetate molecule, which can bind to another acetyl-CoA to repeat the process.

Oxidative Phosphorylation

11. The NADH and FADH$_2$ molecules produced in the previous steps transfer their electrons to a series of proteins in the inner mitochondrial membrane.
12. Here, the electrons pass through an *electron transport chain* (mostly *cytochrome* complexes) that use the energy from the electrons to pump hydrogen ions into the intermembrane space.
13. Oxygen atoms must be available to accept the hydrogen ions at the end of the electron transport chain. If there are no oxygen atoms to receive the electrons, the electron transport chain stops, and hydrogen ions are not pumped into the intermembrane space.
14. *ATP synthase*, the same enzyme and channel protein found in the thylakoid membrane, is also present in the inner mitochondrial membrane. It uses the potential energy of the high concentration of hydrogen ions passing through it to convert ADP to ATP.

Increased Energy Demands

When an organism needs more ATP, such as when an animal is vigorously active, other molecules can be fed into the cellular respiration pathway. Stored fats, glycogen, and even proteins can be used in various ways to keep cellular respiration going. However, the limiting factor is usually oxygen. Cardiovascular fitness refers to the efficiency of the cardiovascular system to deliver adequate oxygen to mitochondria as they attempt to produce enough ATP to meet the demands of the organism.

It is interesting to note that the buildup of carbon dioxide, not the lack of oxygen, is what typically triggers increased breathing rate. Carbon dioxide in water produces carbonic acid. High levels of carbonic acid trigger the respiratory system to breathe harder.

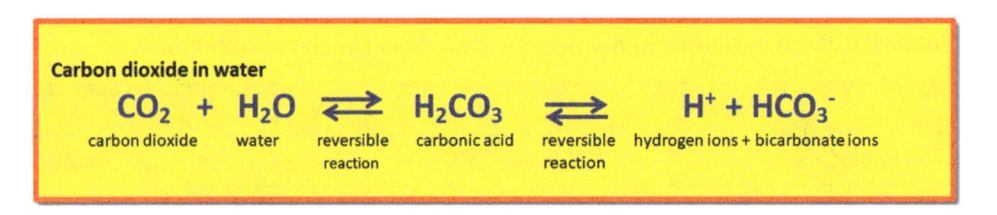

Carbon dioxide in water

$$CO_2 + H_2O \rightleftharpoons H_2CO_3 \rightleftharpoons H^+ + HCO_3^-$$

carbon dioxide — water — reversible reaction — carbonic acid — reversible reaction — hydrogen ions + bicarbonate ions

Lab Skill 8.1 Separating Compounds Using Chromatography

Biologists often need to isolate different compounds, such as proteins or nucleic acids from cells. Gel electrophoresis is one method for separating compounds in a solution. But another common method in biology is chromatography. **Chromatography** is a lab technique that separates different types of solute particles based on their distribution in a fluid stream. The fluid, called the *mobile phase*, may be a liquid or a gas. The particles are usually separated in the fluid based on their molecular weight, polarity, solubility in the solvent, or affinity to the *stationary phase* on which they are placed. The stationary phase can be a strip of paper, as will be used here, or a gel, silicone beads, or other materials.

Pigments are compounds that absorb light at different wavelengths. The ink in a black magic marker absorbs all of the wavelengths of light, which is why it appears black. But the ink particles themselves are often not actually black. The ink is usually a combination of several different pigments. If enough pigments are mixed together so that all wavelengths of light are absorbed, the ink will appear black. This is also why mixing different colors of paint or dye will produce darker and darker colors until it becomes black.

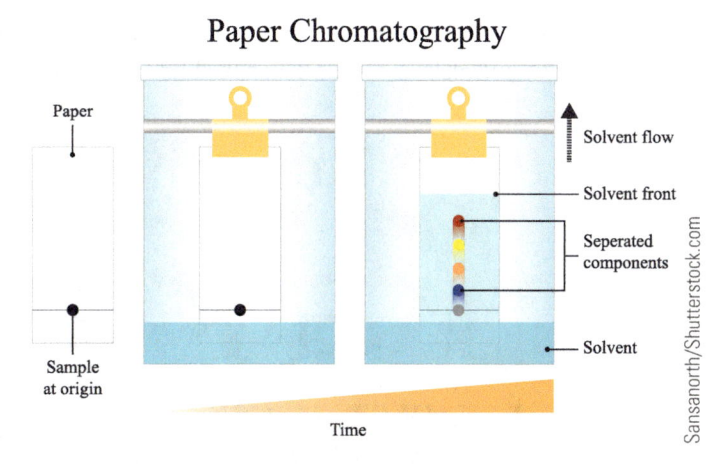

This can easily be demonstrated by paper chromatography. Pigments such as those in ink are usually soluble in a solvent such as acetone or alcohol. The solvent will wick up a strip of paper by capillary action if the end of the strip is placed in it. If some of the ink is placed on the paper, just above the solvent, ink particles will be swept along as the solvent wicks up the paper. In Figure 8.11, the illustration on the left shows black ink at the bottom of the paper at the beginning of a chromatography procedure. As the solvent rises up the paper, the different pigments in the ink became visible. The result is called a **chromatogram**.

Figure 8.11 Paper chromatography. The chromatogram (right) shows the individual components of the original black dot (left).

The topmost point where the solvent wicked up the paper is called the "solvent front." Regardless of how far the solvent has *actually* moved, specific types of molecules will travel a specific distance *compared to the solvent front* depending on the mobile and stationary phases used. This *ratio* of the distance a pigment moved compared to the distance the solvent moved is called the retardation, or **R_f factor**.

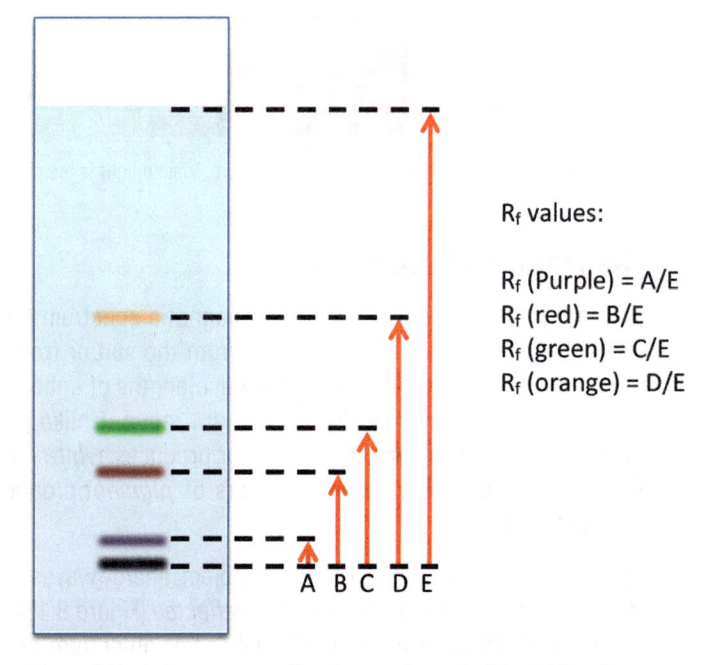

Figure 8.12 illustrates a hypothetical chromatogram produced by paper chromatography. The black line at the bottom of the strip of paper is the application point, or where the sample solution was originally placed. The point of application must be above the level of the solvent when running the chromatography procedure

R_f values:

R_f (Purple) = A/E
R_f (red) = B/E
R_f (green) = C/E
R_f (orange) = D/E

Figure 8.12 A chromatogram. R_f values are determined by dividing the distance a particular solute traveled by the distance the solvent traveled.

or the sample will diffuse into the solution. "A" represents the distance from the application to the first band of pigment. B represents the distance from the application point to the second band, and so forth.

The R_f factor for a solute depends on its properties (such as solubility, polarity, molecular weight) but they are also specific to the stationary phase (type of paper) used as well as the mobile phase (type of solvent used.). Tables of R_f values are available for many different compounds separated by many different mobile and stationary phases. If an unknown compound has been separated by paper chromatography, it may be identified by looking up its R_f value on a standard table of R_f values for that chromatography paper and solvent.

Lab Skill 8.2 Using a Spectrophotometer

A **spectrophotometer** is a common piece of laboratory equipment that measures how much a solution absorbs or transmits certain wavelengths of electromagnetic energy.

Electromagnetic energy is a form of energy that travels through empty space in the form of waves and photons. Photons have no mass and move at the speed of light. The wave patterns of different forms of electromagnetic energy determine the type of energy. Visible light is only one small part of the electromagnetic spectrum (Figure 8.13). Many common household devices use different parts of the electromagnetic spectrum. Cell phones, microwaves, radio waves, remote control units, night vision goggles, and security cameras all use parts of the electromagnetic spectrum we can't detect with our eyes. Wavelength (the distance between peaks of waves) is inversely related to frequency (number of peaks per second). So a type of electromagnetic energy with a high frequency will have a low wavelength, and vice versa.

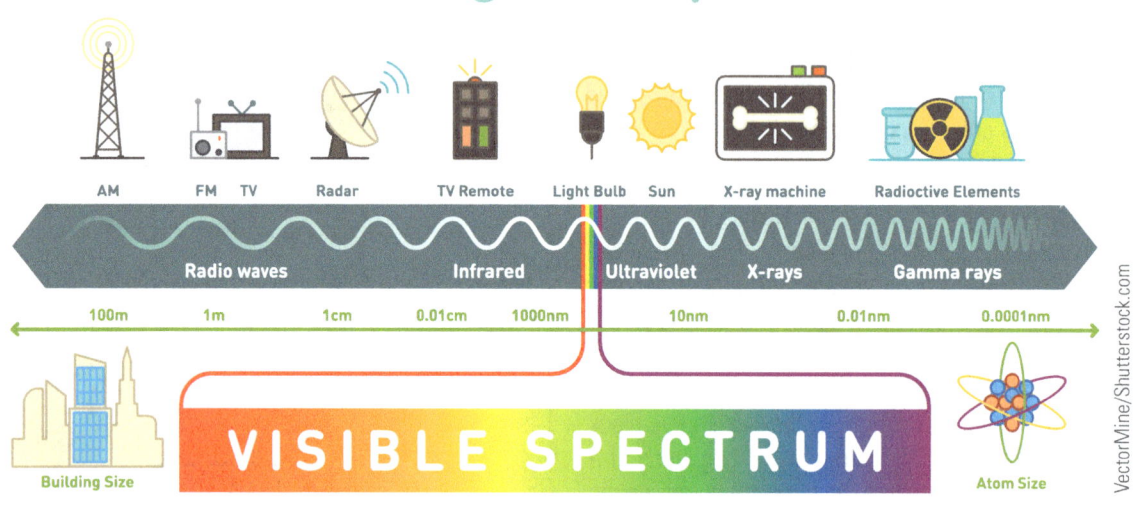

Figure 8.13 The electromagnetic spectrum. Visible light is the part of the spectrum human eyes can detect.

The Nature of Light

Visible light is the part of the electromagnetic spectrum that can be detected by human eyes. Light from the sun or from a standard light bulb contains all of the wavelengths of light and some wavelengths outside the visible light range. Unlike pigments, combining primary colors of *light* produces *white light* (Figure 8.14). (Combining primary colors of *pigment* produce *black pigment*.)

When electromagnetic energy strikes an object, energy waves can be *absorbed, transmitted, refracted,* or *reflected* (Figure 8.15). If the energy is absorbed, it is converted into some other form, such as heat, or, as in photosynthesis, excites electrons. If the energy is reflected, it bounces off of the surface of the object. If the energy is transmitted, the energy passes through the object. And if the energy is refracted, the waves changes direction as they pass through an object.

The color of an object is determined by the wavelengths of light that are NOT absorbed by its pigments. A red object absorbs other

Figure 8.14 Overlapping circles of light. The primary colors of light are red, blue and green. The color we perceive for any object is determined by the combination of the intensities of these wavelengths as they strike the retina of the eye.

wavelengths and reflects red wavelengths. A white object reflects all wavelengths, and a black object absorbs all wavelengths.

The wavelengths of visible light are, for human eyes, between about 740 nm (red) and 380 nm (violet) (Figure 8.16). Wavelengths longer than 740, and at a lower frequency than red wavelengths, are generally undetectable by human eyes. They are described as *infrared* waves. (*Infra* = lower.) Digital cameras, night-vision goggles, and remote controls use infrared beams. Wavelengths shorter than 380 nm, or at a higher frequency than violet, are considered *ultraviolet*. (*Ultra* = above.) A "black light" produces ultraviolet waves. If ultraviolet waves strike a fluorescent pigment, the fluorescent pigment will convert the ultraviolet wave into a longer wavelength that is visible, making the object appear to glow.

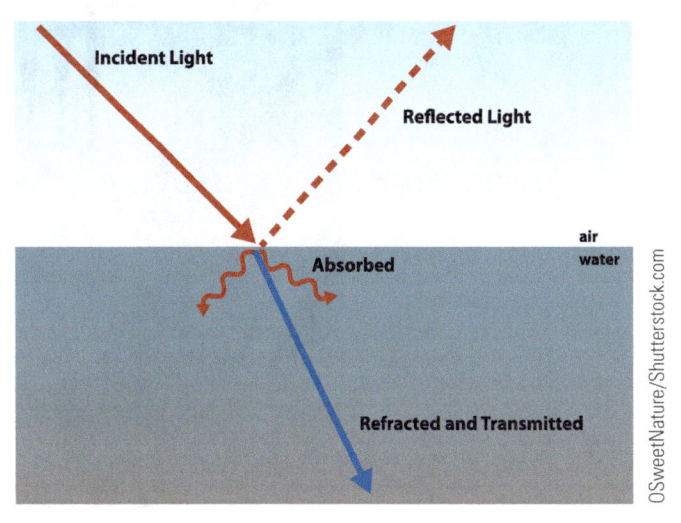

Figure 8.15 Light and other forms of electromagnetic energy can be absorbed, reflected, refracted, and transmitted by objects.

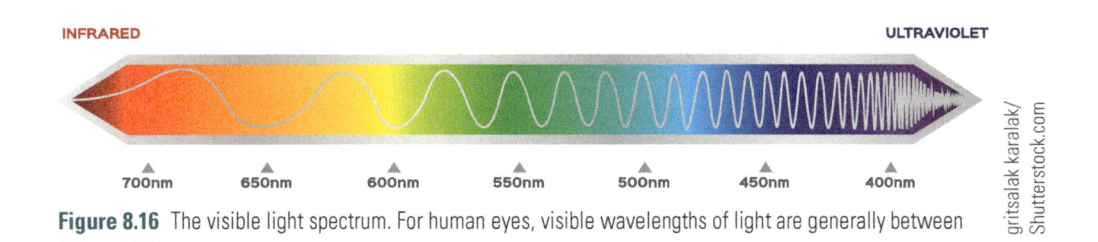

Figure 8.16 The visible light spectrum. For human eyes, visible wavelengths of light are generally between 740 and 380 nm.

Using Spectrophotometry in Biology

Spectrophotometry is an important tool in biochemistry and clinical medicine for determining quantity of specific compounds in a solution. Specific types of molecules absorb certain wavelengths of electromagnetic energy. A spectrophotometer separates light waves by refracting them through a prism. It then detects the amount of that wavelength of light that passes through a solution (Figure 8.17). The result is a reading of % absorbance, or % transmittance. Absorbance and transmittance are inversely related, so 100% absorbance is equivalent to 0% transmittance.

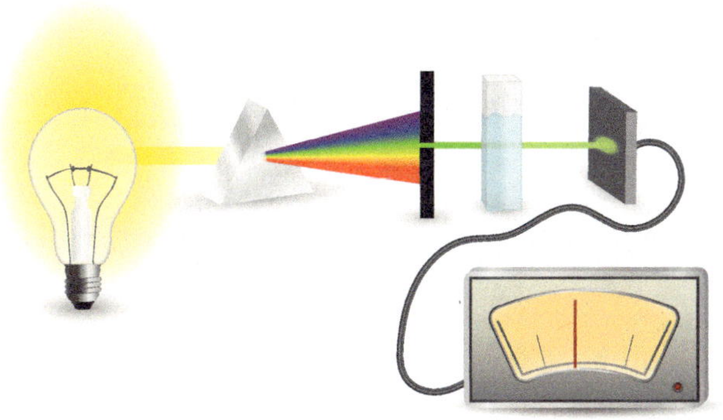

Figure 8.17 A spectrophotometer can detect how much of a particular wavelength is absorbed by molecules in a solution.
extender_01/Shutterstock.com

The **Beer–Lambert law** states that there is a direct linear relationship between the concentration of a solution and its absorbance of electromagnetic energy. If a standard *absorption coefficient* is known for a compound, then a spectrophotometer can be used to determine the concentration of that compound in a solution. If a spectrophotometer is set to emit wavelengths absorbed by a specific compound, the greater the absorbance, the greater the concentration of the compound in that solution. A spectrophotometer can detect the presence and concentration of compounds like DNA, proteins, or pigments in a solution.

A spectrophotometer can also provide an **absorbance spectrum**, a graph that illustrates which wavelengths of electromagnetic energy a particular pigment absorbs. This is useful in biology for studying the pigments involved in photosynthesis, and which wavelengths of light are actually used by the plant.

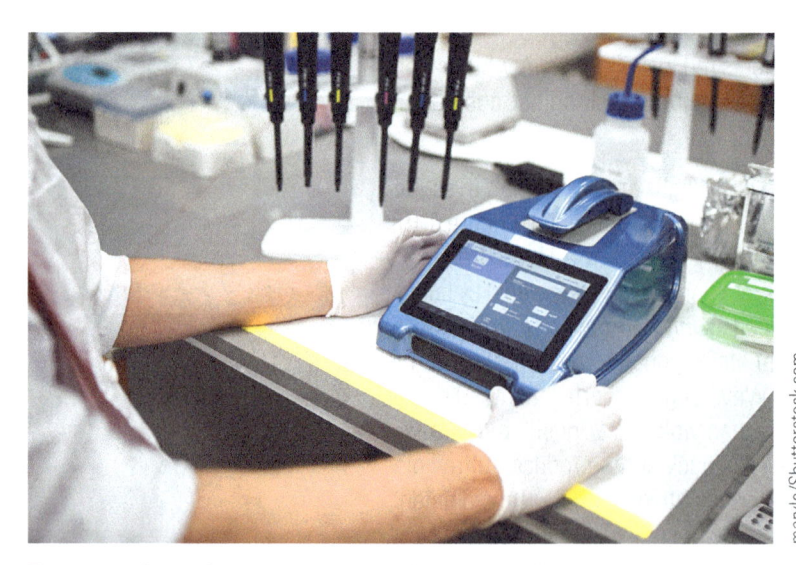

Figure 8.18 A nanodrop spectrophotometer that can pass different wavelengths of ultraviolet and visible light through a specimen, analyze it, and provide a detailed report.

Spectrophotometers come in many shapes and sizes. Over the last few years, new "nanodrop" spectrophotometers have been developed that can analyze a few microliters of a sample (Figure 8.18). They use a laser and an onboard computer to provide rapid and sophisticated analysis of biological compounds.

PRE-LAB QUIZ 8: Are you ready to proceed?

1. Complete the table by indicating if the term given applies to photosynthesis, cellular respiration, or both.

Compound, process, or structure	Photosynthesis only	Cellular respiration only	Both
Glucose			
Pyruvate			
Cristae			
ATP synthase			
Stroma			
High concentration of hydrogen ions			
Calvin cycle			
NADH			
Electron transport chain			
Granum			
FADH$_2$			
Cytochrome complex			
ATP is generated			
Acetyl-CoA			
Citric acid cycle			
NADPH			

2. Compare and contrast the structure of a chloroplast with the structure of a mitochondrion.

3. What is the difference between *absorbing* light and *transmitting* light? If a spectrophotometer indicated that a sample's absorbance is 0.21 (or 21%), what is the transmittance?

4. How many turns of the Calvin cycle are needed to produce one G3P molecule? How many G3Ps are needed to make one glucose? How many turns of the citric acid cycle are needed to break down one glucose molecule?

5. What exactly is a redox reaction? Define *reduction* and *oxidation* as they apply to redox reactions.

6. What is a chromatogram? How is it useful in biology?

7. What is an R_f value, and how is it calculated?

8. What is the electromagnetic spectrum? What range within the spectrum is visible?

9. Give two uses for a spectrophotometer in biology.

10. How many molecules of ATP can be generated from one glucose molecule?

Problem to Solve 8.1 Can I Observe Cellular Respiration and Photosynthesis in a Plant?

Since this activity takes about an hour to run, set it up first. Then complete the other activities while waiting for your results.

Since cellular respiration and photosynthesis are opposite reactions, it should be easy to determine if a plant is doing one or the other at any one time by measuring which compounds are being taken up or released. We could theoretically measure the amount of oxygen being released if photosynthesis is occurring, or we could measure the amount of oxygen being consumed if it is doing cellular respiration. We could also measure the water being used for photosynthesis, or produced by cellular respiration. However, the most convenient way to measure cellular respiration or photosynthetic activity is with carbon dioxide (Figure 8.19).

Figure 8.19 Elodea is a common aquatic plant.

Recall that carbon dioxide in water can bind to water molecules and form carbonic acid, H_2CO_3, which then dissociates into hydrogen ions and bicarbonate ions. These are reversible reactions, but the greater the concentration of carbon dioxide in the solution, the more likely carbonic acid and bicarbonate will be present in the solution. The presence of carbonic acid can be detected with a pH probe or pH indicators.

You will use sprigs of elodea, a small, aquatic plant, for this experiment. One sprig will be kept in a dark environment for an hour, and the other in a bright environment for an hour. You will need to create a hypothesis and prediction for how the presence or absence of light will affect carbon dioxide concentrations in the solution, and the subsequent effect that will have on the pH of the solution.

This experiment requires accurate pH measurements. If using a pH probe, ensure it is calibrated, rinse it in clean water between uses, and ensure the tip is fully submerged in the solution when taking a reading. If using pH test strips or universal indicator, try to get an accurate reading as possible.

Title

Title: Inferring photosynthesis or cellular respiration from changes in pH

Purpose

Goal: Place sprigs of an aquatic plant in light and dark conditions to determine if carbon dioxide levels change. Use changes in the pH of the solution to infer changes in CO_2 levels.

Hypothesis: If the plant exposed to light, the pH of the solution will _____ because _____. If the plant is in the dark, the pH will _____ because _____.

Protocol

Materials:

(Per group)
- 4 large test tubes
- 2 test tube racks
- pH probe, test strips, or pH indicator solution
- 5 beakers (100-150 ml)
- 2 pieces of foil
- parafilm
- mass scale and weigh boat
- tape and marker for labeling test tubes
- 2 transfer pipettes (3 ml)
- Forceps

(Per class)
- large bowl of water containing elodea sprigs
- LED grow light (or other suitable light source)

Protocol:

1. Label the four test tubes with group number and LE (light: experimental), LC (light: control), DE (dark: experimental), and DC (dark: control). Label four clean beakers with the same labels and set them aside for later.

2. Obtain two sprigs of elodea about 8-10 cm each in length. Try to obtain sprigs that are close to the same size.

3. Measure the mass of the elodea sprigs. Trim longer sprigs if necessary until they have close to the same mass. (Sprigs will be reused, so return any larger cut sections back to the container. Very small pieces can be discarded.)

4. Determine which sprigs will go in the light and dark conditions and record their masses in the appropriate boxes of the data table.

Continued from previous page.

Protocol

5. Fill all 4 test tubes three-fourths of the way full with water from the bowl of elodea (making sure they are equal). Place one elodea sprig in the test tube labeled LE, and the other in the test tube labeled DE.

6. Half-fill one clean beaker with distilled water. Rinse the pH probe in the distilled water and ensure it is calibrated. (Buffer solutions may be provided for checking calibration. Calibrate the probe as directed by your instructor.)

7. Use the pH probe (or other pH indicator as directed by your instructor) to measure the initial pH of the water in the bowl of elodea. Record it in the data table.

8. Cover the tops of all test tubes with parafilm.

9. Wrap the tubes labeled DE and DC tightly with foil, place them in a test tube rack, and place the rack in a cabinet.

10. Place the tubes labeled LE and LC in a test tube rack and place them under the grow light (or other light source as directed).

11. Record the start time in the data table.

12. Wait at least one hour. (Complete other activities in this lab exercise.)

13. Collect the test tube racks, and remove the foil and parafilm from the tubes.

14. Pour the liquid from each test tube into the appropriately labeled beakers.

15. Use the pH probe (or other pH indicator) to measure the pH of each solution and record it in the data table.

16. Return the elodea to the bowl, rinse and cap the pH probe, and clean glassware.

Purpose

Continued from previous page.

Protocol

Data:

Test tube	Condition	Mass of elodea (g)	Initial pH (pH_i)	Final pH (pH_f)	ΔpH ($pH_f - pH_i$)	Adjusted pH $pH_{adj} = \Delta pH$ of experimental $- \Delta pH$ of control	Normalized change in pH (pH_{adj}/mass)
LE	Light: elodea						
LC	Light: control	X					
DE	Dark: elodea						
DC	Dark: control	X					

Discussion and Conclusions:

Why did you need to include tubes LC and DC in this experiment?

Did the control tubes show a change in pH? If so, what may explain this?

Continued from previous page.

Why was any change in pH in the tubes without elodea subtracted from the tubes with elodea?

Why was the adjusted pH then "normalized" by dividing by the mass of the elodea? How does this allow you to compare your data to that of other groups?

Did your data support your hypothesis? Explain why or why not.

If your results are not what you expected, what might explain those results? Unexpected results may be due to experimental error, but they may also be indications that there are factors in the experiment that you did not consider. What other factors may have affected your results?

Was the elodea doing cellular respiration in the light condition? Does your data show this? Explain why or why not.

Problem to Solve 8.2 Can I Observe Differences in Human CO_2 Production?

Limewater is a common name used to describe a dilute calcium hydroxide, $Ca(OH)_2$, solution. It has nothing to do with the lime fruit. It is a solution of powdered limestone or chalk. Limewater is normally clear, colorless, and basic, with a pH around 12.

When carbon dioxide is bubbled through limewater, the carbon dioxide reacts with the calcium hydroxide to form an insoluble precipitate, calcium carbonate, $CaCO_3$. This makes the solution appear milky white (Figure 8.20).

If excessive CO_2 is bubbled through the solution, the calcium carbonate is converted to calcium bicarbonate, $Ca(HCO_3)_2$. Calcium bicarbonate is soluble in water, so the solution will appear clear and colorless again.

$$CO_2 + Ca(OH)_2 \longrightarrow CaCO_3 + H_2O$$

$$CO_2 + CaCO_3 + H_2O \longrightarrow Ca(HCO_3)_2$$

In this simple activity, you will observe differences in CO_2 concentration of your exhaled breath. You will not be able to quantitatively measure the amount of carbon dioxide being released, but you will be able to observe differences in the limewater as you blow bubbles through it.

Figure 8.20 Blowing carbon dioxide through limewater causes it to become cloudy.

You will compare the effects of a normal exhalation on limewater, an exhalation after holding your breath, and an exhalation after a short burst of vigorous exercise. It is important that for each trial you exhale roughly equal volumes of air through the limewater. The differences in the appearance of the limewater solutions should allow you to infer relative levels of carbon dioxide in your exhalations, and from this the level of cellular respiration occurring in your body.

Title: Detecting levels of carbon dioxide in exhaled breath under different conditions.

Goal: Observe the effects of different activities on human cellular respiration by observing the relative amount of carbon dioxide exhaled.

Hypothesis: The limewater will appear _____ after blowing one normal exhalation through it. The limewater will appear _____ after holding my breath for one minute and then exhaling through it. The limewater will appear _____ after doing twenty jumping jacks and then exhaling through it.

Materials:
(Per student)
Three large test tubes
Test tube rack
Tape and marker for labeling test tubes
Straw
Limewater solution
Paper towel
Clock or stopwatch

Protocol:

1. Label the test tubes 1, 2, and 3.

2. Fill the test tubes *half full* of limewater. The exact amount is not important, but they should each contain about the same amount of limewater.

3. Place the straw in test tube 1. Wrap the paper towel around the top of the tube to prevent spraying limewater.

4. Take in a deep breath, and then *slowly* blow through the straw until you have fully exhaled that one breath. (Do not inhale through the straw or drink the limewater!)

5. Observe the limewater for changes and record observations in the data table.

6. Transfer the paper towel and straw to test tube 2.

7. Then use a stopwatch to time yourself. Take in a deep breath, hold it for a full minute (or as long as you can), and then slowly blow your exhalation through the limewater in test tube 2.

8. Observe the results and record observations in the data table.

9. Transfer the straw and paper towel to test tube 3.

Date

Title

Purpose

Continued from previous page.

Protocol

10. Do twenty quick jumping jacks (or as many as you can). You should be breathing fairly hard. Take in a deep breath and slowly blow your exhalation through the limewater in test tube 3.

11. Observe the results and record observations in the data table.

12. Discard limewater in the sink, dispose of straw and paper towels, and clean the test tubes.

Data and Observations:

Test Tube	Condition	Appearance of limewater	Implications
1	Normal exhalation		
2	Exhalation after holding breath		
3	Exhalation after vigorous activity		

Discussion and Conclusions:

Did the data support your hypothesis?

Under which condition did it appear that your breath contained the most carbon dioxide? Under which did your breath contain the least carbon dioxide?

Problem to Solve 8.3 Can I Determine the Pigments in a Leaf?

Plants contain a variety of pigments. Chlorophyll is needed for photosynthesis, but other pigments can absorb light at other wavelengths and transfer that energy to chlorophyll. The other pigments in a leaf may not be visible until the leaf stops photosynthesizing and the pigments begin to break down. This is why leaves change color in the fall. As chlorophyll begins to degrade, other pigments become visible.

Pigments give fruits and vegetables their color. Many plants (and other organisms) contain many pigments resulting in the variety of colors we see in nature (Table 8.3). If a human or other animal consumes enough pigments, they can even start turning that color. (Human babies fed too much carrot baby food can turn orange, and white flamingos that eat a diet rich in beta-carotene turn pink.)

Pigments have many functions. One of the major reasons why plants produce pigments is to attract pollinators. Since early humans began painting on the walls of caves, people have used plant pigments to paint, dye, and stain many things. An interesting feature of anthocyanins is that the color of the pigment varies with pH, making these compounds useful indicators for measuring the pH of solutions. In this activity, you will separate pigments from different types of leaves. You will also separate pigments in black ink.

Review Lab Skill 8.2 on paper chromatography. The procedure for this lab is straightforward, but you will be handling volatile chemicals. Wear gloves and safety glasses. *Avoid touching the chromatography paper with your bare hands.* Oils from the skin will block the movement of the chromatography solvent through the paper. You will be preparing a leaf extract. Save the extracted pigments you obtain in this lab activity to use in the next one.

Table 8.3 Examples of plant pigments and their colors

Major category	Major subtypes	Main colors	Examples
Chlorophylls	Chlorophyll a	Blue-green	All land plants and green algae
	Chlorophyll b	Yellow-green	
Carotenoids	Xanthophylls	Yellow	Yellow squash, lemon, banana peel, yellow bell pepper, flowers
	Carotenes	Orange and red	Carrot, pumpkin, squash, flowers
Anthocyanins		Deep red, blue, and purple	Blueberry, eggplant, cherry, flowers
Betalains	Betacyanins	Deep red to purple	Beets, flowers
	Betaxanthins	Yellow and orange	Beets, flowers

Title: Separation of plant pigments in different plant tissues

Goal: To use paper chromatography to demonstrate the variety of pigments in plant tissues, including leaves that appear to just contain green pigments.

Materials:

(Per pair of students)
- leaves (spinach leaves, other dark colored leaves, fruits, or vegetables)
- scissors
- forceps
- small ruler
- mortar and pestle
- 2 beakers (100-150 ml)
- 2 transfer pipettes (3 ml)
- 1 transfer pipette (1 ml)
- 1 centrifuge tube (1.5 ml)
- chromatography paper
- black pen or marker (any kind)
- pencil
- tape
- parafilm
- hotplate (may be shared between groups)
- microcentrifuge (may be shared between groups)

(Per class)
- chromatography solvent (e.g., 9:1 petroleum ether and acetone) approximately 50 ml ethyl alcohol in a beaker, covered with parafilm to prevent evaporation

Protocol:

1. Obtain plant matter of interest, preferably with dark colors. (leaves, fruits, vegetables, flowers, etc.)

2. Cut out two strips (2 cm × 10 cm) of chromatography paper. **Do not touch paper with bare hands! Oil from the skin will block migration of the chromatography solution.**

3. Use a ruler to lightly draw a line in pencil (*NOT PEN*) across each strip of paper 1 cm from the bottom.

Laboratory Exercise 8 Cell Biology II: Photosynthesis and Cellular Respiration

Continued from previous page.

Protocol

4. Half-fill a small beaker with tap water and place it on the hot plate. Bring to a boil.

5. Boil the plant matter in the beaker for 5 minutes. This denatures enzymes that may degrade the pigments.

6. Using forceps, carefully remove the plant matter from the boiling water and place it on a paper towel to cool and dry.

7. Use a transfer pipette (3 ml) to withdraw about 2-3 ml of ethyl alcohol provided for class use.

8. Once the plant matter has cooled, use the scissors to cut it into very tiny pieces, putting the pieces into the mortar (bowl). (See Figure 8.21.)

9. Add a few drops of ethyl alcohol to the mortar. Grind the plant matter with the pestle until a paste forms. Do not add too much ethyl alcohol. The paste should be a dark color. The more concentrated it is, the better.

10. Use a clean transfer pipette to transfer as much of the plant sample as possible to a 1.5 ml centrifuge tube.

11. Centrifuge for 5 minutes at maximum speed. (If necessary, review Lab Skill 4.1 on how to use a centrifuge.)

12. Without disturbing the pellet, transfer the supernatant to the clean centrifuge tube. It should be a dark color and viscous (thick). If not, repeat steps 7 through 10.

13. Discard the centrifuge tube containing the pellet.

Figure 8.21 Mortar and pestle. The mortar is the bowl. The pestle is used to grind materials into a fine powder or paste.

Date | Researcher name

Title

Purpose

Continued from previous page.

Protocol

14. Using a small transfer pipette (1 ml), transfer the supernatant to one of the strips of chromatography paper. "Paint" the pencil line with the plant solution until the line appears dark. See Figure 8.22. Save the excess solution.

15. Use a black ink pen or marker to draw a heavy, dark, black line where the pencil line marks the other strip of chromatography paper.

16. Tape the other ends of the strips of paper to a pencil so that they are next to each other.

17. Place the pencil over a beaker so the strips hang down into the beaker. Ensure the strips almost but not quite touch the bottom of the beaker. Adjust the tape as necessary.

18. Obtain about 5 ml chromatography solution in a small beaker. Use a 3 ml transfer pipette to carefully add the chromatography solution to the beaker until the bottom of the strips are submerged in the solution, but the lines of plant extract and ink are not submerged.

19. Cover the beaker with parafilm to reduce evaporation of the solvent.

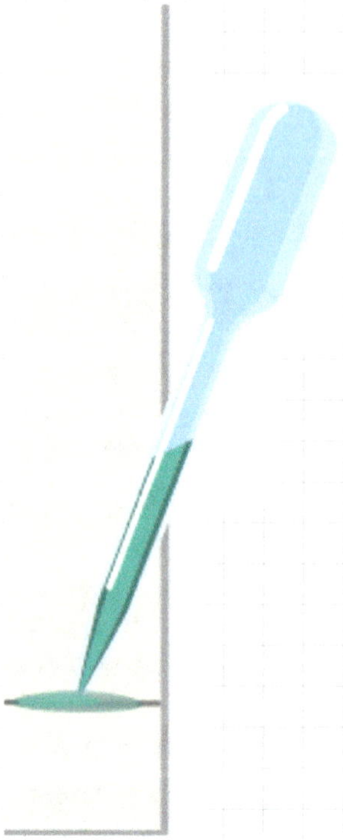

Paul Luyster; Pipette: Nsit/Shutterstock.com

Figure 8.22 Applying solution to chromatography paper.

20. Allow the solvent to wick up the chromatography paper. This will take several minutes.

21. When the solvent front is within 3 cm of the top, withdraw the papers and immediately mark the solvent front with a pencil.

22. Mark the distance each pigment moved.

Researcher signature | Date | Supervisor/Instructor signature | Date

Laboratory Exercise 8 Cell Biology II: Photosynthesis and Cellular Respiration

Continued from previous page.

Protocol

23. Measure the distance from the application point to the front edge of each band of pigment. Enter the distances on the data table.

24. Calculate the R_f values for each pigment. If you can match the R_f values to any values in Table 8.4, identify the pigments in the plant sample.

25. Once the strips are dry, place clear tape over the areas that have shown separation of pigments (to keep the pigments from rubbing off) and tape the strips into the lab manual.

26. Dispose of solvents as directed by your instructor. Clean and put away glassware and equipment.

Data:

Plant Pigments			
Total number of pigments separated from original extract:			
Total distance solvent traveled (between application point and solvent front):			
Pigment color	Distance pigment moved	R_f value	Possible name of pigment

Tape plant pigment chromatogram here:

Title

Purpose

Continued from previous page.

Protocol

Ink Pigments		
Total number of pigments separated from ink:		
Total distance solvent traveled:		
Pigment color	Distance Pigment moved	R_f Value

Tape ink chromatogram here:

Table 8.4 R_f values for selected pigments

Pigments	R_f value in 9:1 acetone:ether	R_f value in 9:1 propanol:ether
Beta-carotene *	0.99	0.94
Xanthophyll *	unknown	0.89
Lutein *	0.68	unknown
Violaxanthin *	0.40	unknown
Chlorophyll a	0.30	0.46
Chlorophyll b	0.13	0.22

* Carotenes, xanthophyll, lutein, and violaxanthin are all carotenoids.

Conclusions:

What was the original color of the plant extract?

How many different pigments were present in the plant extract?

How many different pigments were in the black ink?

Problem to Solve 8.4 Can I Create an Absorbance Spectrum for Different Plant Pigment Combinations?

This simple activity will allow you to observe the absorbance spectrum of your plant extract. Note that it is likely a combination of pigments unique to that plant, so it will not match the absorbance spectrum in the textbook. If you could analyze just one pigment, such as chlorophyll a, the absorbance spectrum of your chlorophyll sample would look like the absorbance graphs in your textbook.

You will be using a nanodrop spectrophotometer. For this activity, the only materials you will need are your centrifuge tube of plant extract from the previous lab activity, a small beaker containing about 1 ml of clean chromatography solution, an adjustable micropipette set at 3 µl, and two white or yellow tips.

1. As directed by your instructor, turn on the spectrophotometer and touch the application for *UV-Vis* (ultraviolet and visible light).
2. Then in the main application screen, enter the wavelengths from the table below. The device will not allow you to enter all of them at once, so you will have to run your sample in the spectrophotometer twice, changing the wavelengths as needed.
3. Use the micropipette to place 3 µl of the clean chromatography solution on the small pedestal of the spectrophotometer. Close the top, and press "blank." It will calibrate the spectrophotometer.
4. Then open the top, wipe the pedestal and lid to clean them, and place 3 µl of the plant extract on the pedestal. Close the lid and press "run."
5. Record the absorbance for each wavelength in the data table. If you get an error message, dilute your extract by adding a small amount of chromatography solution to your centrifuge tube and try again.
6. Change the wavelengths in the settings screen and press run.
7. Record the absorbance for each wavelength in the data table.
8. Clean the pedestal and exit the program.
9. Graph the data.

Data:

Plant extract was produced from:	
Wavelength	Absorbance reading
300	
350	
400	
450	
500	
550	
600	
650	
700	
750	

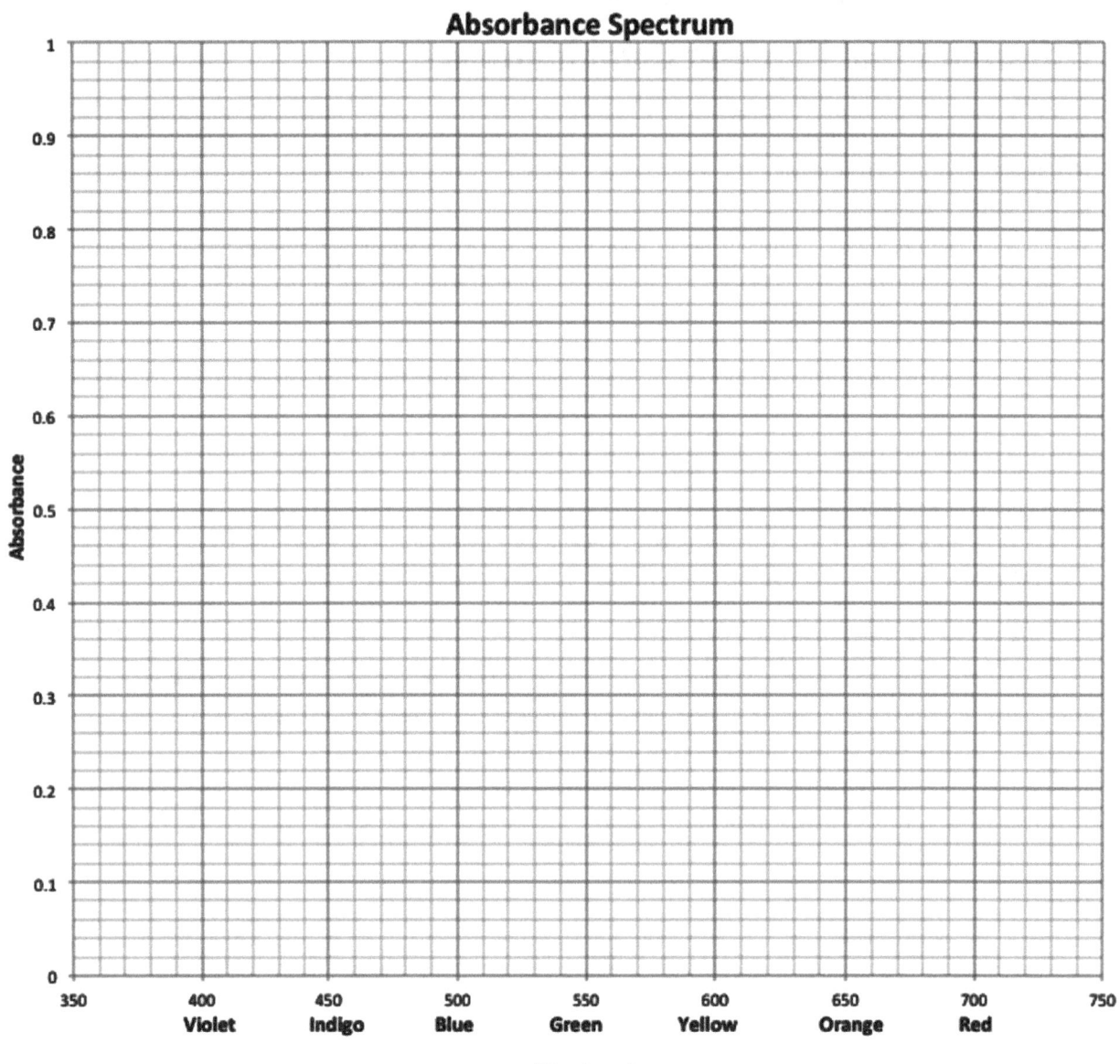

Absorbance Spectrum

Continued from previous page.

Summary and Conclusions:

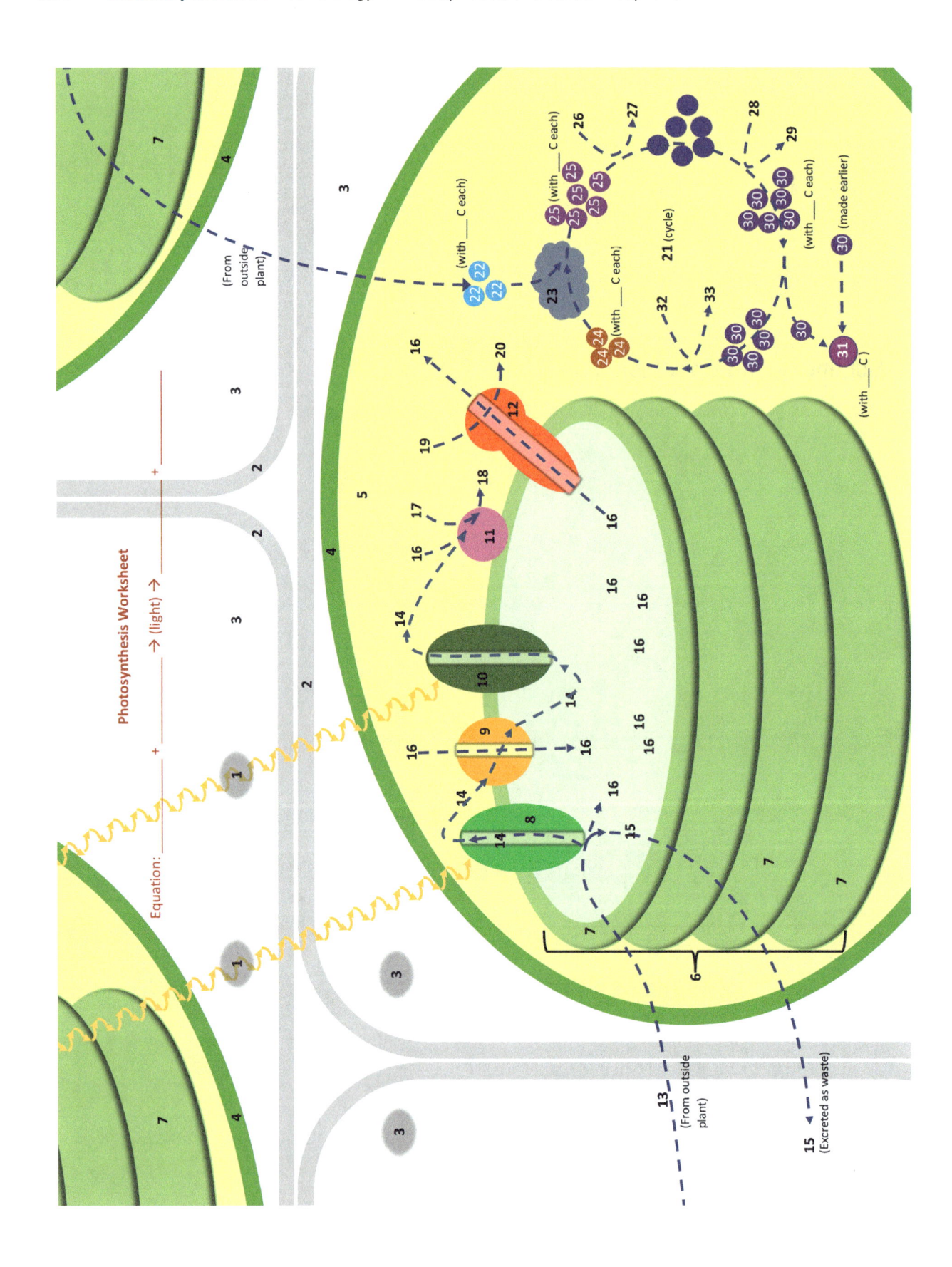

236 Laboratory Exercise 8 Cell Biology II: Photosynthesis and Cellular Respiration

Photosynthesis Worksheet

Equation: _____ + _____ + _____ (light) → _____ + _____

Problem to Solve 8.5 Can I Explain the Chemistry of Photosynthesis and Cellular Respiration?

Photosynthesis and cellular respiration are complex chemical processes. They are difficult to remember, and easy to confuse. The best way to learn them is to visualize what is happening at the cellular and organelle level, and to practice working through the processes.

Use the terms list and answer sheet below to identify the numbered parts of the photosynthesis and cellular respiration diagrams on the next page. (Some terms are used more than once for different numbers on the diagram.) Use the summary in the background information of this lab exercise, your textbook, and other resources to help you.

Terms List

- 3-phosphoglyceraldehyde (G3P)
- 3-phosphoglycerate
- ADP
- ATP
- ATP synthase
- Calvin cycle
- carbon dioxide (CO_2)
- cell wall & cell membrane
- chloroplast
- cytoplasm
- electron transport chain and cytochrome complex
- electrons (e-)
- glucose ($C_6H_{12}O_6$)
- granum
- hydrogen ions (H+)
- light
- NADP+
- NADP+ reductase
- NADPH
- oxygen (O_2)
- photosystem I
- photosystem II
- rubisco
- RuBP
- stroma
- thylakoid
- water (H_2O)

General Orientation	Chemical Process of Light Reactions	Chemical Process of Dark Reactions
1	13	21
2	14	22
3	15	23
4	16	24
5	17	25
6	18	26
7	19	27
Proteins of the Light Reactions	20	28
8		29
9		30
10		31
11		32
12		33

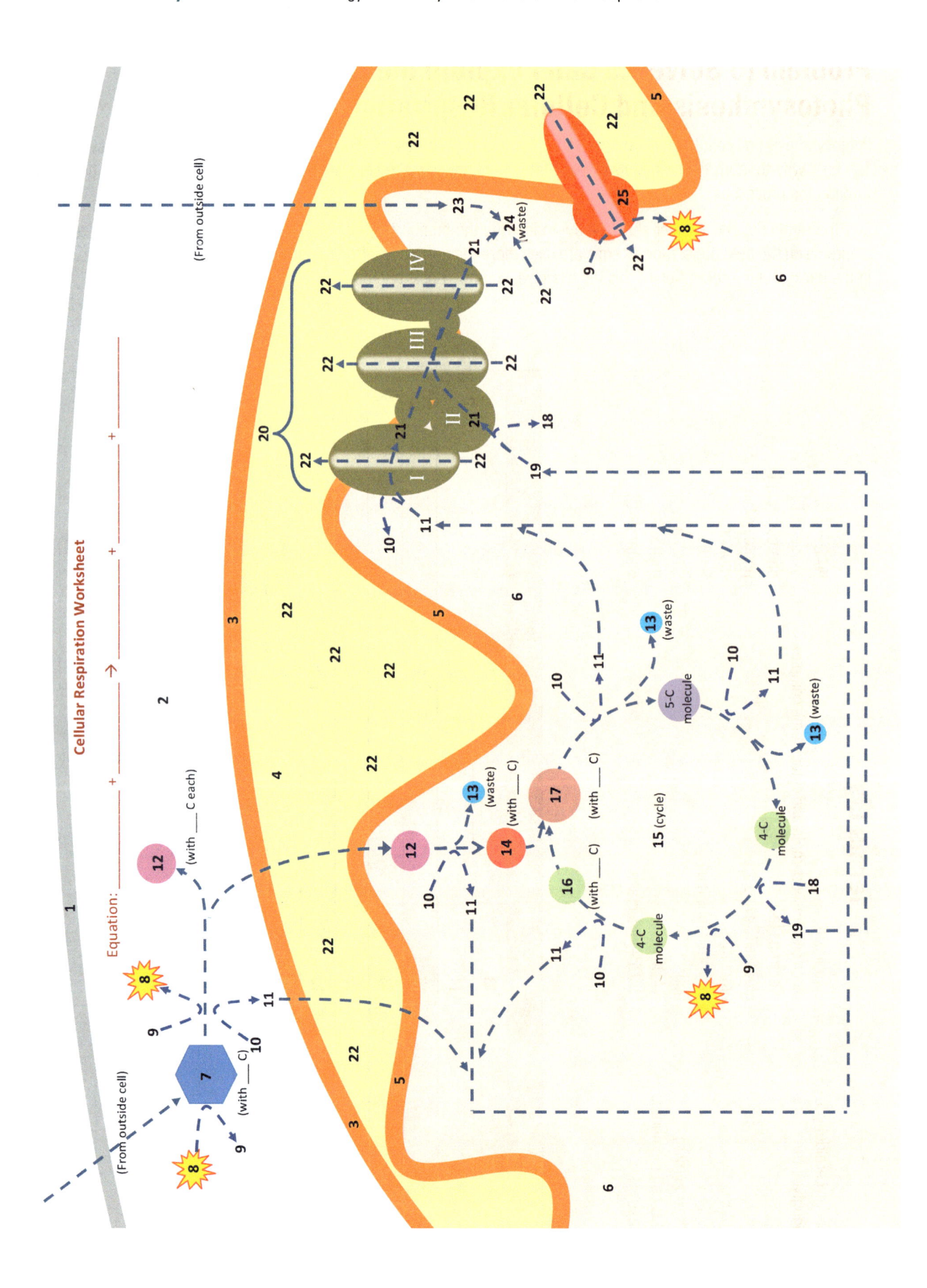

238 Laboratory Exercise 8 Cell Biology II: Photosynthesis and Cellular Respiration

Cellular Respiration Worksheet

Equation: _____ 1 _____ + _____ + _____ → _____ 2 _____ + _____ + _____

Laboratory Exercise 8 Cell Biology II: Photosynthesis and Cellular Respiration   239

Terms List

Acetyl CoA
ADP
ATP
ATP synthase
citric acid/citrate
citric acid/Kreb's cycle
carbon dioxide (CO₂)
cytopplasm
electron transport chain

electrons (e⁻)
FAD
FADH₂
glucose (C₆H₁₂O₆)
hydrogen ions (H+)
inner mitochondrial membrane
intermembrane space
mitochondrial matrix
NAD+

NADH
oxygen (O₂)
outer mitochondrial membrane
oxaloacetate
plasma membrane
pyruvic acid/pyruvate
water (H₂O)

General Orientation	Molecules in Glycolysis	Pyruvate Oxidation	Oxidative Phosphorylation
1	7	13	20
2	8	14	21
3	9	**Citric Acid Cycle**	22
4	10	15	23
5	11	16	24
6	12	17	25
		18	
		19	

Name: _____   Date: _____

POST-LAB QUIZ 8: What have you learned?

1. What compounds are produced when carbon dioxide is mixed with water?

2. If a test tube containing water and an aquatic plant is sealed and placed in the dark for a few hours, will the pH of the water increase, decrease, or remain the same? Why?

3. If a test tube containing water and heterotrophic aquatic organisms is sealed and placed in the dark for a few hours, will the pH of the water increase, decrease, or remain the same? Why?

4. What was the control in the elodea experiment? Why might the control demonstrate a change in pH? Why was this change subtracted from the experimental tubes?

5. This graph represents the absorbance spectrum for a hypothetical pigment. What color is the pigment? Why?

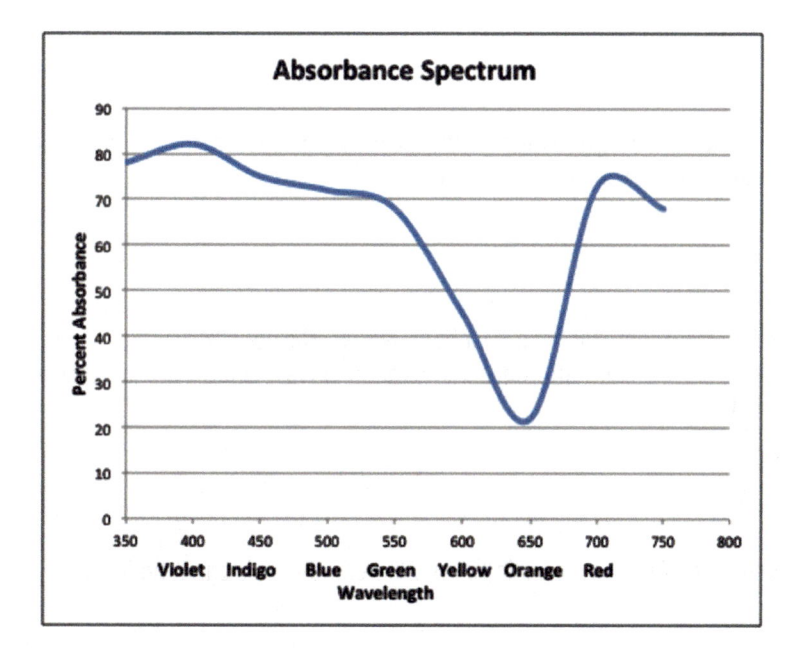

Absorbance Spectrum

(Graph: Percent Absorbance vs. Wavelength from 350 to 800 nm. The curve starts near 78% at 350, peaks around 82% near 400 (Violet), gradually decreases to ~70% through Indigo, Blue, Green, drops sharply to a minimum of ~21% around 650 (Orange), then rises to ~75% around 725 (Red), ending near 68% at 800.)

Wavelength labels: Violet (400), Indigo (450), Blue (500), Green (550), Yellow (600), Orange (650), Red (700)

Laboratory Exercise 8   Cell Biology II: Photosynthesis and Cellular Respiration

This figure to the right represents a chromatogram.
Hypothetical standard R_f values for pigments separated by the mobile and stationary phases used in this chromatogram:

β-carotene	0.99
chlorophyll a	0.30
chlorophyll b	0.13
violaxanthin	0.40
lutein	0.68
xanthophyll	0.20
phaeophytin	0.85

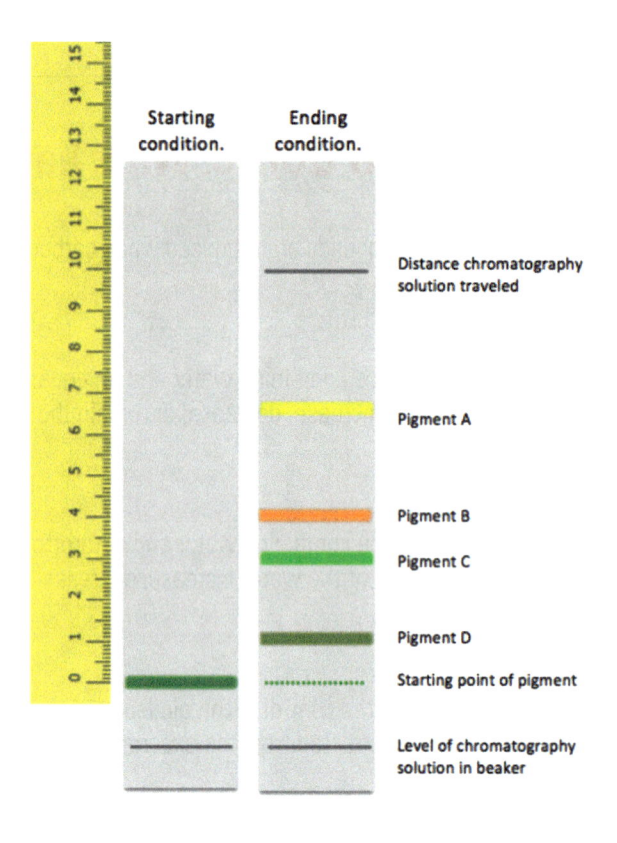

6. Using their R_f values, identify pigments A, B, C, and D.

7. When plants are in direct sunlight, which of the following processes are occurring: light reactions, dark reactions, and/or cellular respiration? Explain your answer.

8. Jillian purchased a "grow light" for her indoor herb garden. She noticed that the light it provides is unusual compared to other lamps. What is the likely difference in the range of wavelengths produced by a "grow light" versus a typical light? (Be specific.)

9. What enzyme generates ATP in eukaryotic cells? Where in the cells can it be found?

10. Why exactly do plants typically appear green? If a plant has purple leaves, which wavelengths of light is the plant probably using for photosynthesis?

Laboratory Exercise 9
Cell Biology III: Fermenting Sugar to Produce Alcohol

Concepts to Understand

After completing this lab, the student should be able to:

1. Describe yeast and explain how yeast fermentation is used in making alcohol.
2. Explain the concept of a null hypothesis.
3. Explain the process of running an experiment.
4. Differentiate between inductive and deductive reasoning.
5. Describe the structure of a scientific report.

Skills to Learn

After completing this lab, the student should be able to:

1. Run a controlled experiment.
2. Collect and represent quantitative data in appropriate tables and graphs.
3. Analyze the data using basic statistics.
4. Write a formal scientific report.

SEE ALSO

Campbell Biology—Chapter 1, Section 1.3; Chapter 9, section 9.5

Appendix E:—How to Write a Scientific Paper

Appendix F:—An Application of Stoichiometry and Dimensional Analysis

Others:

Important Terms to Know

- Abstract
- ABV
- Aerobic
- Alcohol fermentation
- Anaerobic
- Bar chart
- Continuous data
- Deductive reasoning
- Discrete data
- Facultative anaerobe
- Histogram
- Inductive reasoning
- Lactic acid fermentation
- Line graph
- Literature review
- Normalized
- Null hypothesis
- p-value
- Pathogenic
- Pie chart
- Statistical significance
- t-Test
- Water displacement method
- Yeast

Pre-lab Preparation and Background Information

People with training in biology can work in many different fields. Biologists may work in many different areas, including healthcare, animal care, agriculture, ecology, resource management, biomedical research, forensics, genetics, pharmaceutical development, and academia. All of these fields have a research component. Therefore, all biologists need to understand how to do research and how to write scientific papers or reports that explain their experiment and their findings.

Scientific research is simply trying to answer a question using the scientific method (See Lab Exercise 1, Figure 1.2.) Scientific research can be simplified to the following:

- Observe a phenomenon that is poorly understood.
- Develop a hypothesis that may explain that phenomenon.
- Identify an independent variable to test and a dependent variable to measure.
- Test the hypothesis with a controlled experiment by manipulating the independent variable and collecting data on the dependent variable.
- Analyze it to see if there is a *significant* correlation between the variables.
- Determine if the hypothesis was supported by the data.

Biologists (and scientists in general) use two types of logic in scientific research—inductive and deductive reasoning. Both are a necessary part of scientific inquiry, but both have limitations.

Inductive reasoning occurs when we draw general conclusions (or create "rules") from observing many specific events. For example, if I observe that all foods with a natural yellow color, like lemons and grapefruit, taste sour, I might infer that yellow foods always have a sour flavor. This "rule" is only useful as long as the data supports it. If I discover a yellow food that is not sour, like a banana, corn, or pineapple, then my inferred principle must be discarded or revised. Humans are hardwired to look for patterns. For example, if you try an ethnic food and don't like the flavor, you may try another dish from that same country. If you don't like that one either, you might be daring and try one more. If you don't like that one either, it would be quite natural to *infer* that you don't like ANY foods of that particular ethnicity. You obviously have not tasted all of the foods from that part of the world, but you have made a rule, or an *inference* based on the data that you do have. It may be right, but the more data you have, the more confident you can be that your inference is a valid one. And as soon as new data refutes that inference (you discover a food of that ethnicity that you like) then the rule must be discarded.

Deductive reasoning occurs when we apply previously accepted "rules" to specific situations. If you already know that a certain type of ethnic food is spicy, then it is natural to *deduce* that if a certain dish is of that ethnicity, then it must be spicy. Again, the value of deductive reasoning rests on the validity of the rule being used. Rules are generalizations based inductive reasoning, and inductive reasoning requires having complete data.

This interplay between inductive and deductive reasoning is why science is never finished. Every time scientists make new discoveries, previous rules (also known as theories or scientific principles) may need to be abandoned. Today, we are revising much of what we understand about biology as we learn more about how proteins and DNA work within cells. This is not to say that science is never *right*. Instead, science becomes more *accurate* as new data forces us to revise and improve old theories to account for more data.

This lab exercise is designed to help you develop and test a hypothesis, gather data, make an inference from that data. If other researchers find additional supporting data, then that inference may eventually become a theory or scientific principle. All scientific research is built upon the research of others. This lab exercise will require you to review previous material including how enzymes work (Lab Exercise 3) and the process of cellular respiration (Lab Exercise 8). You will also need to remember how to use measuring equipment, such as a graduated cylinder, pipette, and hydrometer (Lab Exercise 1), and an incubator (Lab Exercise 3). Biology research becomes easier as you develop a deeper understanding of biological principles, gain experience working with different types of lab equipment, and practice scientific writing.

Review of Aerobic and Anaerobic Cellular Respiration

In the previous lab you studied the processes of photosynthesis and cellular respiration. **Aerobic** cellular respiration occurs in the presence of oxygen. However, some cells in a multicellular body (like a human) may run out of oxygen during vigorous activity. Some organisms live in places that have no oxygen. **Anaerobic** cellular respiration is an alternative pathway for obtaining energy from food molecules in the absence of oxygen. It is not very efficient, which is why most organisms that normally require oxygen will quickly die without it. Recall that oxidative (or aerobic) cellular respiration generates over 30 ATP molecules from a single glucose molecule. But without oxygen, cells can only produce 2 ATP per glucose molecule. Cells of complex animals will not survive long without oxygen and the ATP produced by aerobic cellular respiration. But some simple organisms are **facultative anaerobes**. They can switch from aerobic cellular respiration to anaerobic cellular respiration and still produce enough ATP to survive.

Anaerobic cellular respiration is also called **lactic acid fermentation** in animals and humans, or **alcohol fermentation** in plants and fungi. In animals and humans, if muscle cells run out of oxygen (due to overtaxing the respiratory or cardio-vascular systems), they switch to a fermentation process that produces lactic acid. This is why muscles "burn" during a workout. Lactic acid creates a burning sensation. (However, lactic acid is metabolized quickly, so muscle soreness the day after a workout is NOT due to lactic acid.) In fungi and plants, the metabolic pathway produces a slightly different compound. Instead of lactic acid, alcohol is formed as a by-product. Humans have used this metabolic pathway for centuries to produce alcoholic beverages.

$$\text{Fermentation: } C_6H_{12}O_6 \text{ (glucose)} \rightarrow 2\ C_2H_5OH \text{ (ethanol)} + 2\ CO_2 \text{ (carbon dioxide)}$$

Yeast

Yeasts are a group of unicellular fungi. There are a number of species of yeast. Some are **pathogenic** (disease causing), while others are routinely used in making bread, Kombucha tea, keifer, and many other foods. When bread "rises," it is because living yeast cells are producing bubbles of carbon dioxide in the dough that makes it puff up.

Yeast can be purchased in multiple forms, or even cultured at home. Yeast is usually sold as freeze-dried cells in small foil packages (Figure 9.1). The cells are dormant, but alive, which is why a package like this will have an expiration date. Eventually, if the yeast cells are not provided water and sugar, they will die. One gram of dry yeast contains approximately 6 billion yeast cells. Therefore, each 7 g packet contains approximately 42 billion yeast cells. In this lab exercise, you will be using a solution containing two packets of yeast per liter of water. That will produce a mixture of approximately 84 billion yeast cells per liter, or *82 million yeast cells per ml!*

Yeasts are facultative anaerobes. In this lab exercise, living yeast will be used to metabolize sugar in an anaerobic environment in much the same way it would be used to make beer, wine, hard cider, or other alcoholic beverages. You will be expected to look up more details on yeast for your paper over this lab.

It is important to state that this lab will not actually make hard cider—you will be testing one variable that may affect the alcohol production and fermentation rate of yeast if someone was making hard cider. You can use the principles covered

YEAST CELL

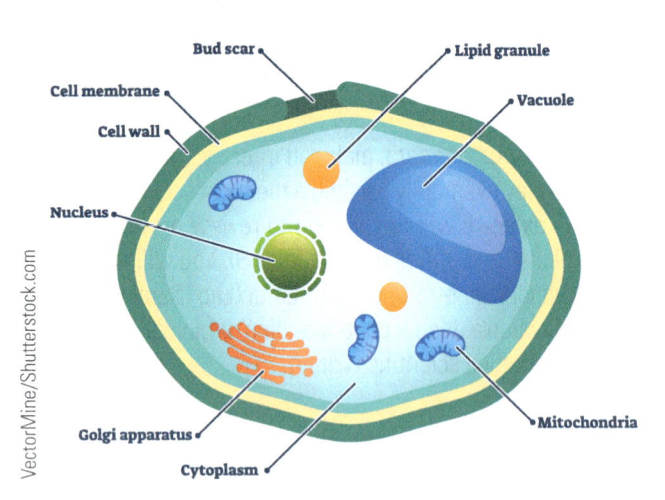

Figure 9.1 Yeast cell (left) and packages of freeze-dried yeast (right).

in this lab exercise to make your own hard cider. Home brewing cider or beer is a popular hobby and a lucrative business. But you should do more research on how to brew beer or hard cider to ensure it will actually taste good. The solution made here will produce very little alcohol, will not taste good, and, as it is a lab experiment, should *not* be consumed.

Properties of Ethyl Alcohol (Ethanol) C_2H_5OH

When creating an alcoholic beverage, the alcohol content of the final product is of prime importance. Ethyl alcohol has a density less than water. A sugar solution has a density greater than pure water. (Review Lab Exercise 2 if necessary.) Remember that a hydrometer can be used to measure the *specific gravity* of a solution—its density as compared to water. Pure water has a specific gravity of 1.000. The specific gravity of ethanol is 0.794. The specific gravity of a mixture of yeast and sugar will be greater than 1.000 depending on the amount of sugar and yeast in the solution.

Yeast in an anaerobic sugar solution will consume sugar and produce carbon dioxide and ethyl alcohol as by-products. The carbon dioxide will bubble out of the solution, and the replacement of sugar with ethyl alcohol will lower the specific gravity of the solution. Brewers use the change in density of their yeast/sugar mixture to determine how much alcohol has been produced. Brewers measure the specific gravity of their brew before and after fermentation. The change in the specific gravity can be used to calculate the amount of alcohol produced.

To estimate the approximate *percentage of alcohol* (alcohol by volume, or **ABV**) in the final product, brewers multiply the change in specific gravity by the conversion factor 131.25. (If you are interested in how this conversion factor was obtained, see Appendix F: An Application of Stoichiometry and Dimensional Analysis.)

$$ABV = (S.G._{initial} - S.G._{final}) \times 131.25$$

If you are interested in seeing how this conversion factor was determined, see *Appenix F: An Application of Stoichiometry and Dimensional Analysis*. The *rate of cellular respiration* can also be estimated by determining the amount of carbon dioxide being produced by the yeast over time. It can be inferred that the metabolic rate of the yeast is proportional to the rate of carbon dioxide produced.

Manipulating pH in food

When reading ingredients on a food label, it may be a bit frightening to see lots of complicated chemical names listed. However, many of those chemicals are simply compounds that slightly alter the chemistry of the food to enhance flavor, reduce spoilage, inhibit microbe growth, add color, change texture, or adjust pH. This project involves manipulating yeast, sugar, and pH. If making real cider, a fruit juice would be used. Acid and base levels in foods are manipulated using chemicals such as ascorbic acid (Vitamin C) and calcium hydroxide $Ca(OH)_2$ (pickling lime). You will measure pH by using Universal Indicator, a compound that produces a different color at each pH level. (You used Universal Indicator in Lab Exercise 2.)

Formally Analyzing Lab Results

Many students mistakenly assume that "to do research" simply means looking things up online or in books. Looking for information that has already been published is a kind of *library* research. *Experimental* research is quite different. In experimental research, you state your problem, develop a hypothesis, test it with an experiment, and analyze your results. *New* data is collected and analyzed, not just copied from other sources. Experimental research should always start from a neutral perspective. Your hypothesis may seem like a great explanation for what you have observed. But when you conduct your unbiased, empirical, and objective scientific experiment, you should begin with a null hypothesis (*null* = without meaning or value).

A **null hypothesis** is a hypothesis that assumes *nothing will happen*. For example, imagine a biologist is testing the effect of a plant hormone on the growth rate of corn plants. Typically, a researcher wants to immediately hypothesize that the treatment will cause the desired outcome. But it is easy to make such a hypothesis and then "look for" evidence to prove it right. A much better practice in empirical science is to state a null hypothesis and then an alternative hypothesis. A null hypothesis in this case would state that the growth rates, with and without the hormone treatment, should be exactly the same. Then the *alternative* hypothesis will state that if the null hypothesis is not supported, then the alternative hypothesis must be accepted. Notice how this puts greater emphasis on disproving the null hypothesis instead of looking for confirming evidence.

This use of a null and alternative hypothesis is important when doing statistical analysis. Fortunately, the statistics for this experiment are built into an automated spreadsheet that your instructor will provide in the lab, so you will not need to calculate any of this by hand. However, the spreadsheet will *normalize* the data, run a *t-test*, and calculate *p-values*.

Normalizing the data simply means to account for discrepancies between groups. In this case, since multiple groups will be doing the same experiment, they may have slight differences in their experimental setup. When different replications of an experiment are not exactly the same, such as the exact starting pH level, various mathematical adjustments can be done so that comparisons between trials are more accurate. Normalizing the data "calibrates" it so that the results from different groups can be better compared.

A **t-test** is a standard statistical test to determine if two averages are *significantly different* from each other. The *t*-test can be used to determine the *p*-value.

A **p-value** is score between zero and one. A very small *p*-value (typically ≤ 0.05) indicates strong evidence *against* the null hypothesis. When the *p*-value is less than 0.05, you can *accept the alternative hypothesis that the groups are significantly different*. A larger *p*-value (> 0.05) indicates weak evidence *against* the null hypothesis, so you *fail to reject* the null hypothesis. This means there is no significant difference between the compared groups.

Writing it all up

Brilliant discoveries mean nothing if they are not written and shared. Your final task in this lab exercise will be to write up your research according to the standard format for scientific papers. The following Lab Skills will help guide you through the process. Consult *Appendix E How to Write a Good Scientific Paper*.

Lab Skill 9.1 Preparing for an Experiment

The following steps describe what biologists must do *before* starting a lab experiment. A common misconception is thinking that the work begins on the lab bench. Much has to happen first, such as defining the problem, figuring out what will be needed for an experiment, and determining which variables will be manipulated and measured.

Clearly state the question or problem: What are you trying to find out? Write it down in your lab notebook. Be as specific and clear as possible. Also write down any secondary questions you might need to answer in order to solve this problem.

Conduct a *literature review*: Has this question already been explored and answered by someone else? If so, what did they discover? Are the results of previous studies conclusive? Does their work already answer your question? If not, what more needs to be learned to find a clear answer to your question?

If no one has published any scientific papers that answer your question, what background information do you need in order to *start* answering this question? List additional areas of investigation that come up as you look through related studies.

As you search the scientific literature for related information, record relevant facts *and where they came* from in your notebook. Do not use any information from an unreliable source, such as non-scientific studies, personal stories, uncontrolled experiments, or opinion pieces. Reliable sources are textbooks, articles in peer-reviewed journals, university websites, or government institution websites. General Google searches will not provide these sources. See Appendix E, *How to Write a Scientific Paper*, for more information about finding *good* sources.

State a hypothesis: Based on your literature review and your understanding of the relevant concepts, what do you *think* is the answer to this question? State your hypothesis as clearly as possible. Usually, it will be a statement explaining how you think the *independent and dependent variables* are related, including your reasons. Record this in your notebook.

Make a prediction: If your hypothesis is right, what specific evidence can you obtain to demonstrate that it is right? Predictions are specific statements that set up the experiment. They are usually "if… then…" statements that indicate how you will manipulate the independent variable, and exactly how this will affect the dependent variable(s).

Design an experiment: Write out an experiment in your lab notebook based on your prediction. Ensure your experiment has multiple trials and a *control*. Include all steps and details so that you know exactly what you are going to do.

  a. List all of the materials and equipment you will need.
  b. Write out a timeline of what you will do when, especially if it will take multiple days.
  c. Write out the protocols you plan to use. It helps to break it into sections. Your protocol in your notebook should be complete enough that anyone could understand it and exactly replicate your experiment if they followed it.

Determine how you will collect your data: Ensure that you know what *quantitative* and/or *qualitative* data you will need to collect. Ensure that your data will include applicable units, is accurate, and is clearly organized. (You don't want to come back later and not know what your numbers mean!) Create tables to record the data as needed.

Lab Skill 9.2 Conducting an Experiment

As you have learned and practiced in previous labs, experiments require planning and preparation. When you enter the lab, you already know what you will do—or at least have a good plan. This also means having a fairly complete list of materials so that you can quickly gather them before you begin. Many experiments require particular timing, and you don't want your experiment stopped or delayed because you can't find a needed piece of equipment in the middle of your experiment.

Part of the preparation process also includes making any solutions that will be needed during the experiment. Many laboratory reagents are sold in solid or concentrated form. Prepare solutions and dilutions as needed. (Refer back to Lab Skills 2.2 and 2.3.)

Then follow the protocol you wrote in your lab notebook. Collect the data you planned to collect. Carefully record any additions or changes to your protocol. Record data you collect in well-organized tables. Make sure all numerical data include appropriate units.

Note any changes, problems, or mistakes you encounter in your lab notebook. Your notebook should be written in such a way that there is plenty of room to write in the margins or on facing pages as needed. The more crowded it is, the harder it will be for you to decipher your notes later on. A good, organized, and clearly written notebook is an invaluable resource when writing up your findings or doing follow-up research (Figure 9.2).

If the experiment fails due to an experimental error, note the problem and adjust the protocol as needed. Then start the experiment over. Sometimes, experiments need to be run several times before all of the potential problems in the experimental design are worked out.

Figure 9.2 A well-organized notebook makes writing scientific reports much easier.

Lab Skill 9.3 Graphically Representing Data

The fundamental point of a scientific study is to show the relationship between the *independent* variable and the *dependent* variable. Tables, charts, and graphs are used to present data in a way that helps a reader clearly identify trends or relationships. Sometimes the data may need to be converted into different forms, and tables can help show how the data were converted or used to obtain statistical measures. It is much easier to present and understand numerical data when they are displayed in a well-designed graph or chart. The most common types are pie charts, line graphs, bar charts, and histograms. The type of data being presented will determine which chart to use.

Pie Chart. A pie chart is always used to represent percentages or fractions of a whole. For example, if you wanted to show relative number of people in a biology class with different eye colors, a pie graph would represent the percentage of brown, blue, green, or hazel eyes as proportional "slices" of the pie graph (Figure 9.3). Pie charts describe qualitative data (such as eye colors) but in a way that shows the relative percentages of those observations.

Line Graphs. Line graphs are used to represent **continuous** data where each point on the X-axis has a corresponding value plotted on the Y-axis. The X-axis often, though not always, refers to a unit of time, and the data points are plotted to show the change in a variable over time. An example might be the height of a fertilized plant over time compared to the height of an unfertilized plant over time (Figure 9.4). Even though plant height measurements were taken weekly, the height of the plant can be inferred because the plant's height would not have changed drastically between measurements.

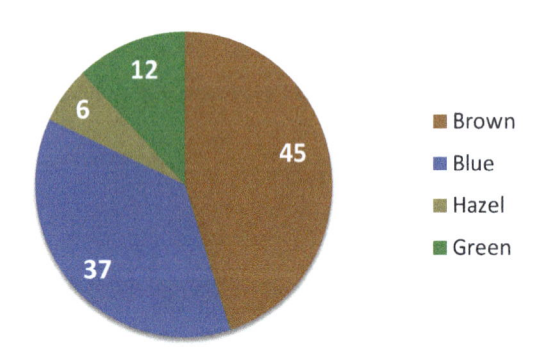

Figure 9.3 Example of a pie chart.

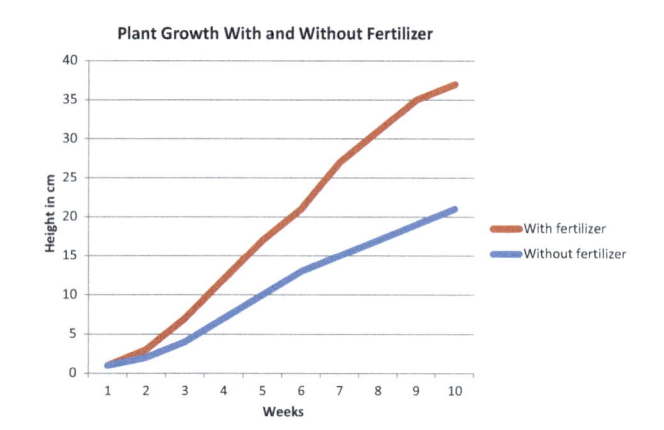

Figure 9.4 Example of a line graph.

Bar charts. Like pie graphs, bar charts are typically used to represent qualitative data. The bars may be oriented horizontally or vertically. Each bar represents a different category. The categories may be ordered so that the longest bars are on top to easily show trends. For example, if you wanted to show how many people purchased a particular flavor of ice cream at an ice cream shop over the course of a day, it could be presented as a bar graph (Figure 9.5). You may be wondering why this data would not be presented as a pie chart. Since some customers may have bought more than one scoop and flavor of ice cream, a bar graph is more useful.

Histograms. A histogram is a very common type of bar graph used to represent quantitative data, but where the independent variable is **discrete**, not continuous. Discrete data refers to data collected under a certain condition rather than over a continuous range. A researcher may study the effects of different doses of medication, or the effect of an enzyme at several different temperatures. Since the researcher does not have data for every possible dose, or every temperature, a histogram should be used. When making a histogram, the independent variable should be placed on the X axis, and dependent variable on the Y-axis. For example, if you wanted to show the results of different amounts of fertilizer on plant

Number of Customers Buying Ice Cream by Flavor

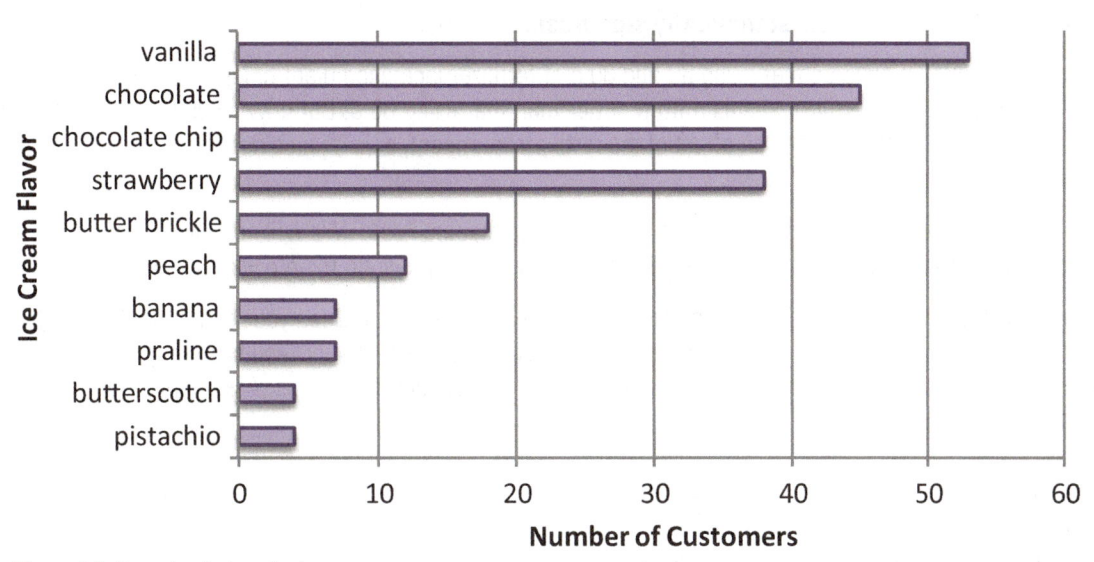

Figure 9.5 Example of a bar chart.

Plant Growth After 4 Weeks With Different Fertilizer Treatments

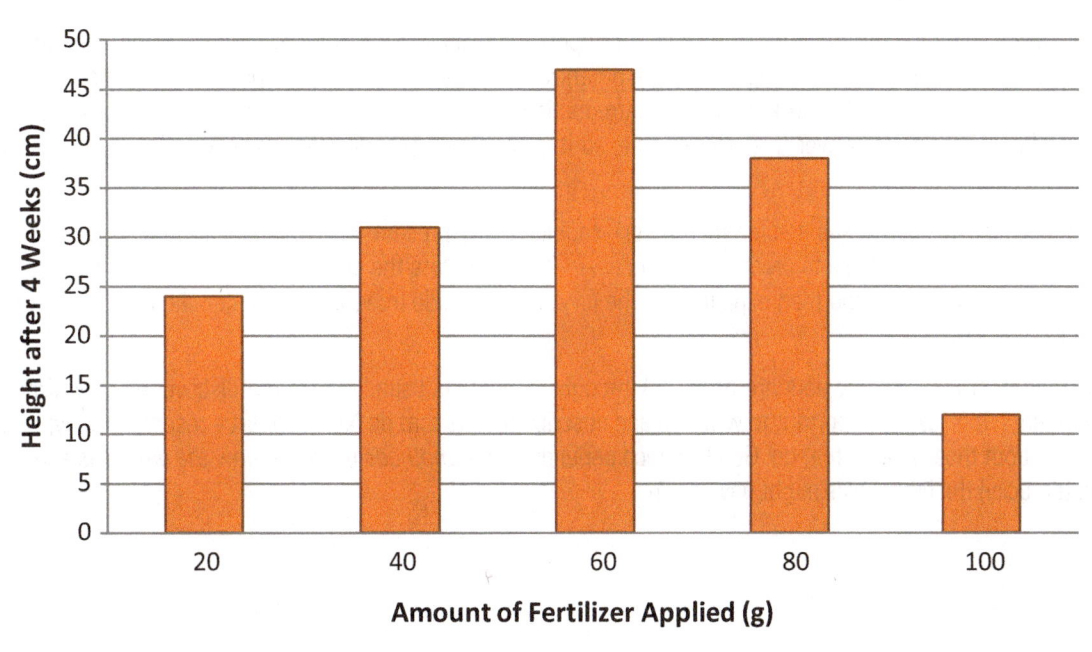

Figure 9.6 Example of a histogram.

growth after 4 weeks, a histogram would be more appropriate than a line graph. The independent variable (in this case, the amount of fertilizer) should be oriented on the X-axis, and the dependent variable, (in this case plant height after 4 weeks) plotted on the Y-axis (Figure 9.6). No data were collected for amounts of fertilizer other than the five shown, so a line graph should not be used.

Lab Skill 9.4 Analyzing Data with Statistics

What does the data mean? Do the numbers you collected actually tell you what you need to know? Use statistical analysis to determine if your findings are **statistically significant**.

Consider Figure 9.6. It appears that using 60 and 80 g of fertilizer increased plant height the most after four weeks. And the growth with 60 g of fertilizer seems a little better than that using 80 g. But how much better is it? Is the difference significant enough to attribute it to an effect of the fertilizer treatment, or is it just a natural difference due to the randomness of the population of plants?

This is where statistics play an important role in interpreting research data. We will not cover all of the methods used for statistical analysis in this class. An automated spreadsheet will be provided to help you run some basic statistical tests.

Draw your conclusions: Consider again your hypothesis and prediction. Does your data (and the analysis of it) match your prediction? Does it support your hypothesis? If so, you *can* state that your results *DO* support your hypothesis. If it doesn't, you *must* acknowledge that your results *DO NOT* support your hypothesis. Ask yourself the following questions and record your thoughts in your notebook:

a. How confident am I that my results are truly representative of how the world works? Will I (or others) get the same result if they repeat this experiment?
b. Did anything go wrong that might have affected my results? If so, what?
c. If I did this experiment again, would I change my experimental design? If so, how?
d. Do these results indicate that more research needs to be done? Are there unanswered questions or new questions to deal with?
e. What is the practical value of this research?
f. Does this research raise any ethical concerns?

Finally, after analyzing the data, write a scientific report. See *Appendix E How to Write a Scientific Paper* for detailed instructions for how a scientific paper should be organized and prepared. Scientific papers must follow a specific format including sections for an **abstract** (summary), introduction, materials and methods, results, conclusion, and references. You will be writing a paper over this lab exercise, so it is very important you look over Appendix E *before* doing the following lab activities.

Your report, (or in some cases a scientific poster), should clearly present your research and findings. It should be much like what a lawyer would use to present a court case—it should state the charge (your hypothesis), provide the evidence (your experimental data), and then explain how the jury should decide (or what the reader should conclude) based on the analysis of the evidence.

Note that most research projects are not breakthroughs. Scientists usually make big discoveries only after many failed experiments. It is just as important to write up and describe results that do NOT support a researcher's hypothesis, as it is to write about those results that do. Good science depends on the *methods used and how the data was analyzed*, not how good (or lucky) the hypothesis might have been.

Name: _____   Date: _____

PRE-LAB QUIZ 9: Are you ready to proceed?

1. What are independent variables in scientific experiments? What are dependent variables?

2. What is the difference between *discrete* and *continuous* data? Will an experiment measuring fermentation rates and changes in specific gravity at five different pH values provide discrete or continuous data? Explain your answer.

3. What is a null hypothesis? Why is establishing a null hypothesis a normal part of scientific research?

4. What is a "review of the literature" in scientific writing? Why is it done?

5. What is an abstract in a scientific paper? What is usually included in an abstract?

6. Even though the abstract is the first part of a scientific paper, it should be written last. Why? (See Appendix E if necessary)

7. What is the difference between inductive reasoning and deductive reasoning?

8. When you compare an independent variable to a dependent variable and draw conclusions about their relationship, what type of reasoning are you using?

9. What does the p-value represent or indicate in a statistical analysis? What would a p-value of 0.1 actually mean?

10. The following hypothetical data was collected. Which type of graph would be most appropriate for presenting this data? Why?

Age range	Number of hours spent using a cell phone per day
1-4	2.1
5-9	2.9
10-14	4.7
15-19	5.9
20-24	7.1
25-29	7.0
30-34	6.2
35-39	3.5
40-44	4.1
45-49	3.1
50-54	3.7
55-59	2.1
60-64	2.9
65-69	1.6
70 and over	1.9

Problem to Solve 9.1 Can I Conduct an Experiment and Write a Scientific Report?

A Case Study: Microbrewery Biology

Bill Brewer owns and operates a small microbrewery. He wants to expand his product line to include hard apple cider. But as a business owner, he wants to maximize production speed and minimize waste. Bill has offered to pay your team a consultant fee to help him set up his hard apple cider production.

Bill already knows how to make beer, so he understands that yeast can ferment sugar to produce alcohol. He also knows that he tends to get the best results when he allows fermentation to occur at about 45°C (113°F). Bill also knows that beer is made with hops and grains. The hops are flavoring compounds, and the grain is broken down into malt sugar to activate the fermentation.

Bill also knows that *cider* is made with fruit juices, like pressed apples, not hops and grains. Fruit juices are often acidic. He also knows that apples contain starch, simple sugars, fiber, and other compounds that give apples their flavor. He has pressed them to obtain the juice. He has yeast ready to add to the juice. But he is not sure what to do to *maximize the fermentation rate and alcohol content* of his cider (Figure 9.7).

Figure 9.7 Alcoholic beverages are produced by living organisms. Brewers must understand, and sometimes research, the biological processes involved.

Bill wants to know… "Does the acidic nature of fruit juice affect fermentation? If so, what starting pH will best maximize fermentation - both the rate of the reaction and the amount of alcohol produced?"

Background information to look up, write about, and cite in your paper.

- What is *fermentation*? What is the chemical equation for it?
- What are *yeasts*? What activates them?
- What are enzymes? How are they involved in the fermentation process?
- Why might different pH levels affect fermentation?
- How can the pH of cider (or other foods) be safely adjusted?
- Why do brewers use a hydrometer in the brewing process?
- What *exactly* do the two dependent variables in this exercise actually measure?

Your group will work out a hypothesis, conduct an experiment, and each <u>individually</u> write up a formal scientific report for Bill's company.

Title

(Title) Determining best pH for yeast fermentation

Purpose

(Goal) Test rate of yeast fermentation in a 10% sugar solution at five different pH levels. Determine if any differences in fermentation rate are statistically significant. Present the study and results in a formal scientific report.

Protocol

Topics to look up and discuss in my introduction and literature review:

1.

2.

3.

4.

5.

Hypothesis:_____.

Prediction:_____.

ate

itle

urpose

Continued from previous page.

rotocol

Materials:

(per team of 3-4 students)

1 medium or large beaker (approximately 250–300 ml capacity)
5 small beakers labeled 1 through 5 (80–120 ml capacity)
5 fermentation tubes labeled 1 through 5 (Figure 9.8)
1 medium graduated cylinder (50 ml)
1 small graduated cylinder (10 ml)
1 hydrometer and hydrometer tube
5 glass stirring rods
Wax pencil
2.0 M Citric acid solution (dropper bottle)
0.3 M Calcium hydroxide $Ca(OH)_2$ solution (dropper bottle)
Universal Indicator and color chart
Safety glasses and gloves
mass balance and 5 weigh boats
(Per class)
1.5 liters of 10% sucrose solution
Yeast packets (2 per class)
Incubator set at 45°C

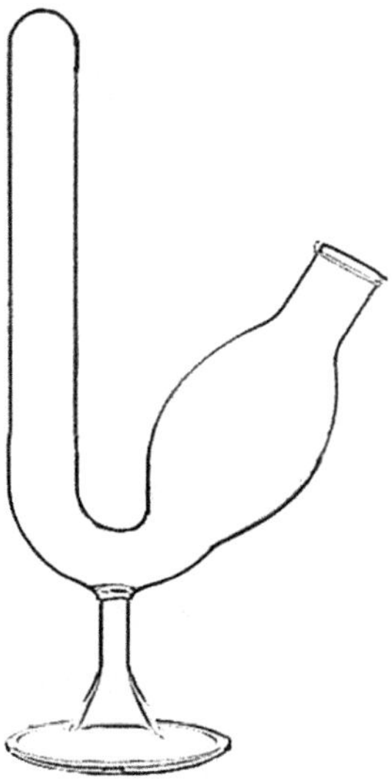

Figure 9.8 Fermentation tube.

Protocol:

1. Ensure all glassware is clean and dry. Remove any wax pencil markings on fermentation tubes.
2. Label the five small beakers and five fermentation tubes with numbers 1-5 using tape. Also include group number on labels for fermentation tubes.
3. Using a mass balance (accurate to 0.01g) and five small weigh boats, measure out five samples of yeast, 0.50 g each.
4. Obtain 200 ml of the 10% sucrose solution in the medium beaker.
5. Using a small graduated cylinder obtain 5 ml of universal indicator and add it to the 200 ml of sucrose solution. Swirl to mix.

esearcher signature | Date | Supervisor/Instructor signature | Date

Purpose

Continued from previous page.

Protocol

6. Pour 40 ml of the sucrose/universal indicator solution into each of the labeled beakers. Place a glass stirring rod in each beaker. (Take care not to mix up stirring rods during the remaining steps.)
7. Place the beakers on a white background while adding acid or base solutions. Use the universal indicator color chart to determine the pH of the solution.
8. Add citric acid one drop at a time, stirring between drops, to beaker #2 until the pH is 5 or 6. Add citric acid to beaker #1 one drop at a time, stirring between drops, until the pH is about 4. Record actual pH values in the data table.
9. Add CaOH to beaker #4 one drop at a time, stirring between drops, until pH is about 8 or 9. Add CaOH to beaker #5 one drop at a time, stirring between drops, until the pH is about 10. Record the actual pH values in the data table.
10. Add 0.5 g yeast to each beaker and stir well.
11. Quickly use the hydrometer to determine the specific gravity of the solution in beaker #3. Record the starting specific gravity in the data table. (It will be the same for all five solutions.) Then quickly transfer all of the solution from the hydrometer (and any left in beaker #3) to the fermentation tube labeled #3.
12. Immediately pour all of the solutions from the other beakers into the appropriately numbered fermentation tubes. Ensure no yeast remain in the beakers. (If some yeast remains, pour some of the solution back into the beaker, swirl to dislodge the yeast, and pour the solution back into the fermentation tube.)
13. Cover the opening and carefully tip each fermentation tube to remove air from the vertical part of the tube. Make sure no bubbles remain.

ate

itle

urpose

Continued from previous page.

rotocol

14. Place the fermentation tubes in the incubator at 45°C. **Record the start time.**

15. Check every 10 minutes to ensure no tube collects too much gas. If gas almost fills the inverted tube (see Figure 9.9), remove the tube and record the **stop time**. Immediately mark the level of the gas in the tube as shown in Figure 9.9. If any tubes are still incubating after 1 hour, remove them from the incubator, record the stop time, and go to the next step.

While waiting for fermentation to proceed, go over Appendix E How to Write a Good Scientific Paper.

Proceed with the following steps after a fermentation tube has been removed from incubator.

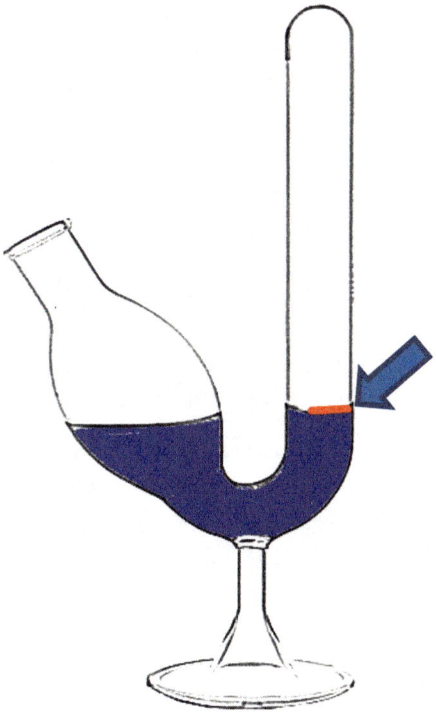

Figure 9.9 Measuring volume of gas produced. Mark the level of liquid with wax pencil (as indicated by arrow).

16. Immediately after removing a fermentation tube from the incubator, mark the gas level on the fermentation tube with a wax pencil. Be as accurate as possible. Do this while the tube is sitting flat on the table, NOT while holding it! (See Figure 9.9.)

17. Determine the number of minutes the tube was incubated and record in data table.

18. Pour the contents of the fermentation tube into the hydrometer tube and measure the final specific gravity. Record the final specific gravity in the data table. Then discard the solution in a waste container as directed.

19. Using the **water displacement method** below, determine the volume of gas collected in the fermentation tube:

    a. Add a small amount of water to the tube **and invert.** Compare the level of water to the mark.

    a. Add or remove water until the quantity of water in tube is exactly at level of the mark **when inverted.** (See Figure 9.10.)

esearcher signature | Date | Supervisor/Instructor signature | Date

Date | Researcher Name

Title

Purpose

Continued from previous page.

Protocol

b. Pour water from fermentation tube into an appropriately sized graduated cylinder to determine volume. Make sure all water in fermentation tube is transferred to graduated cylinder. Accuracy is critical. This volume of water represents the volume of CO_2 produced.

c. Repeat for all fermentation tubes. Ensure graduated cylinder is dry before each use. Carefully record all measurements in data table.

20. When finished, remove labels and clean all glassware. Ensure all yeast residue is removed from inside fermentation tubes and wax pencil is wiped off.

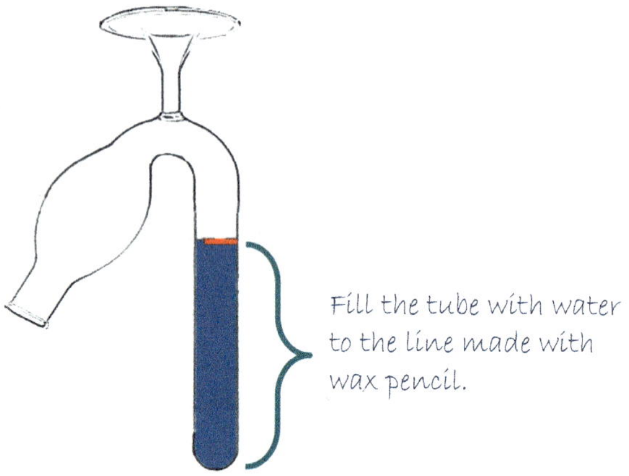
Fill the tube with water to the line made with wax pencil.

Figure 9.10 Water displacement. The volume of gas produced can be determined by measuring the volume of water that occupies the same space.

Researcher signature | Date | Supervisor/Instructor signature | Date

Table 9.1 Data for testing effect of pH on fermentation rate

Beaker or Ferment. Tube	Actual Start pH	Starting Specific Gravity (S.G. initial)	Start Time	End Time	Time Incubated (minutes)	Final Specific Gravity (S.G. final)	Change in Specific Gravity (S.G. initial - S.G. final)	Volume of CO_2 produced (ml)	Rate of CO_2 produced (ml/min.)
1									
2									
3									
4									
5									

Laboratory Exercise 9 Cell Biology III: Fermenting Sugar to Produce Alcohol

Statistical Analysis Worksheet

Your Group's Data (Record your data in yellow boxes.)

Statistical Analysis Worksheet	Ideal pH	Actual pH	Volume of CO_2 (ml)	Minutes in bath (min)	Rate of CO_2 production (ml/min)	Normalized Rate to adjust for differences from ideal pH
1	2					
2	4					
3	6					
4	8					
5	10					

Tube	Initial specific gravity	Final specific gravity	Change in specific gravity	Normalized Change in SG to adjust for differences from ideal pH
1				
2				
3				
4				
5				

CLASS Data (Transfer your data from green boxes above and data from rest of class to yellow boxes.)

Tube	Rate of CO_2 production (ml/min) Group 1	Rate of CO_2 production (ml/min) Group 2	Rate of CO_2 production (ml/min) Group 3	Rate of CO_2 production (ml/min) Group 4	Rate of CO_2 production (ml/min) Group 5	Rate of CO_2 production (ml/min) Group 6	Average rate of CO_2 production (total/n)	Standard Deviation
1 (pH2)								
2 (pH4)								
3 (pH6)								
4 (pH8)								
5 pH(10)								

Tube	Change in Specific Gravity Group 1	Change in Specific Gravity Group 2	Change in Specific Gravity Group 3	Change in Specific Gravity Group 4	Change in Specific Gravity Group 5	Average Change in Specific Gravity	Percent Alcohol
1							
2							
3							
4							
5							

t-test to Reject Null Hypothesis (H_0) (T-test p value is calculated for you. If p value is less than .05, reject the null hypothesis - meaning there is a significant difference between your rate and the control.)

Tube	t-test p value	Decision (Reject H_0 or Do Not Reject H_0)
1 (pH2)		
2 (pH4)		
3 (pH6)		CONTROL
4 (pH8)		
5 pH(10)		

Analysis

Your instructor will provide an electronic spreadsheet that you can use to calculate the statistics for your data in class. Record your group's data and the class data as directed. Your instructor may also provide an electronic copy of the class data. Your paper should refer to both your group's data and the class data.

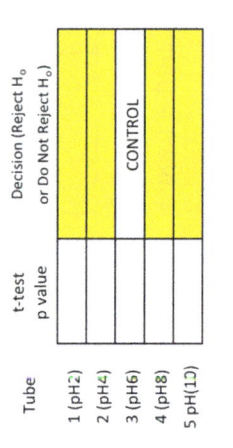

Affect of pH on Fermentation Rate (normalized, all groups)

(Graph: Carbon Dioxide Production (ml/min) vs pH Value, y-axis 0.000–1.200, x-axis 0–10)

Continued from previous page.

Conclusions

What was your one independent variable? What were your two dependent variables?

Did the dependent variables indicate the same results? If not, which variable is more reliable? Why?

How did you determine alcohol content? What assumptions are inherent in this calculation? (See Appendix F for more information on how the equation for determining alcohol content was derived.)

Did your data support your hypothesis? How do you know?

If someone not in this class asked you to explain the purpose of this experiment, how would you explain it to them?

Bill Brewer is funding your research. He wants to know your preliminary findings (before you prepare your formal paper). Summarize in a sentence or two what you have found out.

These questions are to help you write the Discussion and Conclusions section of your paper. The questions themselves should not appear in your paper, but this information should be discussed.

Name: _____   Date: _____

POST-LAB QUIZ 9: What have you learned?

1. What were your null and alternative hypotheses for this experiment?

2. What is the difference between measuring the *volume* of carbon dioxide produced by yeast versus the *rate* of carbon dioxide produced by yeast?

3. How does the *rate of CO_2 produced* by the yeast relate to fermentation? How does the *change in specific gravity* relate to fermentation? (Explain clearly.)

4. The *t*-test compared two means (averages). Which averages were compared in order to get the *t*-test *p*-value?

5. What were your *p*-values? Why did you not have a *p*-value for tube 3? What did your *p*-values for tubes 1, 2, 4, and 5 indicate?

6. What exactly is yeast? <u>Why</u> exactly did differences in pH affect the behavior of the yeast?

7. Which pH seemed to promote the best alcohol production? Which pH seemed to produce the highest fermentation rate? If these were not the same, pH values, why not?

8. Which types of graphs will best represent your data from this experiment? What variable will be represented on the X-axis? What variable will be represented on the Y-axis? (If including multiple graphs, describe all of them.)

9. What *inferences* did you make during this experiment? (Give at least 2.)

10. What *deductions* did you make during this experiment? (Give at least 2.)

Laboratory Exercise 10

Genetics I: Meiosis and Promiscuous Alien Sex

Concepts to Understand

After completing this lab, the student should be able to:

1. Compare and contrast sexual and asexual reproduction.
2. Explain the relevance of meiosis to the process of evolution.
3. Explain when, why, and where meiosis occurs in animals.
4. Compare and contrast meiosis with mitosis.
5. Relate the Laws of Segregation and Independent Assortment to the events of meiosis.

Skills to Learn

After completing this lab, the student should be able to:

1. Use biological models to represent a complex biological process
2. Demonstrate the phases of meiosis.
3. Demonstrate the Laws of Segregation and Independent Assortment
4. Demonstrate how sexual reproduction produces variation in a population.

SEE ALSO:

Campbell Biology—Chapter 13, section 13.1-13.4

YouTube—Mitosis and Meiosis Simulation

Others:

Important terms to know

Allele	Genotype	Phenotype
Asexual	Germ line cells	Recessive
Budding	Gonad	Sex chromosomes
Chiasma	Haploid	Sex-linked
Complete dominance	Incomplete dominance	Sexual
Crossing over	Karyotype	Somatic
Diploid	Kleinfelter's syndrome	Spermatogenesis
Dominant	Law of Independent Assortment	Spermatogonia
Down's syndrome	Law of Segregation	Synapsis
Fission	Monosomy	Tetrad
Fragmentation	Nondisjunction	Trisomy
Gametes	Oogenesis	Turner's syndrome
Gene	Oogonia	Vegetative propagation

Pre-lab Preparation and Background Information

Only one type of cell goes through the process of meiosis: **germ-line cells**. Germ-line cells are *diploid* cells capable of producing *haploid* gametes. **Gametes** are sperm or egg cells. Sperm and egg cells can unite to create a new organism. Non-gametic cells are called **somatic** cells. (*Soma* = body.)

Somatic and gametic cells differ in several ways. Somatic cells vary in size, shape, function, and specialization. They are **diploid** (contain a full maternal and a paternal set of chromosomes) and undergo *mitosis*.

Germ-line cells are only found in structures that produce sperm or eggs. In female animals, *ovaries* contain germ-line cells called **oogonia** that divide to produce eggs in a process called **oogenesis**. In male animals, *testes* contain germ-line cells called **spermatogonia** that produce sperm in a process called **spermatogenesis**. Ovaries and testes are **gonads**. Both men and women have gonads, but in humans and mammals, the gonads of a female are internal, while the gonads of a male are external. Sperm and egg cells are **haploid**, with half the number of chromosomes as a diploid somatic cell.

Why Sex?

Organisms may reproduce **sexually** and/or **asexually**. Asexual reproduction occurs when an organism produces cells that grow into a new organism without the need for a mate. When an organism reproduces asexually, the offspring are (usually) produced by mitosis and are clones of the parent.

Some animals, like a *planarian* (a type of flatworm) or a starfish are very good at reproducing by **fragmentation**. If fragmented, large pieces can grow into a new complete organism (Figure 10.1).

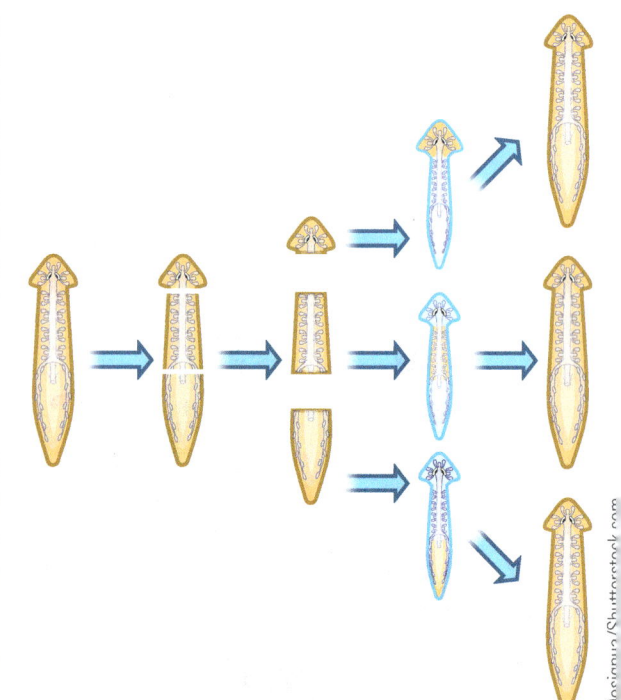

Figure 10.1. Planaria and other simple organisms can often reproduce by regenerating missing parts.

Figure 10.2. A hydra, a simple aquatic animal, in the process of budding. The bud refers to the growth of a new organism out of the parent body.

Figure 10.3. *Kalanchoe diagremontiana*, sometimes called "a pregnant plant," continuously produces baby plants along the edges of the leaves that drop off and start a new plant.

Some organisms can create mini versions of themselves that grow out of the parent organism until mature enough to break off and become independent. This is called **budding**. For example, a simple aquatic animal called a hydra can do this (Figure 10.2).

Some organisms, like bacteria and some protists, simply split into two new organisms by **fission**. And plants may reproduce asexually by **vegetative propagation,** where they send out horizontal stems or aerial roots that lead away from the parent plant to start a new plant. Another version of vegetative propagation occurs when some plants produce tiny plantlets that drop from their leaves (Figure 10.3). All of these strategies are simple and require no fertilization, pollination, finding the right mate, performing complex mating displays, or any of the drama typical of human sexual relationships.

Sexual reproduction, on the other hand, is far more complicated. Sex requires two organisms of the same species to produce special cells that can combine to produce a new viable organism. The gametes must have exactly the right genetic information to produce a unique and healthy individual of the same species. Consider everything that has to occur in exactly the right way for this to happen.

A male must produce a sperm cell that has half the genetic material needed by a new organism. This sperm must be able to swim, survive physical and chemical challenges as it journeys to an egg, and successfully fuse with the egg. The female must be able to produce an egg that also has half the needed genetic material, and is receptive to one, and only one, sperm, and only to a sperm of her species. In terms of evolution, males are most evolutionarily "fit" when they impregnate as many females as possible. But females only want to be fertilized by the *best* males. In animals, this has led to all kinds of complex co-evolved mating rituals, intra-sexual competition over mates, and many other weird reproductive strategies and behaviors. Why go to all the trouble? There are definite pros and cons to sexual versus asexual reproduction. These reproductive strategies are compared in Table 10.1.

One might conclude that asexual reproduction makes more sense. It is easier, simpler, more likely to succeed, and does not require finding a suitable mate. And believe it or not, most species gain no pleasure from sex. It is really only humans and other higher mammals that seem to enjoy it. So, for most organisms it is just something they instinctively do.

What if people reproduced asexually? Imagine if humans could produce babies by budding. A new baby could grow out of a male or female's arm or thigh or back, and once large enough, fall off and become a baby. If that isn't weird enough, realize that the baby would be *you*—or more precisely, your "mini-me" clone!

Table 10.1 Comparison of sexual and asexual reproduction

	Asexual reproduction	**Sexual reproduction**
Types	Fission, budding, fragmentation	Sex
Cell division	Mitosis	Meiosis
Parent	Only requires one	Requires two organisms of opposite sex (usually)
Offspring	Are identical to parent	Are unique from parents
Difficulty	Easy	Complicated
Time required	Fast	Usually slower
Occurs in	Microbes, simple animals, fungi, and plants	Protists, most fungi, almost all plants, almost all animals
Mating	Does not require finding a mate	Must find a mate of same species and opposite sex
Selection	No mate selection required	May have to determine the best mate or fight to have opportunity to mate
Survival of offspring	If parent can survive, cloned offspring will likely survive	Offspring will be unique from parents, for better or worse

The problem with asexual reproduction is just that—the offspring are clones. For species to be able to adapt and evolve to changing environments, there must be variation in the population. And a population made up of clones simply lacks enough variation to evolve. If an environmental factor can affect one member of a clone population, it will affect them all. A disease could wipe out the entire population. So as complex as sexual reproduction is, it is *critical* for evolution. This is why most organisms use some form of sexual reproduction, even though plants, fungi and some simple animals can *also* reproduce asexually if necessary.

A Quick Review of Chromosomes

Review the steps of mitosis in Lab Exercise 4. It is critical that you understand chromosomes and mitosis well before tackling meiosis. Recall that humans have 46 chromosomes in their somatic cells and that these chromosomes occur in pairs. A *homologous pair* of chromosomes contains two chromosomes of the same size, shape, and *type* of genetic information (with the exception of the 23rd pair). One chromosome from each pair came from each parent of that organism (Figure 10.4).

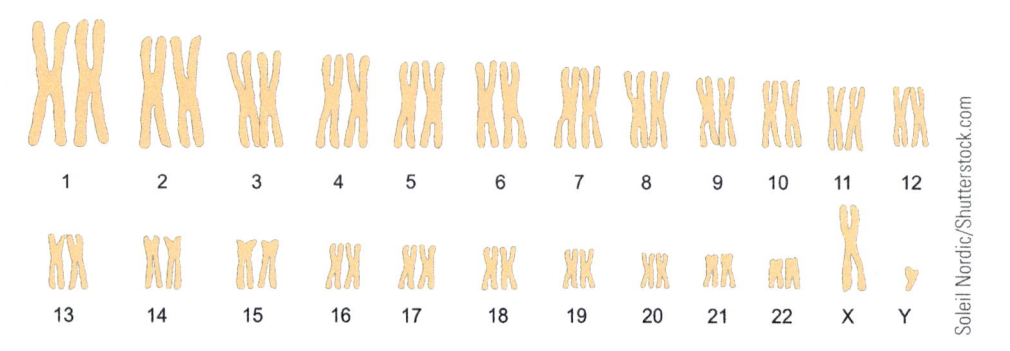

Figure 10.4. The 23 homologous pairs of chromosomes in human cells. The members of a pair contain the same *type* of genetic information, but the chromosomes are not identical. Note that these chromosomes are replicated, and the sister chromatids of any one chromosome *are* identical.

Homologous chromosomes are not *identical*, but they are *similar*. A pair of shoes is like a homologous pair of chromosomes. The shoes can easily be matched by their appearance. They have the same colors, brand name, logos, and other features. But if you've ever tried to put the right shoe on your left foot, you know they are quite different from each other.

Genes and alleles

Recall from Lab Exercise 4 that particular homologous pairs carry specific *genes*. Genes are usually identified as bands on a chromosome. A gene is a "recipe" for the proteins that determine a *trait*. A trait is a characteristic of an organism resulting from its genetic makeup. There may be multiple versions of a gene for a trait, which is why different people have different hair color, skin color, eye color, and so on. These versions of a gene are called **alleles**. For example, chromosomes in a homologous pair both carry the *gene* for hair color, but one chromosome in that pair may carry the *allele* for brown hair, and the other may carry the *allele* for blonde hair (Figure 10.5).

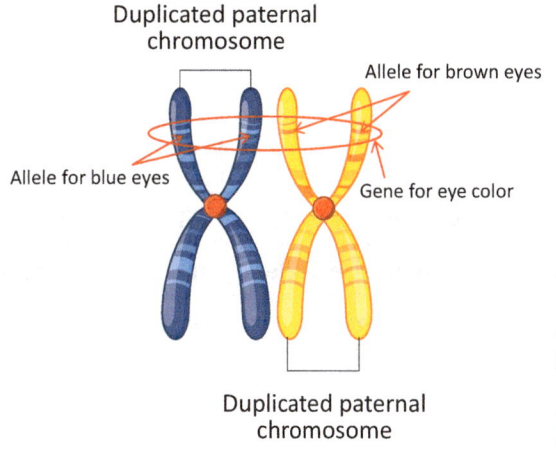

Figure 10.5. Homologous pairs carry *genes* for specific traits but may have different *alleles* for that gene.

It is important to recognize that alleles are not equal. Often, one allele functions to make a protein, but a different version of the allele is less effective. Since cells have two alleles per gene, if a functional allele is present, the cell can usually make the protein and express the trait.

Alleles are usually represented as capital and lowercase letters, where the capital letter represents a **dominant,** or functioning, allele, and the lowercase letter represents a **recessive** allele that is either non-functioning, or functions less effectively. For example, we could describe hair color as being a result of two alleles—one for brown pigment (*B*) and one for lack of brown pigment, or blonde, (*b*). Therefore an organism's genetic makeup can be described with a **genotype** (pair of letters) that represent the alleles, such as *BB*, *Bb*, or *bb*. An individual who has *BB* or *Bb* alleles will have the information for brown proteins, and will have the **phenotype** (expressed trait) of brown hair, whereas an individual with genotype *bb* will not have the information for brown proteins and will have the phenotype of blond hair. In cases of **complete dominance**, the presence of one dominant allele produces the dominant trait. So if the dominant allele for hair color demonstrates *complete* dominance, there could only be two hair phenotypes—brown in the case of either the *BB* or *Bb* genotype, and blond in the case of the *bb* genotype. But in humans this is clearly not the case, as there are many hair colors and shades. Many genes are not expressed according to the simple rules of complete dominance.

The expression of genes is more complex in higher organisms and especially in humans. For example, one allele is not always fully dominant over the other. In some cases, depending on the organism and the gene involved, a mixture of alleles results in a mixture of phenotypes. Consider a plant that has two alleles for the color of the flower. Assume the dominant allele that produces a red pigment is *R*, and the recessive allele that doesn't produce pigment is *r*. Normally, if the dominant allele is present (*RR* or *Rr*), we would expect the plant to produce red flowers. However, in some cases two dominant alleles produce the dominant phenotype (*RR* = red), two recessive alleles produce the recessive phenotype (*rr* = white), but the combination of a dominant and a recessive allele produces a *blending* of the traits (*Rr* = pink). This is called **incomplete dominance.** See Table 10.2 for a comparison of complete and incomplete dominance. There are many other interesting ways in which alleles are expressed. We will consider many of the more complex genetic factors in later lab exercises. For now it is sufficient to understand dominant alleles, recessive alleles, and incomplete dominance.

Comparing Mitosis and Meiosis

Now that we understand alleles and their expression, we can start to see why meiosis is so important. Mitosis produced identical cells. Meiosis, on the other hand, produces sperm and egg cells that contain *half* of the genetic makeup of the parent cell, and only one of the alleles from each homologous pair. The fusion of sperm and egg in fertilization restores the chromosome content in the new organism, but it also produces an entirely new combination of alleles, which produces

Table 10.2 A comparison of genotypes and phenotypes in complete dominance and incomplete dominance

	Example genotype	Dominant allele exhibits *complete* dominance		Dominant allele exhibits *incomplete* dominance	
		Expressed phenotype	Example	Expressed phenotype	Example
Two dominant alleles	RR	Dominant trait	Red flower	Dominant trait	Red flower
Two recessive alleles	rr	Recessive trait	White flower	Recessive trait	White flower
One dominant and one recessive allele	Rr	Dominant trait	Red flower	Blending of traits	Pink flower

variation in the population. Before we explore the implications of splitting and recombining alleles, we need to understand the events of meiosis.

Meiosis, like mitosis, is a form of cellular division, but with a number of very important differences. Mitosis made two identical copies of the nucleus, and the two nuclei were separated by cytokinesis into two new daughter cells. (It is important to note that the term "daughter cell" simply means the cell produced. It should not be interpreted as female.) The number of chromosomes in the parent cell and daughter cells were the same. Daughter cells grow, replicate the DNA, and repeat the cell cycle.

Meiosis produces four completely unique daughter cells. The number of chromosomes in the daughter cells is half that of the parent cell. The daughter cells of meiosis do not contain homologous pairs of chromosomes. Daughter cells are unable to perform either mitosis or meiosis. Daughter cells are gametes.

Recall that mitosis involved four phases: prophase, metaphase, anaphase, and telophase. Meiosis goes through the same phases twice. So meiosis is divided into two parts and eight phases. See Figure 10.6. If you remember the events of mitosis, most of meiosis is similar. In fact, meiosis II occurs exactly the same way as mitosis. However, meiosis I differs from meiosis in a number of ways.

In meiosis I, a diploid cell enters prophase. The nuclear membrane breaks down, and the chromatin condenses into chromosomes. The two centrosomes (each containing two centrioles) begin to migrate to opposite poles. In meiosis I, unlike mitosis, the homologous pairs undergo **synapsis** (find each other) and form *tetrads*. A **tetrad** is a homologous pair of chromosomes that have paired up and are entangled with each other. Where chromosomes touch each other is called a **chiasma**. While they are paired, they trade parts in a process called **crossing over**. A part of one chromatid "crosses over" to the other chromosome, and vice-versa. This produces chromosomes that have been "recombined" from each other. Note that the sister chromatids of one chromosome are no longer identical (Figure 10.7). The figure only shows crossing over occurring once in one homologous pair. During prophase, crossing over happens many times per homologous pair, and between all pairs of chromosomes.

The next major difference between meiosis I and mitosis occurs in metaphase I. Recall that in mitosis, the chromosomes lined up on the *equatorial plate* of the cell. In metaphase I of meiosis the homologous pairs, or tetrads, line up on the equatorial plate (See Figure 10.6). Then in Anaphase I the pairs are separated, and one chromosome from each pair (with its sister chromatids still intact) moves toward one centrosome, and the other chromosome moves toward the other centrosome. Telophase and cytokinesis occurs as in mitosis, but the daughter cells of meiosis I are haploid. There are no pairs of chromosomes in the daughter cells, as one chromosome from each pair ended up in each daughter cell.

Meiosis II in both daughter cells resembles mitosis. The nuclear membrane breaks down and the chromosomes become visible in prophase II. The chromosomes line up on the equator in metaphase II. The sister chromatids are pulled apart in anaphase II. New nuclei are formed in telophase II. Cytokinesis separates the two daughter cells into four daughter cells, all unique and haploid.

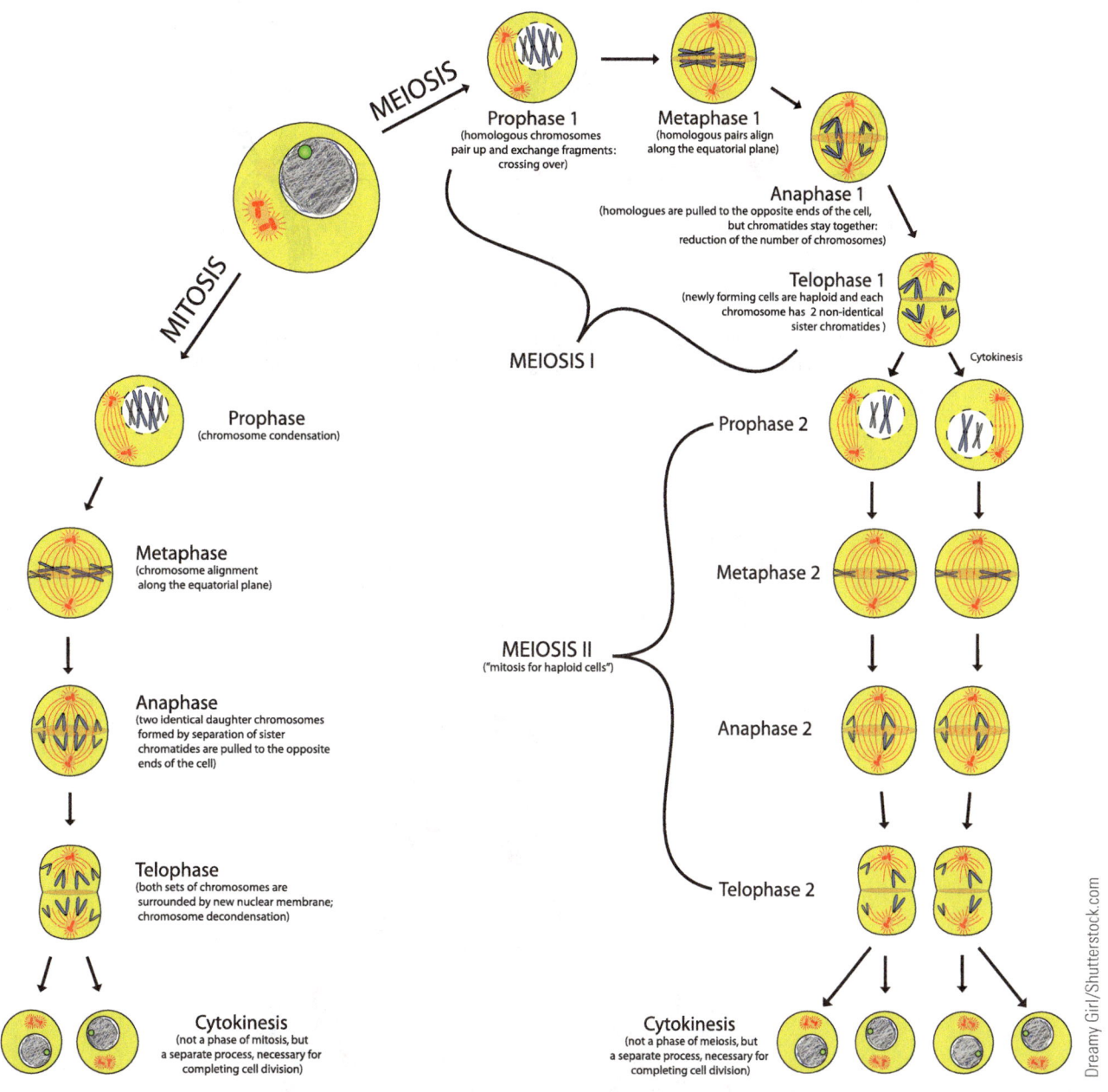

Figure 10.6. Comparison of mitosis and meiosis. Note that meiosis is split into two parts, meiosis I and meiosis II. Like mitosis, each is further divided into prophase, metaphase, anaphase, and telophase.

Meiosis and Variation within Species

Unlike mitosis, meiosis produces four cells unique from each other, and unique from the parent cell. Since one member of each homologous pair was isolated in a daughter cell after meiosis I, both daughter cells had all of the *genes* of the parent cell, but only half of the *alleles*. A cell that is diploid has two alleles per gene. The haploid daughter cells produced by meiosis I have one allele per gene. All of the *genes* are represented in each daughter cell, but the *alleles* for those genes may be different. And since crossing over occurred in prophase I, the sister chromatids are not identical. This means that unlike mitosis, the daughter cells of meiosis II are also different from each other (Figure 10.8). The variations observed in a species results from the trillions of possible combinations of unique gametes produced in sexual preproduction.

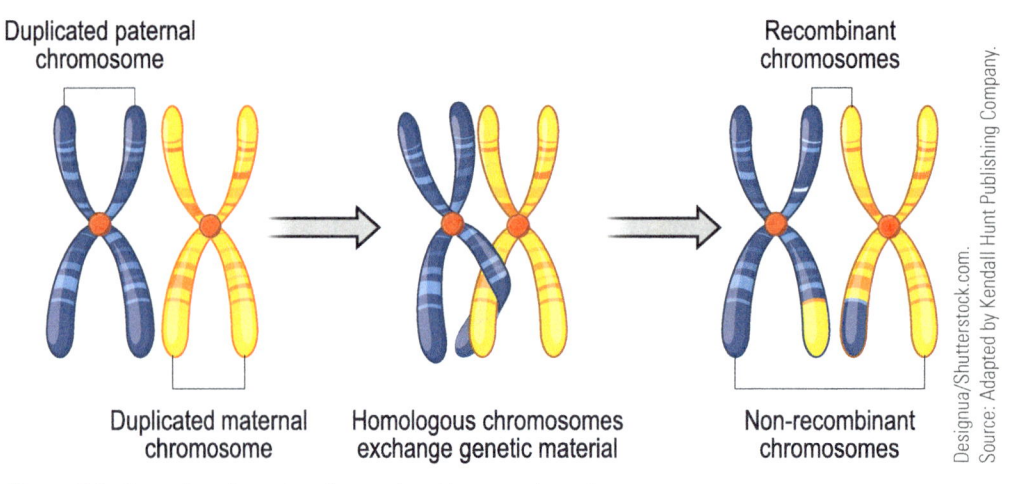

Figure 10.7. Synapsis, or formation of a tetrad, a chiasma, and crossing over.

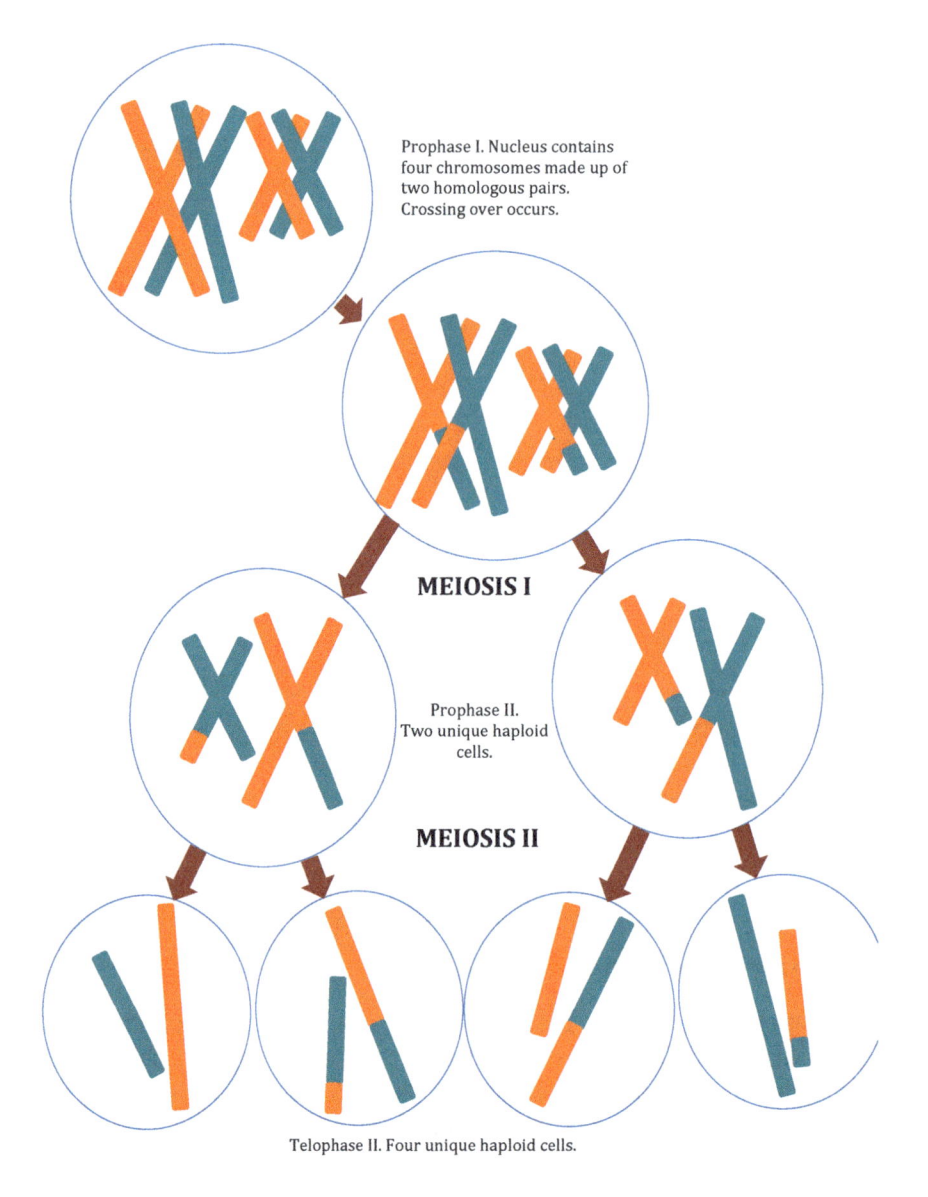

Figure 10.8. Simplified view of meiosis with four chromosomes. This illustrates how daughter cells (gametes) become different.

Laws of Independent Assortment and Segregation

The **Law of Independent Assortment** refers to the way in which chromosomes line up in metaphase I of meiosis. When homologous pairs are pulled to the equatorial plate in the center of the cell, maternal and paternal chromosomes are randomly arranged (or independently sorted). The maternal chromosome will go to one daughter cell and the paternal chromosome will go to the other, but each pair of chromosomes line up independently of the other chromosomes. Some gametes my get an abundance of maternal or paternal chromosomes, but usually, each gamete contains a mixture.

Similarly, the **Law of Segregation** refers to the splitting (segregating) of the homologous pairs of chromosomes in anaphase I and the splitting of the chromatids in anaphase II. The purpose of meiosis is to create haploid cells that contain one chromosome from each homologous pair. If homologous pairs and chromatids are not segregated appropriately, pairs of chromosomes could end up in one daughter cell and neither member of the pair in another daughter cell. If such a gamete were to be fertilized, the resulting offspring would have too many or too few chromosomes, resulting in too many copies of some alleles and not enough of another.

Nondisjunction results from the failure of the Law of Segregation. Nondisjunction is observed in organisms with extra or missing chromosomes. Plants often have extra sets of chromosomes (called **polyploidy**). This not a problem for them. In animal zygotes, however, such a condition often leads to severe developmental problems or death. In some cases, the embryo may survive with this unusual number of chromosomes, but the organism must deal with a range of genetic complications. The likelihood of producing egg and sperm cells with the wrong number of chromosomes increases with age.

Figure 10.9 illustrates the results of nondisjunction in meiosis I and II during oogenesis (egg formation). Note that the figure only shows one homologous pair. The other pairs of chromosomes may segregate normally. If nondisjunction occurs, and an egg with an extra chromosome is fertilized, that fertilized egg's homologous "pair" ends up with three chromosomes instead of two. This is called **trisomy**. If an egg with no chromosomes for that pair is fertilized, the fertilized egg will have one chromosome in that pair (from the sperm). This is called **monosomy**.

Sex Chromosomes

Geneticists can photograph microscopic views of chromosomes during cell division, enlarge them, and use a computer to generate images of isolated chromosomes. These images can be manipulated and sorted based on their size. A **karyotype** is a visual representation of an organism's chromosomes, sorted by size (Figure 10.10). This figure is a stylized representation of a karyotype.

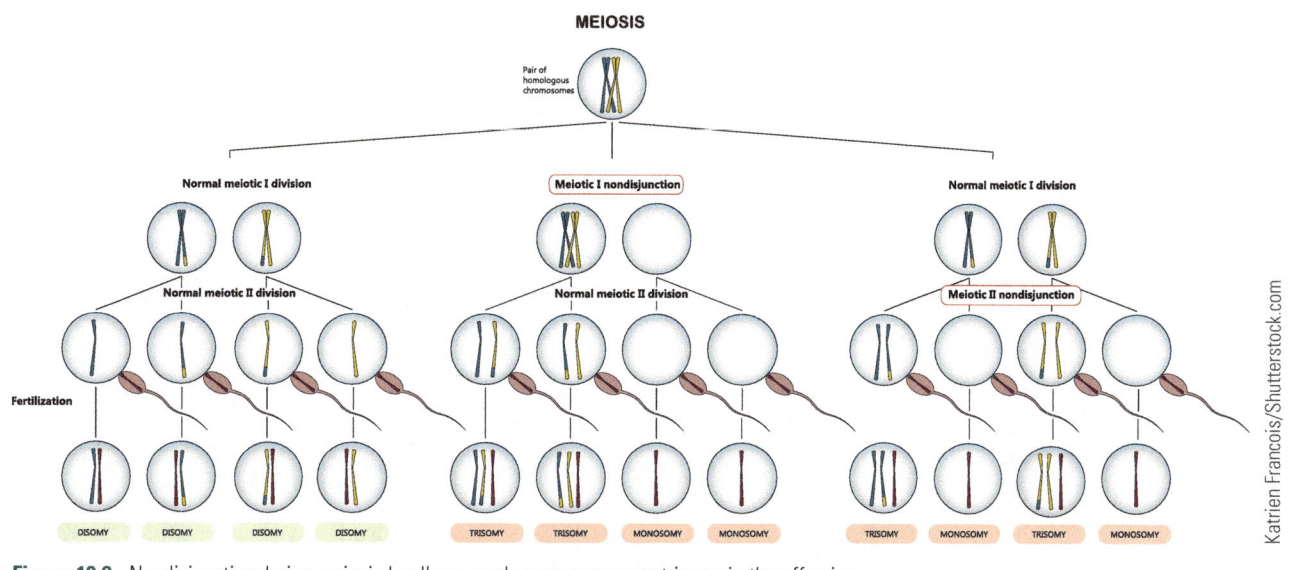

Figure 10.9. Nondisjunction during meiosis I or II can produce monosomy or trisomy in the offspring.

Human karyotype

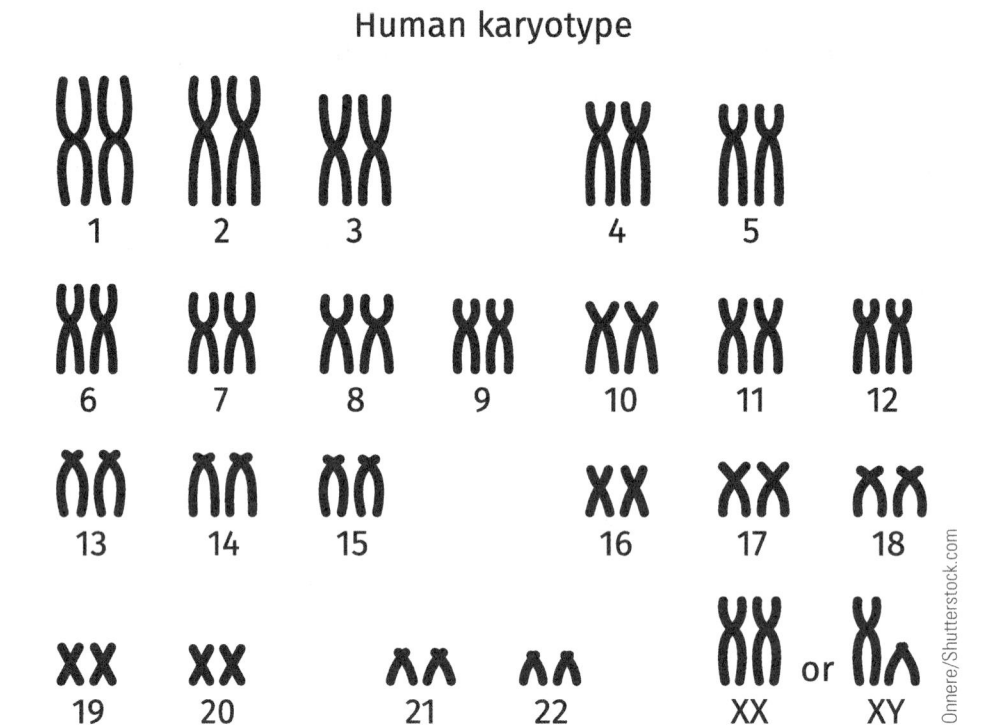

Figure 10.10. A karyotype of human chromosomes. The homologous pairs are labeled 1 through 22. The 23rd pair is composed of two X chromosomes or one X and one Y chromosome. Two X chromosomes produces a female, one X and one Y produce a male.

Note that the last pair of chromosomes in Figure 10.10 are not numbered, but instead are labeled with the letters X or Y. These are the **sex chromosomes** that determine the sex of an organism. Two X chromosomes produce a female organism. An X and a Y chromosome will produce a male organism. Note that the Y chromosome is small. It typically carries very little genetic information, so it is often referred to as a "dud" chromosome. The X chromosome can sometimes carry alleles for certain traits. These traits are considered **sex-linked**. Since males only have one X chromosome, the one allele they have for that trait will determine their trait.

Lab Skill 10.1 Creating and Using Biological Models

The term *model* is often associated with a person who displays designer clothing in fashion shows or with little plastic toys that look like a car, train, plane, or ship. *Biological models*, however, are simplified representations of complex biological phenomena. Models are extremely important for explaining everything from atoms, to cells, to galaxies. Biological structures and processes are complex; models are often employed to help clarify them.

A good biological model is a simplified version of a biological concept (structure, process, phenomenon) that helps explain it, provides predictive power about how it works, and is consistent with related models and experimental data. A plastic model plane such as the one being built in Figure 10.11 may look like a real plane, and if modeling the appearance is the goal, this model serves that purpose. But if it does not actually fly or behave like a plane, it is more a work of art than a *scientific* model. The model plane shown in Figure 10.12 may not look as much like an actual plane, but it may be far more useful as a scientific model if it flies and *behaves* more like a real plane. A scientist studying aerodynamics might find this model far more useful than the one in Figure 10.11.

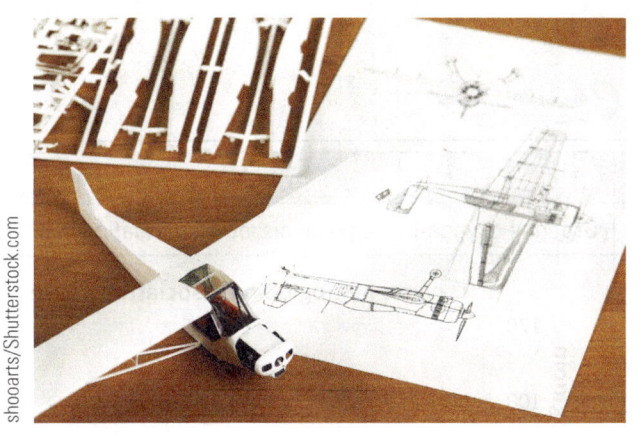

Figure 10.11 A typical model plane.

Figure 10.12. A functional model plane.

You are already familiar with hundreds of biological or scientific models. You used the "planetary model" of the atom to solve chemistry problems. You have studied the structures of a cell using visual or plastic models of cells. Since we often use models in biology, it is worth considering their strengths and weaknesses.

Models are simpler that what they represent. A map is a simple representation of the geography of an area. It would be impractical for the map to be as detailed as the land it represents. Instead, important features are noted on the map, and less important details are left out. With computers, maps can even be interactive. We can have a digital map display the features we want, and we can control the level of detail by increasing or decreasing the scale. Maps are useful because they have predictive power. If you "follow the lines" given on a map, you will (hopefully) end up at the desired location indicated at the end of the line. But the predictive power is only as good as the map. If the map is poorly designed, or if you can't understand it, you will not be able to use it to determine where you are going.

A biological model, like a map, is simpler than the concept it is modeling. But the model conveys the important information in a way that makes it easier to understand. In many cases, the model may be manipulated to show different types of information. But like maps, the predictive power is only as good as the model. The value of models is that they help us create abstract mental frameworks of how things work. For example, navigating a city is challenging without a map. But after studying a map of the city for a few minutes, you can create a *mental* framework of the major roads, buildings, and other landmarks that can help you get around. Of course, such a mental framework cannot include every detail of the city, but your mental model will become more detailed and richer as you gain *experience* navigating the city. The goals of courses like this one are to help you establish richer mental models of biology. Table 10.3 provides some examples of biological models.

Table 10.3 Types of biological models. Examples of models shown here are all used to explain organismal and cellular respiration

Physical model	Functional model	Visual/diagram model
Plastic model of lung	Bell jar that demonstrates mechanical breathing	Diagram showing steps of cellular respiration

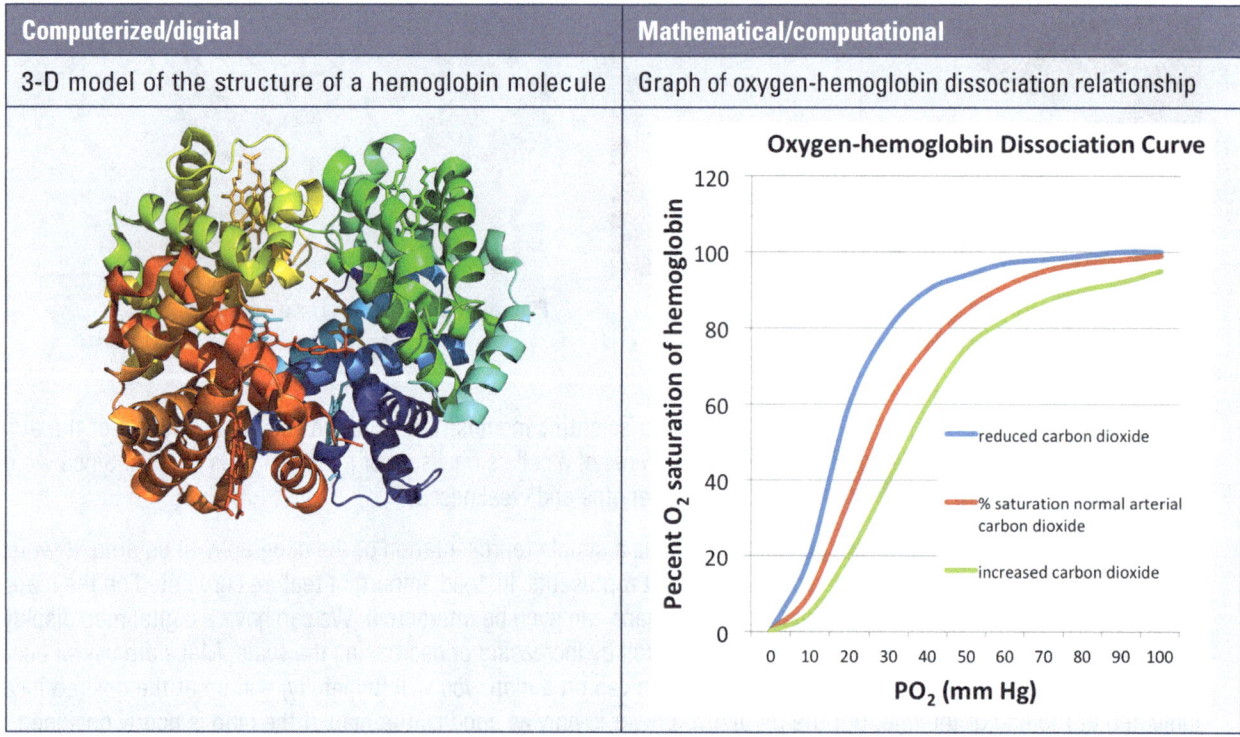

Computerized/digital	Mathematical/computational
3-D model of the structure of a hemoglobin molecule	Graph of oxygen-hemoglobin dissociation relationship

Image credits: left to right: Vladislav Gajic/Shutterstock.com; AlexLMX/Shutterstock.com; VectorMine/Shutterstock.com; ibreakstock/Shutterstock.com; Paul Lusyter.

Biological models have limits. A model of a cell is useful when learning about cells, but many specific cells look nothing like a general cell model. In the following exercises you will use models to represent meiosis. It is important to note that the *model* is not the important part of the lab activity. But the abstract concepts that are represented by the models are very important. Beads are obviously not chromosomes, and you will likely never have "promiscuous alien sex" again. But the *principles* illustrated by these models are foundational to understanding evolution, genetics, and inheritance.

Name: _____   Date: _____

PRE-LAB QUIZ 10: Are you ready to proceed?

1. Complete the table.

	Somatic cell	Germ-line cell	Gamete
Examples		Oogonium/ spermatogonium	
Location			In gonads until released
Number of chromosomes	Diploid		
Type of cell division		Both mitosis and meiosis	

2. Complete the table by placing an X in the appropriate columns. Note that some descriptions apply to multiple columns.

	Mitosis	Meiosis I	Meiosis II
Begins with diploid cell			
Begins with haploid cell			
Ends with diploid cell			
Ends with haploid cell			
Produces genetically identical cells			
Produces genetically unique cells			
Includes crossing over			
Includes separation of chromatids			
Includes separation of homologous pairs			
Produces gametes			
Produces somatic cells			
Is part of the cell cycle			
Goes through PMAT stages			
Produces 4 haploid cells			
Produces 2 diploid cells			
Produces 2 haploid cells			
Includes synapsis and formation of tetrads			
Is preceded by interphase and DNA replication			

3. What are the pros and cons of asexual reproduction?

4. What are the pros and cons of sexual reproduction?

5. A gene for plant height is described as having two alleles, T for tall and t for short. An organism that is tall could have which two *genotypes*?

6. To which phases of meiosis do the Law of Segregation and Independent Assortment apply?

7. In which specific phase(s) of meiosis could nondisjunction occur?

8. Define the terms monosomy and trisomy.

9. What are at least two benefits and two limitations of using biological models?

10. An organism's genetic characteristics for a trait are described as "*Aa.*" What does this mean?

Problem to Solve 10.1 Can I Use a Model to Demonstrate the Phases of Mitosis and Meiosis?

The following activity uses "pop-beads" to represent chromosomes. As you play with the beads, understand that the principle behind this activity is quite important, both in the understanding of meiosis itself and also in understanding how sexual reproduction is a key feature to evolution and the diversity of life on this planet.

You will be expected to use the beads to model a 4-chromosome cell throughout the cell cycle, in interphase, prophase, metaphase, anaphase and telophase of *mitosis*. Then you will be expected to model the steps of *meiosis I and II*, including crossing over.

Pieces of string will represent the cell membrane. A plastic bag will represent the nucleus. Small plastic cylinders will represent centrioles. Magnets will represent the kinetochore proteins and centrioles that link the chromatids to spindle fibers and each other.

The model does not account for all aspects of the cell during cell divisions. There are no organelles represented, though they are certainly present during cell division. Chromosomes move through three dimensions in a cell. A table surface limits this model to a two dimensional plane. And most cells, including human cells, have far more chromosomes than the four being represented in this activity.

Laboratory Exercise 10 Genetics I: Meiosis and Promiscuous Alien Sex

Title

Title: Creating a Model to Represent the Phases of Mitosis and Meiosis

Purpose

Goal: Use pop beads to create a functional model of mitosis and meiosis.

Hypothesis: After showing how a 4-chromosome cell goes through mitosis, the daughter cells will be _____. After showing how a 4-chromosome cell goes through meiosis, the daughter cells will be _____.

Protocol

Materials:

(Per pair of students)
- 4 bags of pop beads (1 each: red, yellow, blue, and green)
- 4 empty bags
- 4 pieces of string
- 4 small white plastic cylinders
- 8 magnetic connectors
- 4 markers or colored pencils, one each in red, blue, green, and yellow

Protocol:

Work with one lab partner.

1. Create models of four chromosomes, one of each color, as shown:

Protocol

Setup

Continued from previous page.

2. Take one strand of beads in each color and place them in one of the bags.
3. Arrange four of the pieces of string into one large circle.
4. Place the four small, white, plastic cylinders anywhere in the circle, but in pairs.
5. Place the bag of beads in the circle.

Demonstrate cell divisions

6. The model as set up represents a 4-chromosome cell in interphase. Work with your lab partner to create a functional model that shows the process of mitosis. Explain how the model represents each phase as you go. You must figure out what to do and explain how it represents all of the phases of mitosis. Refer back to Lab Exercise 4 if necessary.
7. Use a pen and the colored markers to sketch each phase of interphase and mitosis as represented by the model in the Data section on the next page. Label your sketches with appropriate terms, including sister chromatids, homologous chromosomes, centrioles, nucleus, centrosomes, cell membrane, cytokinesis, anaphase, metaphase, prophase, interphase, telophase, replication, and daughter cells.
8. Return all materials to their starting points by following steps 2 through 5 again.
9. Work with your lab partner to create a functional model that shows the process of interphase, meiosis I, and meiosis II. Explain how the model represents each phase as you go. Refer to Figure 10.6 and your textbook if necessary.
10. Use a pen and the colored markers to sketch each phase of interphase and meiosis I and II as represented by the model in the Data section on the following pages. Label your sketches with the appropriate terms such as sister chromatids, homologous chromosomes, centrioles, nucleus, centrosomes, cell membrane, cytokinesis, anaphase I and II, metaphase I and II, prophase I and II, interphase, telophase I and II, replication, daughter cells, crossing over, tetrad, synapsis, recombinant DNA, diploid, and haploid.
11. Demonstrate your model for your instructor.

Laboratory Exercise 10 Genetics I: Meiosis and Promiscuous Alien Sex

Title

Purpose

Continued from previous page.

Protocol

Data:

Mitosis

G1 of Interphase	Anaphase
G2 of Interphase	Telophase
Prophase	Interphase G1 (daughter cells)
Metaphase	

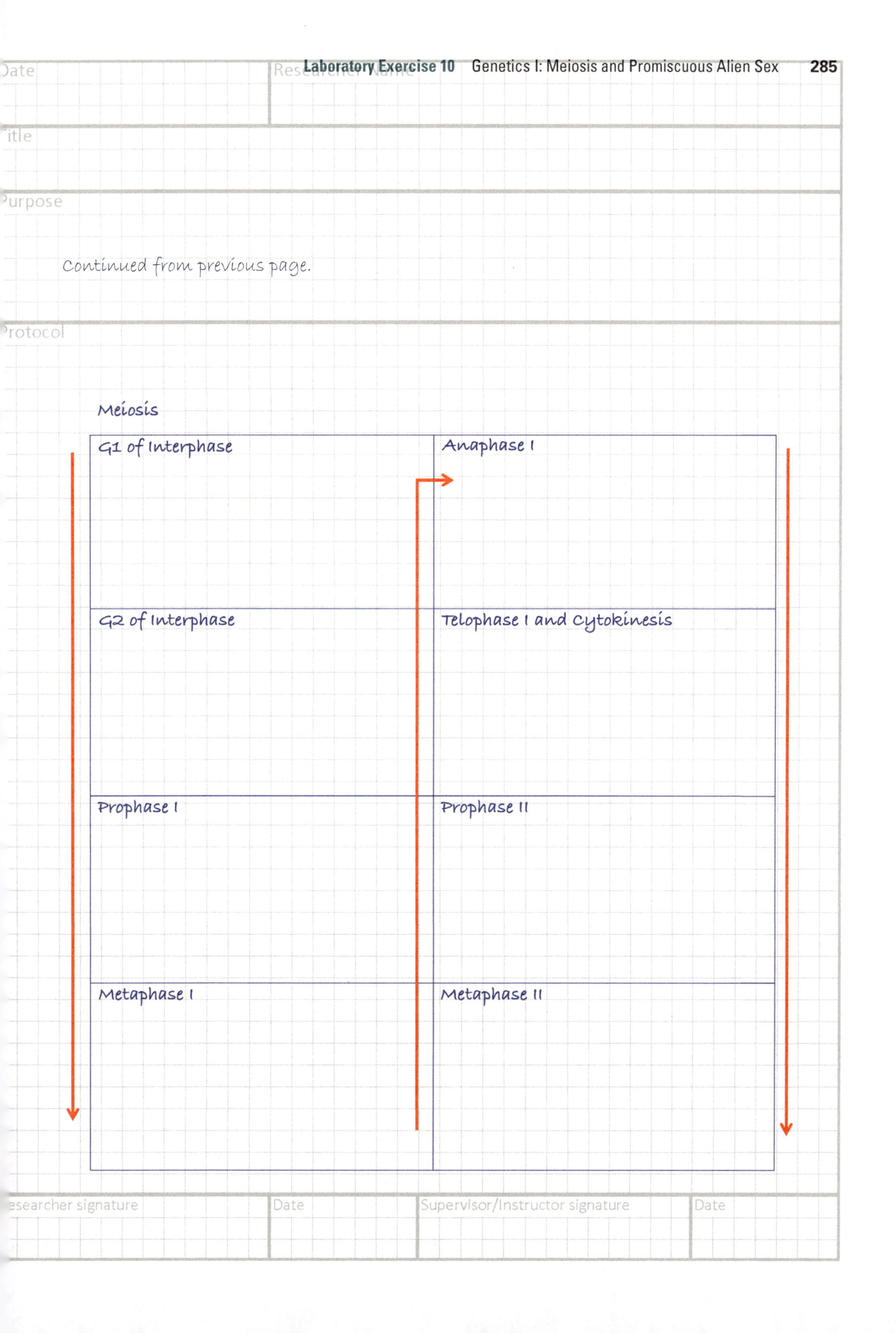

Continued from previous page.

Meiosis

G1 of Interphase	Anaphase I
G2 of Interphase	Telophase I and Cytokinesis
Prophase I	Prophase II
Metaphase I	Metaphase II

Title

Purpose

Continued from previous page.

Protocol

Meiosis continued

Anaphase II

Telophase II

Cytokinesis / Daughter cells

Summary and Conclusions

Problem to Solve 10.2 Can I Model Nondisjunction and Interpret Nondisjunction Karyotypes?

The following human karyotypes represent different nondisjunction disorders.

1. Use the pop-bead model to demonstrate *when and how* nondisjunction occurs.
2. For each of the following syndromes, look up the condition in your textbook or online and describe its major effects. Indicate if the condition is due to monosomy or trisomy and which pair of chromosomes is affected.

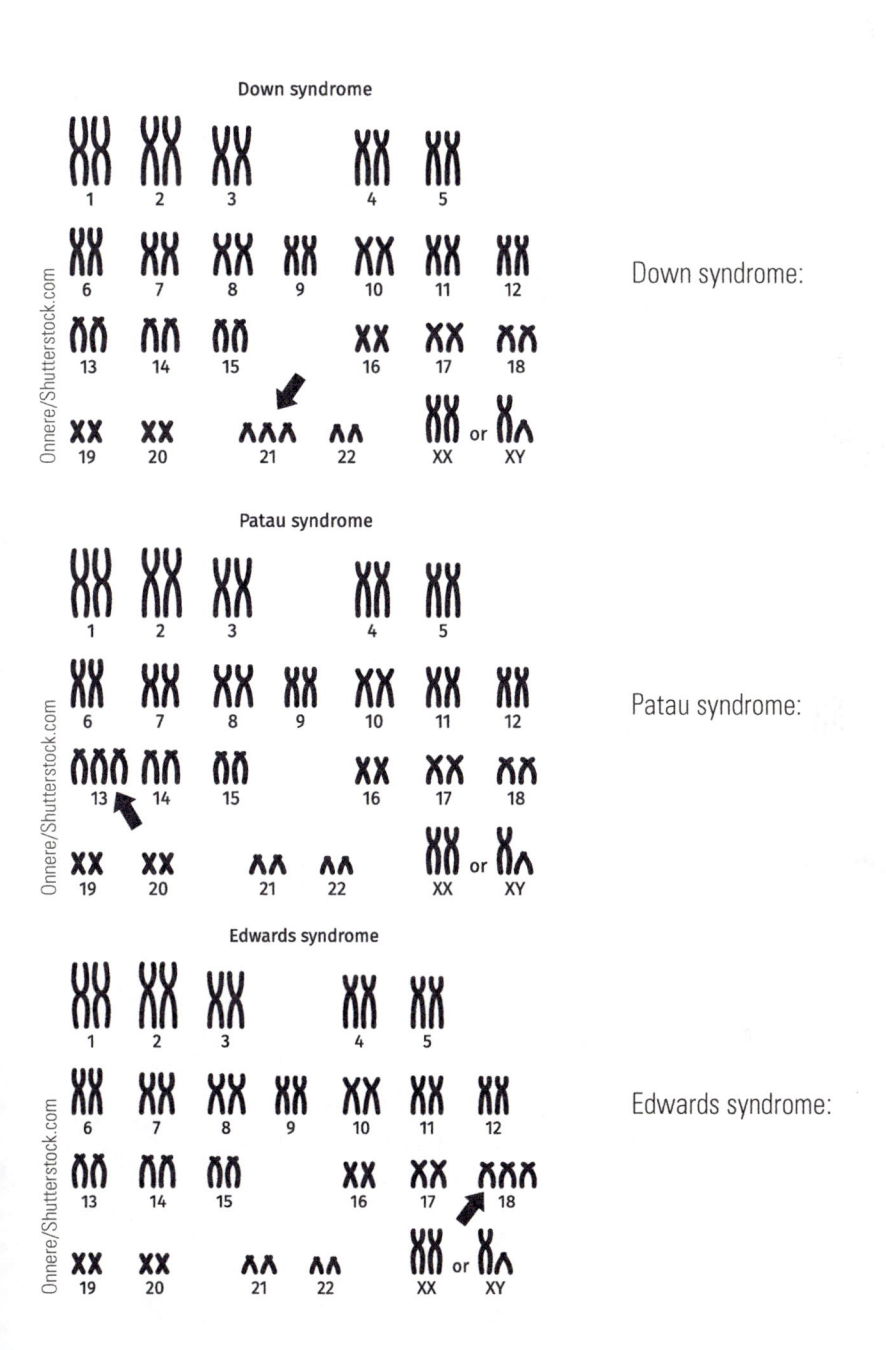

Down syndrome:

Patau syndrome:

Edwards syndrome:

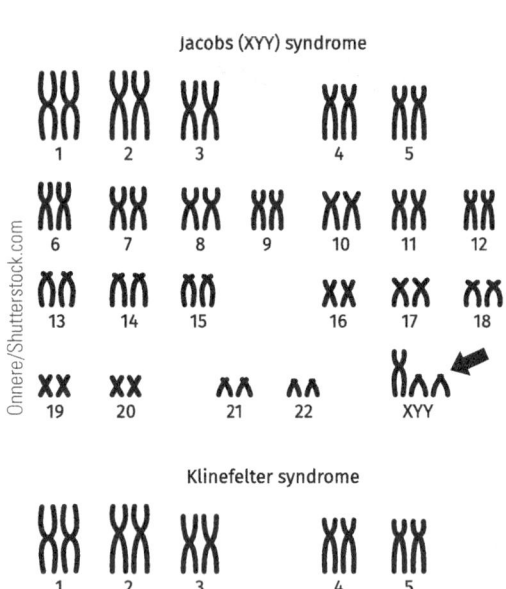

Laboratory Exercise 10 Genetics I: Meiosis and Promiscuous Alien Sex

Jacobs (XYY) syndrome

Jacobs syndrome:

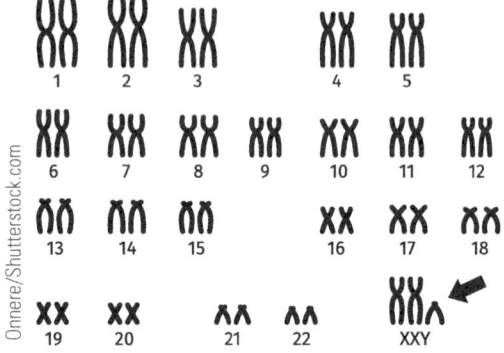

Klinefelter syndrome

Klinefelter syndrome:

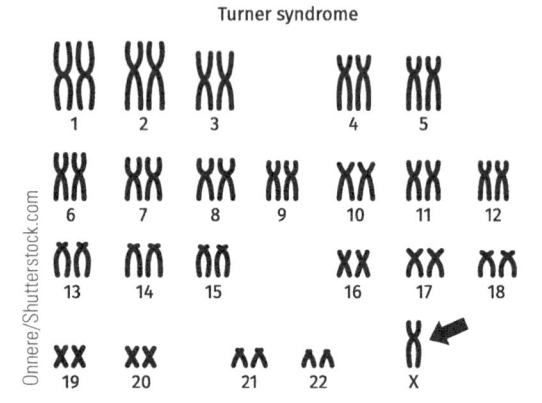

Turner syndrome

Turner syndrome:

Problem to Solve 10.3 Can I Simulate Promiscuous Alien Sex?

Goal: Simulate alien sex and make alien babies!

Materials: (per group of five students)

- The Promiscuous Aliens Data Sheet (on pages following these instructions)
- Deck of cards with words on one side, letters on the other. (Note that in some cases, lowercase letters are underlined for clarity.)
- Four slips of paper, two marked with "SPERM" and two marked with "EGG"

Setup: (Per group of five students)

- Separate the deck of cards into seven piles as indicated by the words on them.
- Find and remove the "Y" card from the SEX deck and give it to the person in the group whose last name starts with the letter closest to Y in the alphabet. (In the case of a tie, flip a coin.)
- Shuffle each pile, keeping the word side up. Place decks in a central location.

How to Have Alien Sex

The person with the Y card is *Femalien-A*. Everyone in the group should write the person's actual name in the Femalien-A box of the data sheet.

Femalien-A must do the following:

1. **Inherit your genetic code**. Take two cards from each deck of traits. (Set the Y card aside until step 4.)
2. **Find out what your sexy alien body looks like!** Read the letters on the cards out loud so everyone can record them in the genotype blanks on the data sheet in the Femalien-A box. Then determine the phenotypes for each trait. Notice that "sex" refers to X and Y chromosomes, but "color" refers to the "C" alleles carried on the X chromosomes. These alleles are "linked" physically since the color allele is on the X chromosome, but the genotypes and phenotypes should be recorded separately.
3. **Ovulate!** With the cards face down, randomly select one card from each pair. Then secretly write the letters from those cards on the slip of paper marked *EGG*. Keep this slip hidden for now. Shuffle your cards (except the Y card) back into the trait decks.
4. **Choose your mate!** Hand the Y card to the "alien" you are considering as a father for your future children.

This person who received the Y card is *Malalien-Y*. Everyone in the group should write that person's actual name in the Malalien-Y box of the data sheet.

Malalien-Y must do the following:

5. **Inherit your genetic code.** Take one card from the SEX pile and two cards from each of the other decks of traits. Add the Y card to the other cards.

6. **Find out how sexy you look!** Read the letters on the cards to the group so everyone can record them on the data sheet in the genotype blanks. Determine your phenotype for each trait.

7. **Make a sperm!** With the cards face down, randomly select one card from each pair. Then secretly write the letters from those cards on the slip of paper marked *SPERM 1*. Keep this slip hidden for now. **(Keep all of your cards.)**

8. **Impress the girl!** Alien courtship rituals are weird, and you must convince your prospective mate that you are the right daddy for her babies. Impress Femalien-A until she agrees to make a baby with you. (Who knows what counts for sexy on your planet? You might need to get creative. But keep it PG please.)

9. **The Big Moment!** Femalien-A and Malalien-Y must decide who in the group will be their baby alien. They must hand their EGG and SPERM 1 slips to that person.

The person who received the EGG and SPERM 1 slips is *Minimalien-I*. Everyone in the group should write the person's actual name in the Minimalien-I box of the data sheet.

Minimalien-I must do the following:

10. **Find out what you inherited.** Read the letters on the EGG slip of paper out loud so everyone can record the letters in the appropriate circle marked EGG on the data sheet. Repeat with the SPERM 1 slip.

11. **Find out if you take after mom or dad or both!** Use the EGG and SPERM 1 information to record your genotypes and phenotypes in the appropriate blanks. Determine if you are a *malalien* or a *femalien*. When finished, shuffle your cards back into the trait decks.

Malalien-Y is promiscuous. He wants to pass on his genes to as many females as possible to increase his "fitness."

Malalien-Y must now do the following:

12. **Make another sperm!** Repeat step 7, recording the letters on the paper slip marked *SPERM 2*. Then set aside the Y card and shuffle the rest of your cards back into the trait decks.

13. **Choose a mate!** This time *you* get to choose another member of the group to be the mother of your babies. You might ask her to demonstrate her mothering skills (whatever those might be) to verify that she is the right choice.

The person chosen to be the second mate is *Femalien-B*. Everyone should write this person's actual name in the appropriate box.

Femalien-B must do the following:

14. **Inherit your genetic code, find out how sexy you look, and ovulate!** Repeat steps 1 through 3.
15. **The Big Moment!** Femalien-B and Malalien-Y must give their EGG and SPERM 2 slips of paper to the last person in the group to make another baby alien!

The person who received the EGG and SPERM 2 slips is *Minimalien-II*. Everyone in the group should write that person's name in the Minimalien-II box of the data sheet.

Minimalien-II must do the following:
16. **Find out what you inherited and which parent you most take after**. Repeat steps 10 and 11.

As a group, discuss and answer the questions in the Discussion and Conclusions section.

Promiscuous Aliens Data Sheet

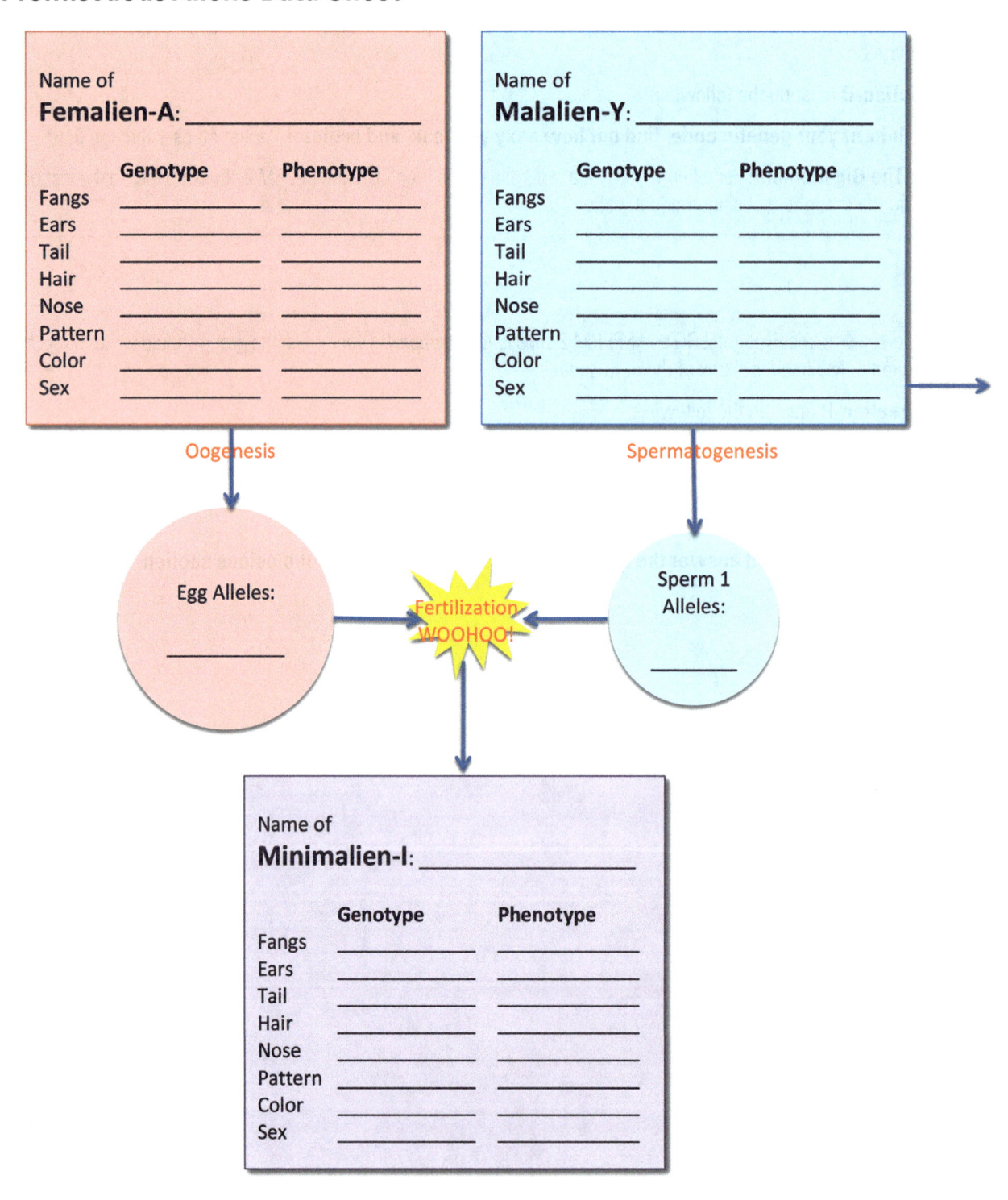

Name of Femalien-A: _____

	Genotype	Phenotype
Fangs	_____	_____
Ears	_____	_____
Tail	_____	_____
Hair	_____	_____
Nose	_____	_____
Pattern	_____	_____
Color	_____	_____
Sex	_____	_____

Name of Malalien-Y: _____

	Genotype	Phenotype
Fangs	_____	_____
Ears	_____	_____
Tail	_____	_____
Hair	_____	_____
Nose	_____	_____
Pattern	_____	_____
Color	_____	_____
Sex	_____	_____

Oogenesis → Egg Alleles: _____

Spermatogenesis → Sperm 1 Alleles: _____

Fertilization WOOHOO!

Name of Minimalien-I: _____

	Genotype	Phenotype
Fangs	_____	_____
Ears	_____	_____
Tail	_____	_____
Hair	_____	_____
Nose	_____	_____
Pattern	_____	_____
Color	_____	_____
Sex	_____	_____

world of vector/Shutterstock.com

Laboratory Exercise 10 Genetics I: Meiosis and Promiscuous Alien Sex

Key to Alien Alleles (Aliealleles)

Traits:	Dominant	Recessive	Type of dominance
Fangs	F = long	f = short	Complete
Ears	E = short pointy	e = long floppy	Complete
Tail	T = tiger	t = bunny	Complete
Hair	H = long	h = short	Incomplete
Nose	N = trunk	n = snout	Complete
Pattern	P = spots	p = no spots	Complete
Color*	C = blue	c = red	Incomplete

sex-linked

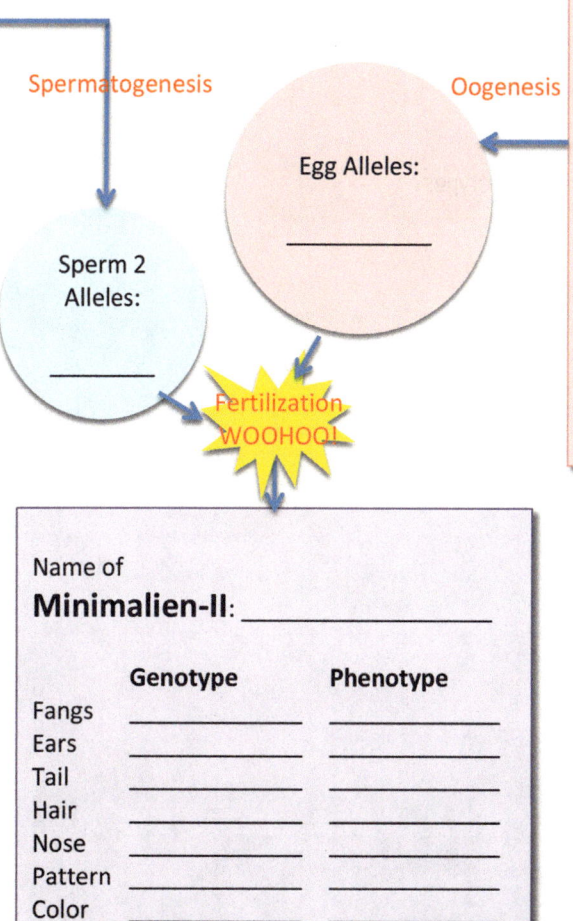

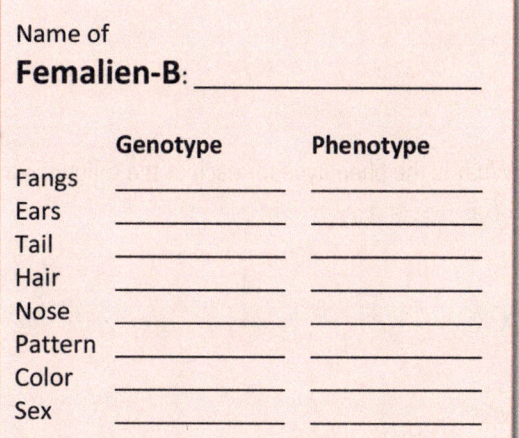

Spermatogenesis

Oogenesis

Egg Alleles: _____

Sperm 2 Alleles: _____

Fertilization WOOHOO!

Name of **Femalien-B**: _____

	Genotype	Phenotype
Fangs	_____	_____
Ears	_____	_____
Tail	_____	_____
Hair	_____	_____
Nose	_____	_____
Pattern	_____	_____
Color	_____	_____
Sex	_____	_____

Name of **Minimalien-II**: _____

	Genotype	Phenotype
Fangs	_____	_____
Ears	_____	_____
Tail	_____	_____
Hair	_____	_____
Nose	_____	_____
Pattern	_____	_____
Color	_____	_____
Sex	_____	_____

world of vector/Shutterstock.com

Discussion and Conclusions:

Could two aliens with long fangs ever produce a minimalien baby with short fangs? Explain.

Could two aliens with long hair ever produce a minimalien baby with short hair? Explain.

Can a malalien ever be purple? Explain.

What is the phenotype for each of the following alien genotypes?

$X^C X^C$

$X^C X^c$

$X^c X^c$

$X^C Y$

$X^c Y$

Name: _____   Date: _____

POST-LAB QUIZ 10: What have you learned?

1. *Why* do higher-level organisms, including almost all plants, animals, and fungi, have sex?

2. What syndrome is caused by trisomy in the 21st pair of chromosomes? What are the effects of this condition?

3. In which step(s) of the Promiscuous Alien activity did the Law of Independent Assortment apply?

4. Why does Klinefelter syndrome cause "feminization"? Why does Turner syndrome cause "masculinization"?

5. An organism has diploid cells containing six chromosomes. You look through a microscope and see a cell that looks like the image below. In what stage(s) of mitosis and/or meiosis is this cell? List all that apply.

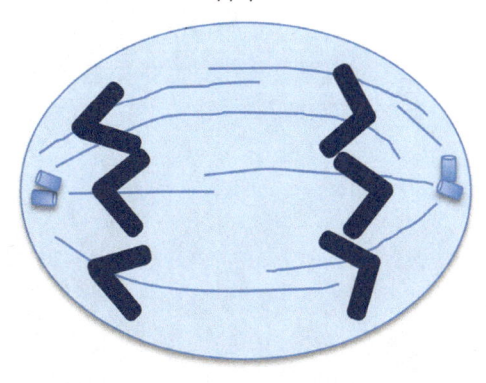

6. When modeling cell division, what did the small, white, plastic cylinders represent? What do these tiny structures actually do in mitosis and meiosis?

7. In the Promiscuous Alien activity, what specific part meiosis was being represented when one of the parent aliens chose a card from each pair of cards?

8. During the alien sex activity, what did the letters on the cards represent? Why was one capitalized and the other lowercase?

9. In the alien sex activity, the color of the alien was "sex-linked." What is a sex-linked trait?

10. Why are lower-case, or recessive, alleles "hidden" by capitalized, or dominant, alleles?

Laboratory Exercise 11

Genetics II: Understanding Inheritance

Concepts to Understand

After completing this lab, the student should be able to:

1. Describe the principles of Mendelian genetics.
2. Use Punnett squares to determine probabilities of genetic outcomes.
3. Interpret a pedigree.

Skills to Learn

After completing this lab, the student should be able to:

1. Apply Mendelian principles to solve genetics problems.
2. Determine the genotypes and phenotypes of Mendelian traits in humans.
3. Construct a pedigree.
4. Use genetic principles to determine blood transfusion compatibility.
5. Determine blood compatibility during pregnancy and appropriate transfusion pairings.
6. Conduct a Chi square test.

SEE ALSO:

Campbell Biology—Chapter 14, section 14.1-14.4; Chapter 15, Sections 15.1-15.2.

Others:

Important terms to know:

Agglutinate	Epistasis	Phenylketonuria
Antibody (Anti-A, Anti-B, Anti-D)	Genotypic ratio	Pleiotropy
Antigen (A, B, and D)	Hemophilia	Polydactyly
Autosomal	Heterozygous	Polygenic
Carrier	Homozygous	Proband
Co-dominant	Huntington's disease	Punnett square
Co-recessive	Monohybrid cross	Rhesus (Rh) factor
Colorblindness	Muscular dystrophy	Sickle cell anemia
Continuous variation	*p* value	Trait
Cystic Fibrosis	P, F1, F2, generations	True breeder
Dihybrid cross	Pedigree	Type 1 diabetes mellitus
Epigenetics	Phenotypic ratio	

Pre-lab Preparation and Background Information

In Lab Exercise 10, you were introduced to some of the basic principles of genetics. The *Promiscuous Aliens* activity demonstrated that genetic **traits** are a result of expressing one or both alleles on homologous chromosomes, and how those alleles are separated and passed on in gametes to the next generation. This lab will take these ideas further, exploring how genes are expressed and inherited.

Gregor Mendel, an Austrian monk who lived in the mid-1800s, discovered foundational principles of genetics still used today. Although he was a monk, he was interested in science, and he took the opportunity to use the monastery garden as a science laboratory. He provides a good example of how a single person can make important discoveries in science without high-tech lab equipment, thousands of dollars in grant funding, or an impressive title or position. He had the most important qualities in a scientist: a questioning and analytical mind, and patience.

Mendel's experiments are worth describing, as this lab exercise will utilize a methodology similar to Mendel's. Mendel made most of his discoveries with peas. Pea plants, shown in Figure 11.1, happened to be a lucky choice for Mendel, since their genetics are simple. The genetic principles for humans and animals are far more complex, and it is unlikely he would have had the same success if he had studied inheritance in mice, rabbits, or humans.

Mendel noted that his pea plants had very specific *binary* characteristics; there are only two variations for most traits. For example, the plants had either purple or white flowers, wrinkled or smooth peas, plump or constricted pods, yellow or green peas, and tall or dwarf plant height. He knew that plant sex is accomplished by pollen traveling from one plant to another. He realized that if he placed the pollen from a stamen of a flower on the pistil of another flower on the *same plant*, the plant would *have sex with itself* (Figure 11.2). This meant that all of the offspring (grown from the peas the flowers produced) would have the same genetic characteristics as the parent plant.

Mendel isolated plants so they could not cross-pollinate. Then he self-pollinated the flowers. He discovered that some

Figure 11.1 Sweet pea plants. A lucky choice for genetics experiments.

plants were **true breeders**, meaning that all of the offspring peas grew into plants that looked exactly like the parent plant. Some true breeders had purple flowers, while other true breeders had white flowers.

He then tried pollinating a true breeder with a true breeder of the opposite characteristic, such as a true-breeding, white-flowering pea plant with a true-breeding, purple-flowering pea plant. He called these the parent plants, or the **P generation**. What surprised him was that *all* of the offspring in the next generation (called the **F$_1$ generation**), looked like only one of the parents. In the case of crossing purple-flowering peas with white-flowering peas, 100% of the peas grew into purple-flowering plants.

He self-pollinated those F$_1$ plants, fully expecting all of *their* offspring to be purple as well. To his surprise, the next generation (called the **F$_2$ generation**) produced a consistent 3:1 ratio of purple flowers to white flowers. (Figure 11.3).

He looked at other characteristics and found the same pattern. Plants that were true breeders for yellow peas crossed with plants that were true breeders for green peas produced an F$_1$ generation of all yellow peas. But if those F$_1$ plants self-pollinated, the F$_2$ generation always produced about 3 yellow peas for each green pea. Table 11.1 shows which traits he observed and the numbers of plants he counted. All of the F$_2$ generations provided a ratio of 3-1 offspring.

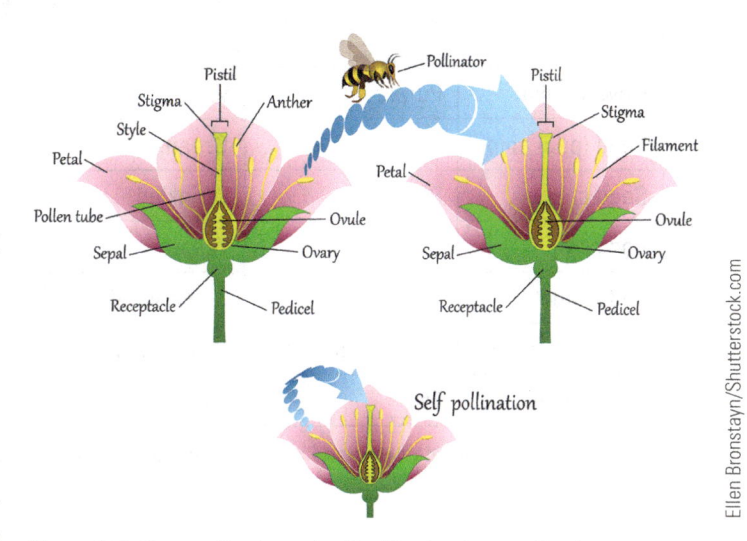

Figure 11.2 Cross-pollination and self-pollination. Cross-pollination occurs when pollen travels from the flower of one plant to the flower of another plant of the same species. Self-pollination occurs when the pollen of a flower pollinates that same flower.

Mendel described the characteristics that showed up in the F$_1$ generation as *dominant*, and the traits that seemed to disappear and then reappear in the F$_2$ generation as *recessive*. In the 1800s, no one knew what DNA was, or how it worked. But Mendel hypothesized that there must be two copies of each characteristic in the plants (or two *alleles*) and that the recessive allele was hidden in the F$_1$ generation. From these observations, he came up with the principles we discussed in Lab Exercise 10: the *Law of Independent Assortment* and the *Law of Segregation*. In short, these laws stated that the alleles for each trait were independently split up and separated in the process of gamete formation. We now know this happens in metaphase and anaphase of meiosis I.

Mendel developed terms and notations to describe the transfer of alleles from generation to generation. We use the same conventions today. Typically, one letter is chosen to represent a trait (such as "P" for pea flower color). The dominant allele is represented by a capital letter (P), and the recessive allele is represented with a lower case letter (p). The term

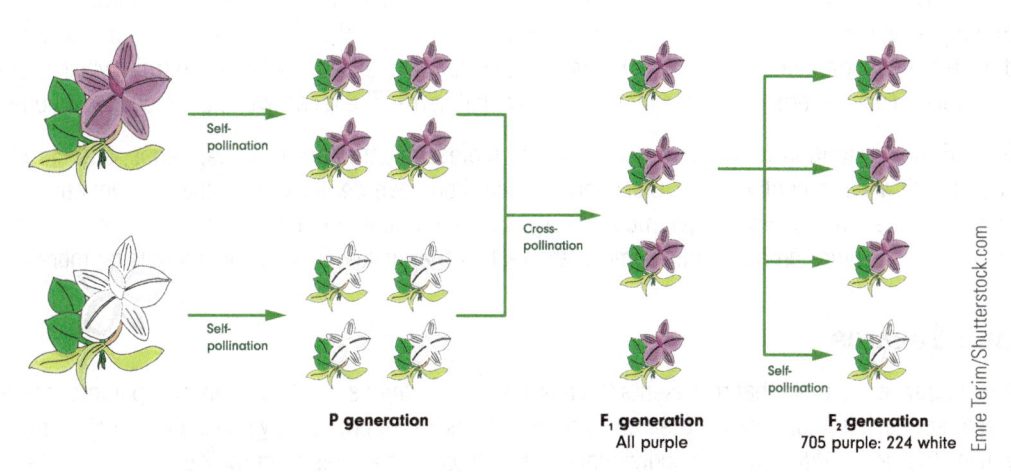

Figure 11.3 Mendel crossed true breeders to produce F$_1$ and F$_2$ generations.

Table 11.1 Mendel's observations

Observed characteristic	Dominant trait	Recessive trait	Observed number of individuals in F_2 generation		Ratio
			Dominant	**Recessive**	
Flower color	Purple	White	705	224	3.15: 1
Seed (pea) color	Yellow	Green	6,022	2,001	3.01: 1
Seed shape	Round	Wrinkled	5,474	1,850	2.96: 1
Pod color	Green	Yellow	428	152	2.82: 1
Pod shape	Inflated	Constricted	882	299	2.95: 1
Flower position	Axial (to the side)	Terminal (at top)	651	207	3.14: 1
Plant height	Tall	Dwarf	787	277	2.84: 1

Note: For every observed characteristic, the F2 generation showed a near 3:1 ratio of dominant to recessive traits.

genotype refers to the pair of letters used to represent the alleles of a trait. The term *phenotype* refers to the expression of the genotype. Not all phenotypes are outwardly visible.

Mendel described a genotype with two identical alleles as **homozygous** (*homo* = same; *zygosity* = genetic makeup). *Homozygous dominant* individuals have two dominant alleles, and the genotype is written as two capital letters. Homozygous recessive individuals have two recessive alleles, written as two lowercase letters. Mendel's purple-flowering true breeders were therefore homozygous dominant (PP), and his true breeder white-flowering plants were homozygous recessive (pp). Mendel described an individual with one of each allele as **heterozygous** (*hetero* = different). He hypothesized that heterozygous individuals would only express the dominant allele.

Mendel's hypothesis, that traits result from pairs of alleles, was supported by the experimental data shown in Table 11.1. Mendel made diagrams to show that crossing a true breeding purple-flowering plant (PP) crossed with a true breeding white-flowering plant (pp) would always produce an F_1 generation of purple heterozygous pea plants (Pp). He could show that self-fertilization of the F_1 plants would always produce a 3:1 ratio of dominant to recessive phenotypes in the F_2 generation (Figure 11.4).

It is important to consider for a moment the concept of dominant and recessive genes. A recessive trait is "hidden" in a heterozygous pair, while a dominant trait is expressed. But alleles are just recipes for proteins. A dominant allele is usually a recipe for a functional protein. A recessive allele is often a variant of the genetic recipe that produces a non-functional, or less functional, version of the protein. Mendel observed this in the purple and white flowers of pea plants. Purple was dominant over white because having the functional "recipe" for purple pigment on either homologous chromosome allowed the plant to make purple pigments (which are just proteins). Having at least one *P* allele produced a purple flower. But if both alleles lacked the proper information for purple pigments (pp), the plant expressed the recessive trait, which was un-pigmented, or white, flowers. In peas, there is no allele for white flower pigments. The recessive allele produced a non-functional protein.

Mendel laid the foundation for modern genetics. The molecular structure of DNA, genes, and chromosomes was not understood until about a century later. Discoveries since then have demonstrated that the genetic expression of a trait is often complicated by many other genetic and *epigenetic* (non-genetic) factors. This exercise will discuss how to solve basic Mendelian genetics problems and some of the factors that make humans and animals far more complex than peas.

Punnett Squares

A **Punnett square** is a table that represents the possible combinations of alleles from two parents' sperm and egg in their resulting offspring. It is important to note that Punnett squares do not predict any one individual offspring's genetic makeup. It only gives the *probabilities* that any individual will inherit certain alleles from their parents.

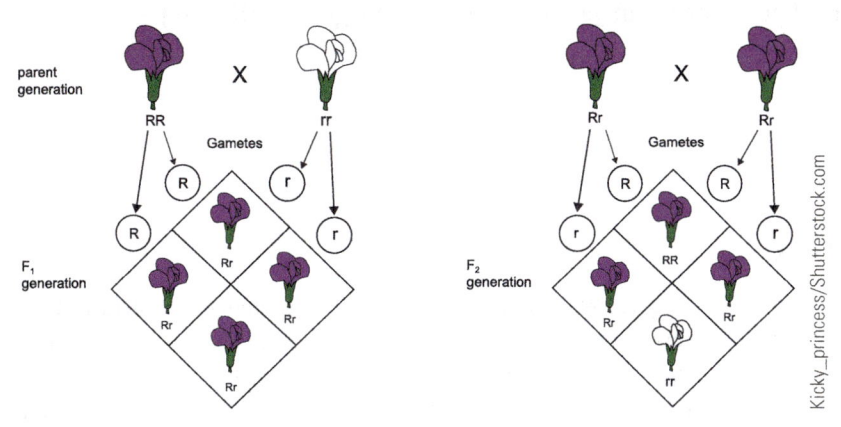

Figure 11.4 Mendel demonstrated how an allele model of genetics explained his observations. True breeders each had two identical alleles for flower color, but purple was dominant over white. All F₁ plants inherited one purple allele and were purple. Self-pollination of the F₁ plants produced a ratio of three purple-flowering plants to one white-flowering plant.

Steps for using a Punnett square

1. Identify traits and determine or define possible alleles of interest.
2. Identify parent phenotypes and genotypes if possible.
3. Determine possible parent *gametes*.
4. Create a table of all possible gamete pairings (a Punnett Square).
5. Determine probabilities of possible genotypes and phenotypes in the offspring (**genotypic** and **phenotypic ratios**).

Creating a Punnett square to predict inheritance for one trait is simple, requiring a 2×2 grid. A **monohybrid cross** is a cross involving one trait. The following example shows how to conduct a monohybrid cross for guinea pig fur color. (The genetics of guinea pig fur color is actually more complex than what is shown here.)

Example of a monohybrid cross:

Two guinea pig parents are known to be heterozygous for fur color. The alleles are defined such that *B* represents the allele for black fur; *b* represents the allele for brown fur.

Step 1 — Identify traits and alleles.
Given: B = dominant black allele, b = recessive brown allele.

Step 2 — Identify parent genotypes and phenotypes.
Given: both are black and heterozygous, Bb.

Step 3 — Determine possible parent gametes.
Each Bb parent has a 50% chance of giving a B or b. Gametes are haploid.

Step 4 — Cross gametes in a Punnett Square.

	sperm B	or b
eggs B	BB	Bb
or b	Bb	bb

Step 5 — Find percentages of genotypes and phenotypes of offspring.

Predicted genotypes: *¼ chance (or 25%) BB; black fur.*
½ chance (or 50%) Bb; black fur.
¼ chance (or 25%) bb, brown fur.

Predicted phenotypes: *¾ (75%) chance black, ¼ (25% chance) brown.*

Dihybrid Crosses

Sometimes we may want to investigate the inheritance of two traits at a time. A **dihybrid cross** is a cross that examines the inheritance of two traits. This requires a larger Punnett square because more alleles are involved and there are more options for gametes. When looking at two traits, each gamete needs one allele from each pair, which results in four possible combinations of alleles. It may be helpful to recall *FOIL* from algebra. To list all of the possible gametes of an organism for *two* traits, you must determine all the ways you can choose one letter (allele) from each pair. Take the *first* letter of each pair, then the *outside* letters, then the *inside* letters, and then the *last* letters.

Example of a dihybrid cross:

Both the male and female guinea pigs are heterozygous for short hair, and black fur. Short hair and black fur are dominant traits. H = short hair, h = long, B = black fur, and b = brown.

Step 1 — Identify traits and alleles.
Given: H = short, h = long, B = black, b = brown.

Step 2 — Identify parent genotypes and phenotypes:
Given: both are heterozygous for both traits: male HhBb, and female HhBb.

Step 3 — Determine possible parent gametes (FOIL parent alleles)
Male sperm: HB, Hb, hB, or hb

Female egg: HB, Hb, hB, or hb

Step 4 — Make a Punnett Square.
Put the letters of a homologous pair together, and write capital letters first in a heterozygous pair.

	HB	***Hb***	***hB***	***hb***
HB	*HHBB*	*HHBb*	*HhBB*	*HhBb*
Hb	*HHBb*	*HHbb*	*HhBb*	*Hhbb*
hB	*HhBB*	*HhBb*	*hhBB*	*hhBb*
hb	*HhBb*	*Hhbb*	*hhBb*	*hhbb*

Zush/Shutterstock.com

greatstockimages/Shutterstock.com

cynoclub/Shutterstock.com

Eric Isselee/Shutterstock.com

Step 5 — find percentages of genotypes and phenotypes of offspring.
genotypes

HHBB 1/16 short black (6.25%)
HHBb 1/8 short black (12.5%)
HHbb 1/16 short brown (6.25%)
HhBB 1/8 short black (12.5%)
HhBb 1/4 short black (25%)
Hhbb 1/8 short brown (12.5%)
hhBB 1/16 long black (6.25%)
hhBb 1/8 long black (12.5%)
hhbb 1/16 long brown (6.25%)

phenotypes

9/16 = short black (56.25%)
3/16 = short brown (18.75%)
3/16 = long black (18.75%)
1/16 = long brown (6.25%)

Sex Chromosomes

In humans, the twenty-third homologous pair contains the sex chromosomes. They are not truly homologous chromosomes, since a Y chromosome is much smaller than an X chromosome. The Y chromosome carries little genetic information. The genotypes for sex chromosomes are XY for male and XX for female. Recall from Lab Exercise 10 that there are rare occurrences of non-disjunction in sex chromosomes that result in trisomy, such as Klinefelters syndrome (XXY), or monosomy, such as Turners syndrome (XO).

What are the chances of a couple having a boy or a girl? Which parent determines the sex of the child? The father can only give an X or a Y chromosome. Any sperm has a 50% chance of containing an X or Y. A woman has two X chromosomes, so her egg has a 50% chance of containing either one of her two X chromosomes, but all eggs contain an X. Therefore, a Punnett Square shows that the chance of having a male or a female child is always 50%. The *father* always determines the sex of the child, since only he can provide a Y chromosome.

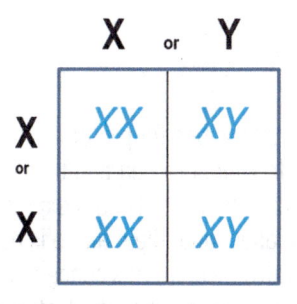

	X	Y
X	XX	XY
X	XX	XY

Consider the following true example. A couple had six boys. (Two were identical twins.) After the youngest son was in high school, the parents experienced a "surprise" pregnancy. If you learned of this family, what chances do you give them of having another son—greater than 50% or less than 50%?

The answer is *exactly* 50%. And it was 50% for *each* pregnancy. It is tempting to give prior outcomes some level of influence over a current pregnancy, but each pregnancy is an isolated event. A common gambler's myth is believing that if you lose enough times in a row, the next turn of the slot machine or roll of the dice *must* have a higher chance of winning. This is not true. Every spin or roll is an isolated event. Likewise, the number of boys or girls previously conceived has NO bearing on future pregnancies. Every pregnancy has a 50% chance of producing a boy or girl. The probability of heads and tails in a coin flip is always 50%. But sometimes you can flip a coin and, just by chance, get six heads in a row.

Complications in Expressing Traits

Sex-linked Traits

Most genes are carried on **autosomes**, the 22 pairs of chromosomes that are not sex chromosomes. Recall that *sex-linked traits* are produced by alleles carried on an X chromosome. In the "Promiscuous Aliens" activity, the color of the alien was sex-linked. Many sex-linked traits of interest are recessive. If a male has a sex-linked trait, he will express it. He has only one X chromosome, so every allele present on it will be expressed. Females may have a recessive sex-linked allele, but the other X chromosome may have the dominant allele and thus hide the sex-linked trait. When this occurs, she is considered a **carrier** of the trait.

Example of a sex-linked trait:

A person with **color blindness** lacks pigments to absorb certain wavelengths of light in the retina of the eye. Color blindness is a recessive and sex-linked trait. Alleles for sex-linked traits are indicated by superscript "tags" on X the chromosomes. (Y-chromosomes contain very few alleles.)

Cross a color-blind male with a normal female. Will any children be color-blind? Will it affect the sons, daughters, or both equally?

- **Step 1 — Determine traits and alleles:**
  - *X^E = X chromosome with color vision allele.*
  - *X^e = X chromosome with colorblind allele.*
  - *Y = chromosome with no alleles.*

- **Step 2 — Determine parent genotypes and phenotypes:**
  - *Colorblind male = $X^e Y$.*
  - *Normal female = $X^E X^E$.*

- **Step 3 — Determine possible parent gametes:**
  - *Male X^e or Y; female X^E or the other X^E*

- **Step 4 — Make a Punnett square.**

	X^e or Y	
	X^e	Y
X^E or	$X^E X^e$	$X^E Y$
X^E	$X^E X^e$	$X^E Y$

- **Step 5 — Find percentages of genotypes and phenotypes of offspring.**
  - *The parents have a 50% chance of having a girl who will be a carrier of colorblindness, and a 50% chance of having a boy who does not have colorblindness. None of their children can actually have colorblindness.*

Incomplete Dominance

Recessive alleles often contain non-functional recipes for proteins. In many cases, one dominant allele is enough to allow an organism to make the needed proteins. In Mendel's pea plants, all it took was one P allele to produce purple flowers. When heterozygous and homozygous dominant individuals both express the same dominant phenotype, the trait is said to express *complete* dominance. But in some cases, one allele alone is not enough to produce the dominant phenotype. This is called *incomplete dominance*. You were introduced to this concept in the Promiscuous Aliens activity in Lab Exercise 10. Color and hair length were incompletely dominant alien traits. Heterozygous individuals with incomplete dominance are unable to fully express the dominant phenotype, resulting in a blending of dominant and recessive traits.

Example of incomplete dominance:

Imagine a flowering plant where red flower pigmentation is incompletely dominant over white. The R allele produces red flower pigments. The r allele does not produce pigment. To be red, the flower needs *both* alleles for red pigment. In this case, the red allele is incompletely dominant, so that heterozygous flowers with only one red allele end up only partially red, or pink.

What is the genotype of a red-flowering plant? What is the genotype for a white-flowering plant? If you cross a red-flowering plant with a white, what will they produce?

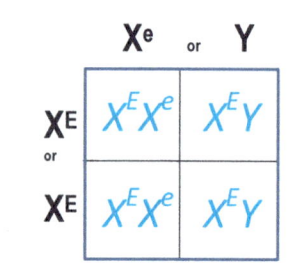

Step 1 — **Determine traits and alleles:**
R = red pigment; r = white color (lack of red pigment).
Red is incompletely dominant over white.

Step 2 — **Determine parent genotypes and phenotypes:**
Red plant must be RR.
White plant must be rr.

Step 3 — **Determine possible parent gametes:**
Red plant can only produce R or R.
White plant can only produce r or r.

Step 4 — **Make a Punnett square.**

Step 5 — **Find percentages of genotypes and phenotypes of offspring.**
If they only pollinate each other, 100% of the offspring produced by these plants will be Rr, and have pink flowers. (Figure 11.5)

	R	R
r	Rr	Rr
r	Rr	Rr

Co-dominance and Co-recessiveness

In some cases, there may be two or more alleles that produce functional proteins that affect a trait. If two dominant alleles can be expressed at the same time, they are **co-dominant**. In some complex situations, there may even be two **co-recessive** genes that may be expressed at the same time. Co-dominance can explain why a brown dog and a black dog can produce black and brown-spotted puppies. Co-dominance is also a factor in many other traits beyond coloration, such as human ABO blood types.

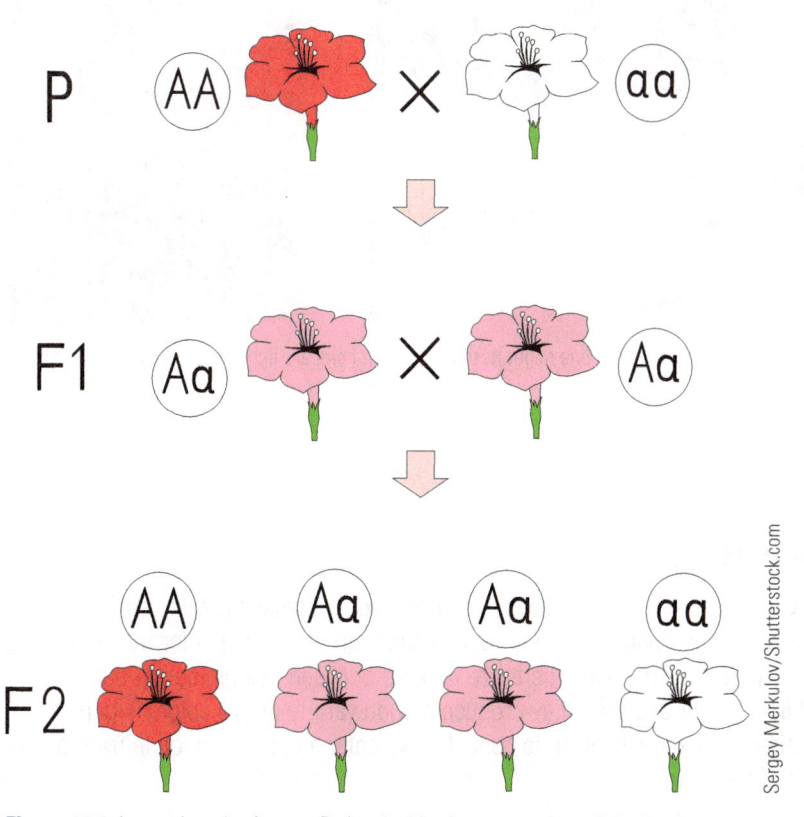

Figure 11.5 Incomplete dominance. Red and white flowers produce pink offspring.

Example of co-dominance:

There are many breeds and colors of horses. However, among the many alleles for horse coat color, brown and white are generally both dominant alleles. A brown horse crossed with a white horse will typically produce a foal with "roan" coloration that has both white and brown hair, *not light brown hair*.

Step 1 — Determine traits and alleles:
H_1 = brown hair; H_2 = white hair.

H_1 and H_2 are co-dominant.

Step 2 — Determine parent genotypes and phenotypes:
Brown horse is H_1H_1.

White horse is H_2H_2.

Step 3 — Determine possible parent gametes:
Brown can only produce H_1 or H_1.

White can only produce H_2 or H_2.

Step 4 — Make a Punnett square.

Step 5 — Find percentages of genotypes and phenotypes of offspring.
All offspring will have the H_1H_2 genotype and will express roan coloration (Figure 11.6).

	H_1 or H_1	
H_2 or	H_1H_2	H_1H_2
H_2	H_1H_2	H_1H_2

Figure 11.6 A roan horse. From a distance one might assume the coat hairs are light brown. On closer inspection, the intermingled white and brown hairs are visible.

Multiple Alleles

Some traits may be expressed by multiple alleles. Eye color in humans is quite complex. To date, fifteen genes have been associated with eye color. There are many factors that can affect eye color including level of pigmentation and other light-reflecting properties of the eye. There are no blue or green eye pigments, but varying levels of melanin (the same brown pigment in dark hair and skin) may produce a wide variety of eye colors (Figure 11.7). Recent research has shown that too many other factors are involved to consider eye color a basic Mendelian trait, but we can oversimplify these factors for the sake of this example.

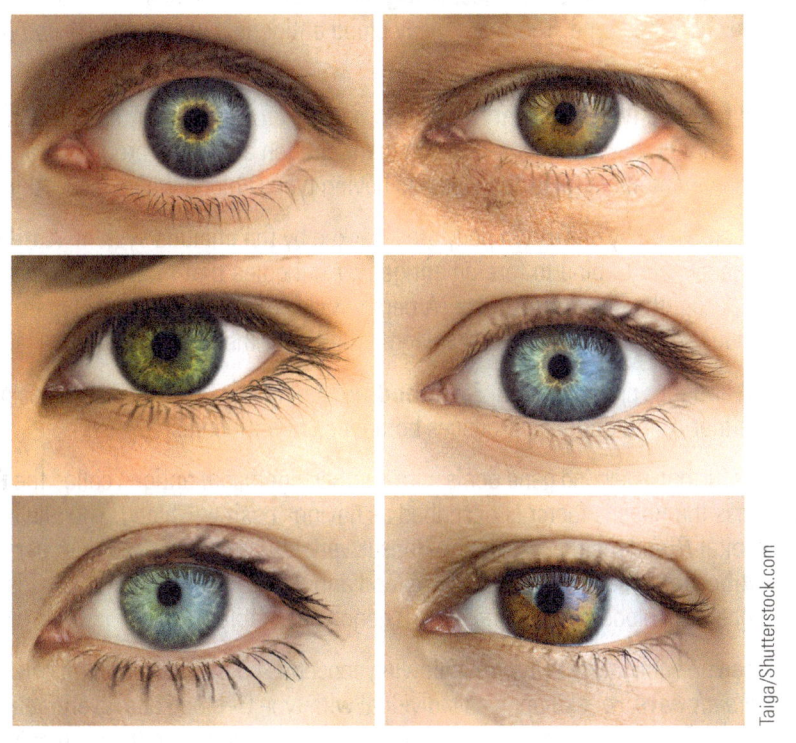

Figure 11.7 Eye color is complex and involves multiple alleles influencing melanin production and other factors that influence light-scattering and reflective properties of the eye.

In cases where there are multiple alleles for a trait, a single letter is used to represent the trait, and subscript or superscript notations indicate specific alleles. Capital letters are used to denote dominance.

An example of multiple alleles:

Imagine that the allele for brown eye color (B) is completely dominant over the blue allele (b_1), but incompletely dominant over green (b_2). Blue and green alleles are also co-recessive. A brown-eyed man, whose mother had blue eyes, married a green-eyed woman. What eye colors might be expressed in their children?

Step 1 — Determine traits and alleles:
B = brown, b_1 = blue, b_2 = green. B is completely dominant over b_1 but incompletely dominant over b_2. Alleles b_1 and b_2 are co-recessive.

Step 2 — Determine parent genotypes and phenotypes:
The man has brown eyes, but his mother had blue eyes. He must have a B allele, but he must also have a b_1 allele from his mother. His genotype must be Bb_1. His wife has green eyes, so her genotype must be b_2b_2.

Step 3 — Determine gametes:
The man can produce B or b_1.

The woman can produce b_2 or the other b_2.

Step 4 — Make a Punnett square.

	B	or	b_1
b_2	Bb_2		b_1b_2
or			
b_2	Bb_2		b_1b_2

Step 5 — Find percentages of genotypes and phenotypes of offspring.
The offspring have a 50/50 chance of inheriting Bb_2 and hazel eyes (a blending of brown and green) or b_1b_2 and blue-green eyes.

Blood types are also a result of multiple alleles. **Antigens** (marker molecules in the glycocalyx) on the surface of blood cells determine a person's blood type. There are many blood antigens, but *antigens A and B* are used to determine one's blood type.

The alleles for blood type are typically identified with a capital *I* for the allele that produces antigens, or a lowercase *i* for the allele that does not produce antigens. The dominant antigen allele can produce A antigens (I^A) or B antigens (I^B). A person with neither dominant allele (*ii*) is described as having type O blood which lacks A and B antigens.

Antigens that are not produced by one's own body are attacked by the immune system. **Antibodies** are proteins produced by cells of the immune system to bind to foreign antigens. Antibodies that bind to antigens on foreign blood cells cause the blood cells to stick together, or **agglutinate**. A person who receives the wrong type of blood in a blood transfusion could experience serious complications or die from agglutination of their blood.

A person who lacks A antigens produces **anti-A antibodies**. A person who lacks B antigens produces **anti-B antibodies**. A person who has neither A or B antigens produces both types of antibodies.

To complicate matters further, a second gene deals with another significant antigen called the **Rhesus factor**, or *D-antigen*. A person who produces the Rhesus factor is described as having "positive" blood. The allele for the Rhesus factor is typically noted with a capital *R*, and the recessive, non-Rhesus variant with a lowercase *r*. A person without the Rhesus factor produces **anti-D antibodies**. There are many other blood antigens, but ABO blood typing is based on antigens A, B, and D. Thus there are eight main blood phenotypes, but 18 genotypes (Table 11.2).

Blood compatibility can be summarized with the following concept: *A person can't receive blood that contains any antigens he or she does not already have.* For example, an individual with type O- blood has no antigens. They can give their blood to *anyone*. Individuals with O- blood are described as *universal donors*. However, they can only receive blood from someone else with no antigens, which would be another O- individual. An individual with type AB+ blood has all of the antigens—A, B, and D. They can receive blood from *anyone*. Individuals with AB+ blood are described as *universal recipients*. However, they can only donate blood to someone with all of those antigens, which would be another AB+ individual.

Table 11.2 Blood type genotypes, phenotypes, and antibodies

TYPE A			TYPE B		
Genotype	Phenotype	Antibodies	Genotype	Phenotype	Antibodies
$I^A I^A$ RR	A+	Anti-B	$I^B I^B$ RR	B+	Anti-A
$I^A i$ RR			$I^B i$ RR		
$I^A I^A$ Rr			$I^B I^B$ Rr		
$I^A i$ Rr			$I^B i$ Rr		
$I^A I^A$ rr	A-	Anti-B, Anti-D	$I^B I^B$ rr	B-	Anti-A, Anti-D
$I^A i$ rr			$I^B i$ rr		
TYPE AB			TYPE O		
Genotype	Phenotype	Antibodies	Genotype	Phenotype	Antibodies
$I^A I^B$ RR	AB+	none	ii RR	O+	Anti A, Anti-B
$I^A I^B$ Rr			ii Rr		
$I^A I^B$ rr	AB-	Anti-D	ii rr	O-	Anti-A, Anti-B, Anti-D

Two parents may have different blood types. When a baby is conceived, the baby's blood type may be incompatible with the mother's. This poses a potential threat, as the mother may produce antibodies that will cross the placenta and mix with the baby's blood. Couples should get their blood tested before or during pregnancy to see if their baby might inherit a blood type that could trigger the mother's immune system (Figure 11.8). The most important is the Rhesus factor. If a baby inherits the Rh factor allele from the father, and the mother lacks this allele (has negative blood), the baby is at risk. This is particularly important for a second pregnancy, as the mother will have likely produced enough antibodies from the first pregnancy that they could attack the embryo at the earliest stages of a second pregnancy.

Figure 11.8 Pregnant couples should check for blood compatibility between mother and baby. Treatments are simple, but without them the baby and mother are at risk.

Blood transfusions are often needed in cases of trauma and major bleeding. If a patient is given a transfusion of blood containing antigens they do not produce, their own antibodies will bind to the transfused blood cells, causing them to clump or rupture, leading to severe reactions including stroke, kidney failure, and death.

Example of a blood genetics problem:

A woman with type A- blood marries a man with B+ blood. They are expecting a child. What is the probability that their child will specifically produce Rh antigens?

Step 1 — Determine traits and alleles:
In this case we really only need to look at the Rh factor for each parent. R = Rh positive, r = Rh negative

Step 2 — Determine parent genotypes and phenotypes:
The mother has A- blood, so her genotype for the Rh factor is rr. We only know the father is positive. We do not know if his genotype is RR or Rr.

Step 3 — Determine possible parent gametes:
If the father is homozygous dominant, he will produce a R or the other R.

If he is heterozygous he will produce a R or a r.

The mother must produce one r or the other r.

Step 4 — Make two Punnett squares, one for each possibility.

	R	R
r	Rr	Rr
r	Rr	Rr

	R	r
r	Rr	rr
r	Rr	rr

Step 5 — Find percentages of genotypes and phenotypes of offspring.
If the father is homozygous, there is a 100% chance the baby will be Rh positive, and the mother should be treated for blood compatibility. If the father is heterozygous, there is still a 50% chance the baby will be Rh positive. Regardless, the baby and mother are at risk if they do not take precautions. Blood tests, monitoring, and possible treatment during the pregnancy can significantly reduce blood compatibility issues.

More Complex Genetic Factors

Many different sets of genes collectively contribute to some phenotypes. Such traits are **polygenic** (*poly* = multiple; *genic* = genes). For example, height, hair color, and skin color can't be described with a simple genotype. Consider a person's height. We tend to think of height as a genetic trait, and to some degree it is, but height is a product of many factors, both genetic and non-genetic, including fetal hormones, fetal developmental rate, hormone levels before and during puberty, timing of puberty, nutrition, exercise, mother's nutritional level during gestation, and more. It is difficult to predict the inheritance of complex traits that are determined by so many genes. When many genes produce a wide range of phenotypes, the trait is said to exhibit **continuous variation** (Figure 11.9).

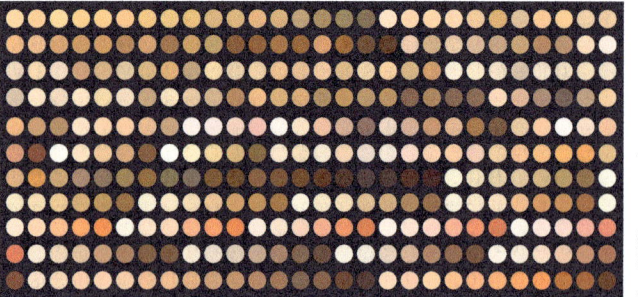

Figure 11.9 Skin color. Many genetic factors contribute to skin coloration.

Environmental Factors and Gene Expression

Further complications arise when we consider the influence of the environment on gene expression. Some genes are turned on or off by external factors. For example, the pH level of the soil or the average environmental temperature can activate or deactivate some plant genes. Environmental factors such as nutrition and stress can turn on or off genes in humans. The relatively new field of **epigenetics** seeks to understand how external factors trigger the expression of specific genes.

However, some traits are genetic, while others are not. For example, a bodybuilder will not produce muscular babies. Nor will a skilled musician give birth to a musical prodigy. When some traits such as athletic ability or musical skill seem to run in families, it is likely due to children growing up in an environment that cultivates those skills.

Behaviors are not directly genetic. Genetic traits arise from the order of nitrogen bases in DNA, and the results of protein synthesis. However, there is interplay between genetics, environmental factors, and behavior. Some genetic factors can indirectly influence the overall neurochemistry of the brain. Thus, some genes may make some people more prone to exhibit certain behaviors. For example, it has been suggested that alcoholism is a genetic trait. This is not supported by biological evidence, since there is no DNA sequence or protein that directly produces alcoholic behavior. But there are genes that make some people release higher or lower levels of neurotransmitters, which may make some more susceptible to addictive behaviors, including, but not limited to, alcoholism. And a continued use of an addictive substance can eventually influence gene expression in neurons, amplifying the effect even further.

Pleiotropy

Another complication called **pleiotropy** occurs when one gene produces a protein that influences many other downstream phenotypes. For example, the gene for the production of melanin can have an influence hair, skin, and eye color. This is particularly true when that gene doesn't work. Consider the importance of channel proteins to cells. If a gene for a certain type of transporter does not work properly, this could affect specific cells throughout the organism, potentially causing many seemingly unrelated problems. Likewise, if a gene for a certain enzyme is mutated, the lack of that enzyme may affect a wide variety of cells in the body.

In some cases, genes even activate other genes. If the activating gene has a mutation or problem, then it doesn't matter what the downstream genes are. They will not get the chance to be expressed.

Epistasis is a type of pleiotropy in which two (or more) genes affect the phenotype of one specific trait. Albinism is an example. Consider a squirrel population in which fur color (brown, black, or gray) can be affected by a different gene that activates the production of fur pigments. If that activating gene is faulty, the squirrel will be white. There is no allele for white fur pigmentation, but the mutated allele produces no fur color at all. And that allele that produces albinism, though recessive, may be passed on to offspring. If two normal-colored squirrels are both carriers for the recessive albinism allele, they may produce an albino offspring regardless of the fur coloration genes they pass on.

Common Genetic Abnormalities

There are many genetic diseases, including some we discussed in Lab Exercise 10, like Downs Syndrome, Klinefelter's Syndrome, and other forms of nondisjunction. The following are some genetic disorders caused by direct mutations in the DNA code. Mutations may spread and persist in the population, particularly if the gene is recessive or the negative consequences do not affect the individual until after childbearing years. Whether or not a gene is dominant or recessive has little to do with the frequency of that allele in the population. Some dominant traits are very rare. Some recessive traits, like blue eyes, are very common.

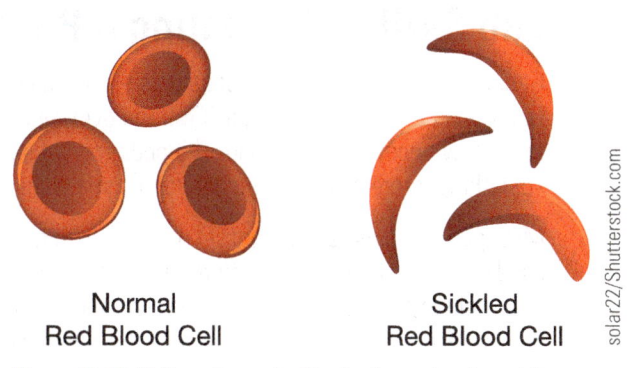

Figure 11.10 Sickle-cell anemia. Blood cells may be shaped like "sickles" or bananas.

Sickle Cell Anemia is recessive and occurs in about 1 out of 500, primarily in the black population. It is caused by a *point mutation*, and it produces irregularly shaped blood cells (Figure 11.10).

Hemophilia is a sex-linked recessive trait that occurs in 1 out of 7,000 people. The condition causes a lack of one or more clotting factors, so the blood doesn't clot properly.

Cystic Fibrosis is recessive and occurs in about 1 in 2,000 people (mainly whites). It causes malfunctions in many organs including the pancreas, lungs, and liver.

Polydactyly is a dominant trait that occurs in about 1 in 400 people. It produces more than 5 fingers or toes on each hand or foot (Figure 11.11).

Phenylketonuria is a recessive trait that occurs in 1 out of 18,000 people (very rare). It causes poor brain development.

Huntington's Disease is a dominant trait that occurs in about 1 out of 10,000 people (very rare). It causes the brain to deteriorate rapidly with age.

Muscular Dystrophy is a sex-linked recessive trait that occurs in about 1 out of 10,000 people (very rare). It causes muscles to deteriorate rapidly with age.

Colorblindness is a common, recessive, sex-linked disorder. It produces an inability to see certain colors, usually red and green, but there are many types of colorblindness.

Type 1 diabetes mellitus results when cells of the pancreas produce a mutated marker that causes the immune system to destroy the pancreas cells that make insulin.

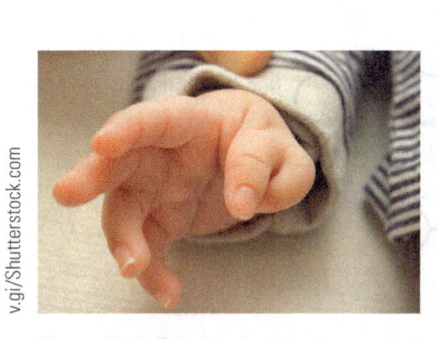

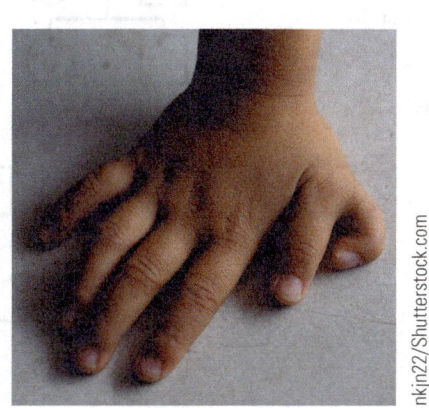

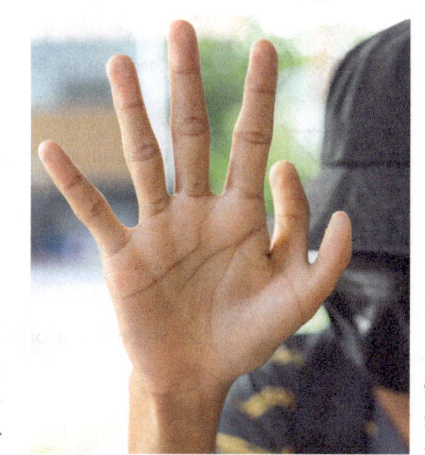

Figure 11.11 Polydactyly. A dominant genetic condition producing extra fingers or toes.

Lab Skill 11.1 Creating a Pedigree

Geneticists, animal breeders, genealogists, and others interested in "family trees" of humans or other animals, use pedigrees to represent inheritance of specific traits over multiple generations. **Pedigrees** are diagrams that represent individuals related by ancestry or breeding. The pedigree is used to indicate which individuals have a specific trait. The trait may be something interesting, like red hair, or a disease, like colorblindness or cystic fibrosis.

Most pedigrees use a standard set of symbols to represent members of the group (Figure 11.12). Some symbols may not be used or may be represented differently in pedigrees where those symbols are not necessary or important. Some pedigrees may include additional notations, such as age, relevant genetic or medical conditions, age at death, and cause of death. Pedigrees of prize animals may include their awards and titles.

Imagine that a young man named Geoff has a family history that includes hemophilia. Geoff knows that hemophilia is a genetic disorder that prevents normal blood clot formation, but that is about all he knows. He doesn't have hemophilia himself, but he is planning to get married in a month. He would like to know if he might be a carrier of hemophilia since members of his family have it. He doesn't know anything about the genetics of hemophilia. He asks you to help him figure this out.

The following is a description of Geoff's family.

- Geoff has a brother and two sisters. His brother has hemophilia, but his parents and sisters do not.
- On his mother's side of the family, he has one aunt and two uncles. One of those uncles has hemophilia. His aunt has passed away.
- Neither grandparent on his mother's side had hemophilia, but his great grandfather, the father of his mother's mother, had hemophilia. His mother's father and his great grandparents have passed away.
- Geoff's father did not have hemophilia. His father's one brother and his father's parents do not have hemophilia.

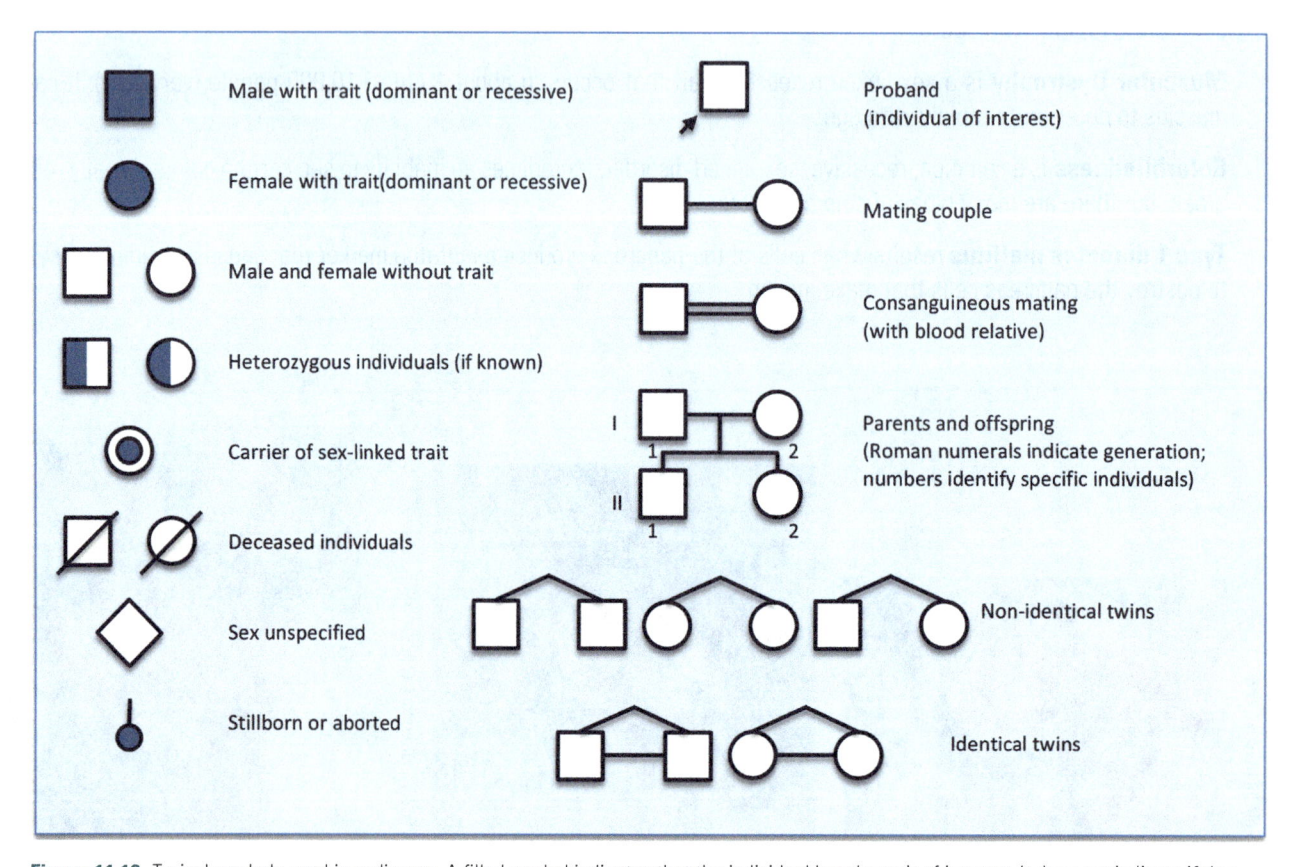

Figure 11.12 Typical symbols used in pedigrees. A filled symbol indicates that the individual has the trait of interest. It does not indicate if that trait is dominant or recessive, or if the individual is homozygous dominant or heterozygous for that trait.

Laboratory Exercise 11 Genetics II: Understanding Inheritance **313**

To create this pedigree, we should start with Geoff, the **proband**, or person of interest. An arrow is used to indicate him. Typically the proband *has* the genetic trait of interest, and the symbol representing that person would be shaded. In this case, he is a male *without* the trait, so he must be represented by an empty square. We can add his brother, who does have hemophilia, and his two sisters.

From here we can add his parents and their siblings. Since they are from a previous generation, they should be positioned above Geoff and his siblings. The slash indicates someone who is deceased.

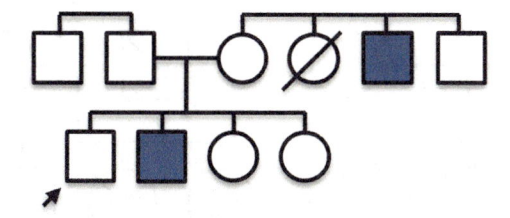

Next we can add the grandparent generation. It should be placed above the others.

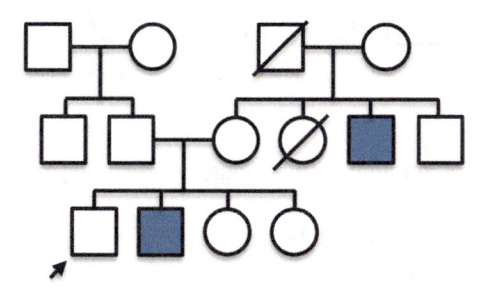

Finally we can add the great-grandparent generation. We can also use Roman numerals to indicate the generations, and numbers to identify individuals of each generation. Since Geoff is interested in whether or not he could be a carrier for hemophilia, we need to determine more information about the genetics of the disease.

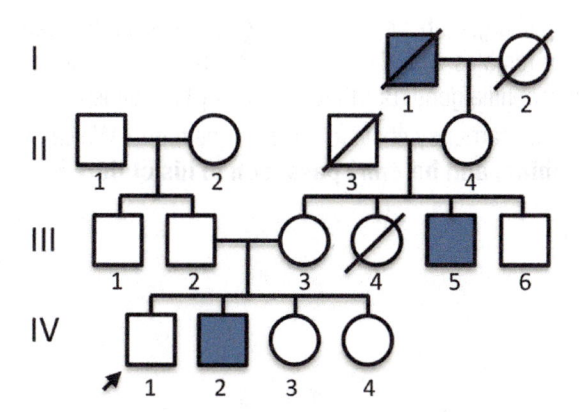

Note that the disease skipped generation II. If a trait is dominant, one or both of the parents of someone with that trait MUST also display it. Dominant traits *must* show up in every previous generation.

But is important to understand that the inverse is not true. A trait that shows up in every generation does not have to be dominant. If many members of a family have a common recessive trait, such as blue eyes, it may be present in every generation.

If the trait is recessive, it may skip generations before it shows up again. When a trait does skip a generation, it *must* be recessive. Since no one in generation II had hemophilia, and since Geoff's parents did not have hemophilia, it *must* be recessive.

Notice also that in this family history, only males have had hemophilia. Recall from the discussion of sex-linked traits that men are more likely than women to express recessive sex-linked traits. Hemophilia is indeed a recessive, sex-linked genetic condition.

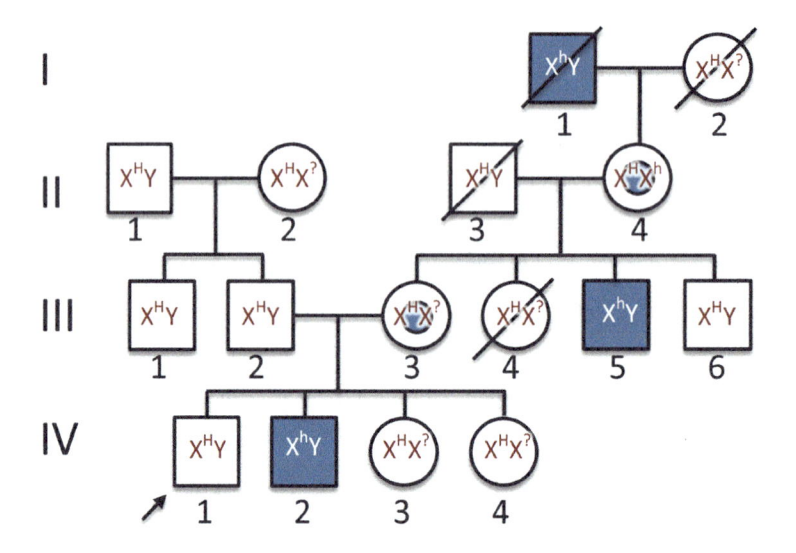

Knowing this, we can now add more information to the pedigree.

- Geoff's brother must have the recessive hemophilia gene on his X chromosome. His genotype *must* be X^hY. In fact, *all* affected males must have this genotype.
- Therefore, all *unaffected* males must have the normal dominant gene X^HY.
- To have a son, a father must pass on his Y chromosome. So no male with hemophilia can pass it on to his son. He will pass it on to his daughter.
- The mother of any male with hemophilia must either *have* hemophilia (X^hX^h) or *carry* it (X^HX^h). No female in this family tree has hemophilia, but II-4 and III-3 had sons with hemophilia. They *must* be carriers and have the genotype X^HX^h.
- Females carrying the hemophilia gene have a 50% chance of passing that gene on to their daughters. Therefore we can't know the genotype for females who did not have children with hemophilia. (It may be tempting to say that I-2 and II-2 do not carry the hemophilia gene. But that would only be a guess.)
- We can add the genotypes and carrier symbols to finish the pedigree. We can also answer the question for Geoff: **He is *not* a carrier of hemophilia, and he *can't* pass it on to his children.**

Lab Skill 11.2 Using a Chi Square Test to Compare Actual Results with Expected Results

A Chi Square test is a calculation that compares observed data with predicted data to see if they are significantly different. For example, Mendelian genetics predicts that out of 20 births, 10 should be male and 10 female. But if you collected data that showed 11 males and 9 females were born, is this close enough to say that the data still agrees with the prediction? A Chi Square test can tell you that.

Recall from Lab Exercise 9 that a *null hypothesis* (H_0) states that there is *no significant difference* between experimental and control variables, or in this case, between the observed and predicted data. Data collected in the real world is unlikely to ever give exactly the same results as a Punnett square. But the *null hypothesis states that any difference is minor and due to random chance*.

The *alternative hypothesis* (H_1) must then state that there *is a significant difference* between the observed and predicted data and that there is some other factor at work that is skewing the data.

Calculating the **p value** (*p* for probability that the null hypothesis is accurate) will indicate which hypothesis to accept.

Generally, *if the p value is greater than 0.05*, the data is *not significantly different* from the expected values. In such a case, the H_0 would be supported and the H_1 rejected.

Generally, *if the p value is less than 0.05*, the data *is statistically different* from the expected values. In such a case, the H_0 would be rejected, and the H_1 would be supported.

Consider a hypothetical scenario where this might be useful. Imagine you had a male gray rabbit (buck) and a female brown rabbit (doe). Gray fur is dominant, and brown is recessive. They mated and produced a litter of twelve baby bunnies (kits). Seven kits were gray, and five were brown. Do the genes for their coloration follow the Mendelian model?

If the gray buck was homozygous dominant (BB) and the brown doe was homozygous recessive (bb), then all of the kits should have been heterozygous and gray (Bb). Since some of the kits were brown, the male must have had a brown allele. If the male was heterozygous (Bb) and the female homozygous recessive (bb), then a Punnett square predicts 50% of the kits should be gray (Bb) and 50% brown (bb). This seems to be a reasonable starting point. Our *expectation* then, according to Mendelian genetics, is that out of 12 kits, 6 should be gray, and 6 should be brown. Our *actual* data is 7 gray kits and 5 brown kits.

Null hypothesis (H_0): The outcome of 7 gray and 5 brown kits is not significantly different from the 6 gray and 6 brown kits predicted by Mendelian genetics. This hypothesis will be accepted if the *p* value is *greater than 0.05*.

Alternative hypothesis (H_1): The outcome of 7 gray and 5 brown kits is significantly different from the 6 gray and 6 brown kits predicted by Mendelian genetics. This would suggest that there may be other genetic factors at work, or the Mendelian model doesn't apply to genes for rabbit fur color. This hypothesis will be accepted if the *p* value is *less than 0.05*.

So how do we calculate the *p* value? First we need to calculate the Chi Square (χ^2) value.

$$\chi^2 = \sum \frac{(o-e)^2}{e}$$

Σ = the sum of
o = observed values
e = expected values

We need to plug in our numbers for each category and add them together. In this case, there are two categories: gray and brown coloration.

For gray, o = 7, and e = 6.

$$\chi^2_{(for\ gray)} = (7-6)^2/6$$
$$= 1^2/6$$
$$= 1/6$$
$$= 1.67$$

For brown, o = 5, and e = 6.

$$\chi^2_{(for\ brown)} = (5-6)^2/6$$
$$= -1^2/6$$
$$= 1/6$$
$$= 1.67$$

$$\chi^2 = 1.67 + 1.67 = 3.34$$

Now we need to look up the Chi Square value on a table to get the *p* value. See Appendix G for a table of Chi Square values. The table contains rows labeled by *degrees of freedom*. The *degrees of freedom (df)* is equal to the number of categories minus 1.

In this example, we had two categories, gray and brown, so the *df* for this data is 1.

We must go across the row until our χ^2 value falls on or between two values in that row. In the partial Chi square table shown in Figure 11.13, 3.33 falls between *p* values of 0.05 and 0.10. *Therefore the p value must be larger than 0.05.* Since the *p* value is greater than 0.05, we do not reject the null hypothesis. There is no *significant* difference between the actual and expected data.

0.10 > P value > 0.05

df	0.995	0.975	0.2	0.1	0.05	0.025	0.02	0.01	0.005
1	0.000	0.001	1.642	2.706	3.841	5.024	5.412	6.635	7.879

$$\chi^2 = 3.33$$

Figure 11.13 Reading the Chi Square table. Note that $\chi^2 = 3.33$, which is between 2.706 and 3.841.

Name: _____ Date: _____

PRE-LAB QUIZ 11: Are you ready to proceed?

1. What exactly is a dominant or recessive trait?

2. Clearly define these three easily confused terms: *homozygous, heterozygous,* and *homologous.*

3. What is the difference between *co-dominance* and *incomplete dominance*?

4. In a certain plant species, red flower color is incompletely dominant over white. Describe the genotypic and phenotypic ratios of offspring produced by crossing two pink-flowered plants.

5. What is a sex-linked trait? Can a male be a carrier of any sex-linked trait?

6. Clearly differentiate between the *antigens* and *antibodies* involved in blood typing. Which antigens and antibodies would a person have if they have A+ blood?

7. In genetics, what causes *continuous variation*?

8. Compare and contrast *pleiotropy* and *epistasis*?

9. What do the Roman numerals and Arabic numbers in a pedigree represent? What do squares and circles represent?

10. What is a *null hypothesis*? If a *p* value for a set of data is 0.1, should you accept or reject the null hypothesis?

Problem to Solve 11.1 Can I Solve Basic Genetics Problems?

For this activity, you will work with a partner. You may be assigned specific questions, or asked to share your answers with the class. You may need extra scratch paper to work out some of the problems.

Basic Genetics

1. S = short hair, s = long hair. Cross parent guinea pigs that are ss (male) and Ss (female). Give the phenotypic and genotypic ratios of the offspring.

2. In the case of the guinea pigs above, what parental genotypes would produce a 3 to 1 ratio of dominant to recessive phenotypes in their offspring?

3. In cattle, the hornless condition is dominant (H), and having horns is recessive. Horn alleles are autosomal. A rancher owns three cows named Apple, Flower, and Ivy. Apple has no horns; Flower and Ivy do have horns. A neighbor's hornless bull named Bubba broke through the fence between pastures and "visited" the three cows. Sometime later, Apple and Flower had horned calves, and Ivy had a hornless calf.

   Give the genotypes. Apple:_____ Flower: _____ Ivy: _____ Bubba: _____

Crossing Two Traits

The autosomal allele for black mouse fur (B) is dominant over brown (b). A separate gene controls pigment production. The dominant allele (A) produces fur pigments, but the recessive allele (a) shuts down fur pigment production altogether, producing an albino.

A black male mouse (BbAa) is crossed with an albino female mouse (Bbaa).

4. What four gametes can the male produce?

5. What four gametes can the female produce?

6. What is the probability that any one of their offspring will be an albino?

7. What is the probability that any one of their offspring will be heterozygous for both traits?

Sex-linked Traits

8. Colorblindness is a sex-linked, recessive trait. A colorblind man named Rob has a non-colorblind daughter named Jane, who herself has had one non-colorblind daughter. Jane is now pregnant with a boy. What are the chances that she will have a colorblind son?

9. Pattern baldness is a recessive, sex-linked trait. Franklin is concerned about going bald. He thinks he is losing his hair. His father went bald on the back of his head. His father's father (paternal grandfather) lost his hair from front to back in a receding hairline. His mother's father (maternal grandfather) never lost his hair. His wife's father (step-father) lost his hair in irregular patches, causing him to shave his head. If Franklin is losing his hair, which relative is he most likely to take after? Why?

10. *Syndactyly* is the condition of having webbed toes, or toes that failed to separate during development (Figure 11.14). The allele that causes it is recessive and sex-linked, but one of the few alleles carried on the Y chromosome. If a man has webbed toes, what is the likelihood that his sons will inherit webbed toes? What is the likelihood that his daughters will?

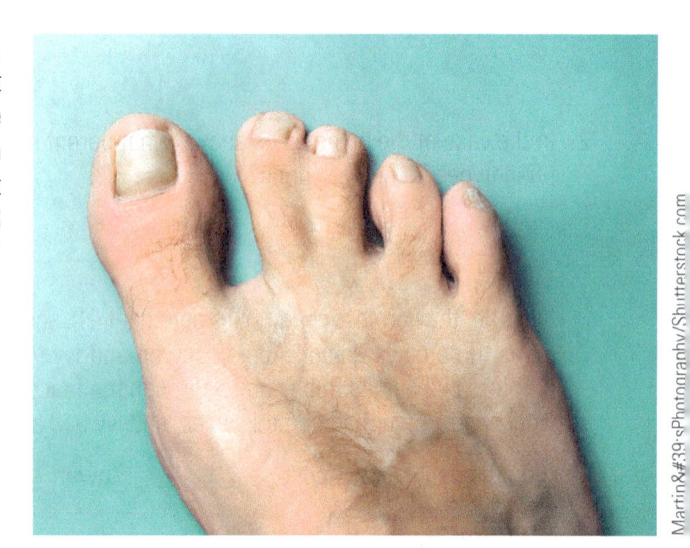

Figure 11.14 Syndactyly, or webbed toes.

Multiple Alleles

11. Mrs. Jones and Mrs. Smith both had babies in the same hospital, on the same day, at almost the same time. You are working in the labor and delivery department when a crisis comes up. Mrs. Smith claims she was told that her child was a girl. Mrs. Jones also claims she was told that her child was a girl. But one newborn is a boy, and the other is a girl. Which baby belongs to which mother? The newborn boy has type B− blood, and the newborn girl has A+ blood.

    Mrs. Jones has AB− blood. Her husband has A− blood.

    Mrs. Smith has A+ blood. Her husband has B− blood.

    Who is the mother of the boy, and who is the mother of the girl? How do you know?

12. Breeds of rabbits often have multiple alleles for coat color. For example, in a simplified version of one breed, a speckled *agouti* gray allele (B) is completely dominant over brown (b_1) and black (b_2). Brown and black alleles are co-recessive.

    Cross a brown rabbit with an agouti rabbit carrying the black allele. What are the possible genotypes and phenotypes of the offspring?

13. I^A and I^B are dominant alleles for A and B antigens, respectively, and i is recessive. R is the dominant allele for Rh factor, and r is recessive.

   What are all of the possible genotypes for a person with type A+ blood?

   A man who has B+ blood ($I^B i ORr$) marries a woman who has A+ blood ($I^A i ORr$). What is the chance that the baby will have O- blood?

Pedigrees

14. In the pedigree below, add numbers for each individual in each generation. Individuals I-2, I-4, II-1, and III-2 have hitchhiker's thumb (a recessive trait where the thumb bends backwards). Indicate them on the pedigree. Use appropriate symbols to indicate all heterozygous individuals.

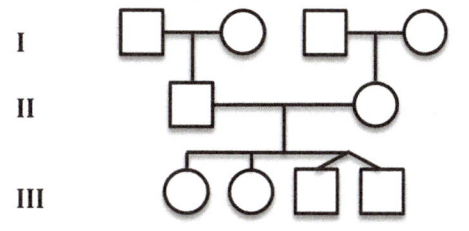

15. What is distinctive about individuals III-3 and III-4?

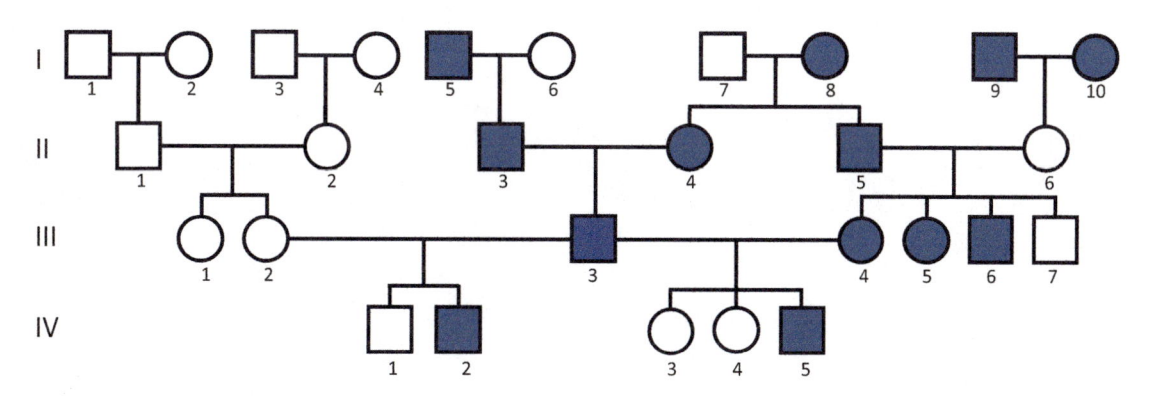

16. Is the trait represented in the pedigree a dominant or recessive trait? Why?

17. Give the genotypes for the following individuals:

    1-6: _____ I-11: _____ II-5: _____ III-3: _____ IV-2: _____

18. What is the relationship between II-2 and IV-2?

19. What is the realtionship between IV-2 and IV-3?

20. Brown eyes are a dominant eye-color allele and blue eyes are recessive. (B=brown, b=blue). Assume for this question that there are no other eye color alleles and that eye color is a simple Mendelian trait.

    A brown-eyed woman, whose father had blue eyes and whose mother had brown eyes, married a brown-eyed man whose parents are also brown-eyed. They have a son who is blue-eyed.

    Draw a pedigree showing brown eyes that includes all four grandparents, the two parents, and the son. Use appropriate symbols. Indicate all possible genotypes.

Problem to Solve 11.2 Can I Describe Genetic Characteristics of Humans?

In this exercise, you will determine your phenotype and genotype for a number of easily observed human traits. For each trait, record your phenotype. If you have a recessive trait, your genotype will be simply two lowercase letters. If you have a dominant trait, you may be either homozygous dominant or heterozygous. You may be able to determine which genotype you have if you consider your family members. For example, if you have curly hair, a dominant allele, your genotype could be KK or Kk. But if you have a parent with straight hair (kk) you MUST be heterozygous. If you can't determine your genotype, use a question mark (K?) to indicate that you have at least one dominant allele, but do not know if you are homozygous or heterozygous.

Mendelian Traits in Humans

	Characteristic	Dominant trait	Alleles	My Pheno-type	My Geno-type
1	Naturally curly or straight hair	Curly is incompletely dominant.	K or k		
2	Freckles	Freckles are dominant.	F or f		
3		Widow's peak is dominant.	W or w		
4	Mid-digital hair—hair on middle section of ring finger	Presence of hair is dominant.	M or m		
5	Handedness	Right handed is dominant over left.	R or r		

	Characteristic	Dominant trait	Alleles	My Pheno-type	My Geno-type
6	Hitchiker's thumb—thumb extends backwards at a near 90° angle	Not having hitch hiker's thumb is dominant.	H or h		
7	Interlocking fingers	Left on top is dominant.	L or l		
8	Cleft chin	Having a cleft is dominant.	C or c		
9	PTC taster	Taste allele is incompletely dominant over non-tasting.	P or p		
10	Thiourea taster	Taste allele is incompletely dominant.	T or t		
11	Sodium benzoate taster	Taste allele is incompletely dominant.	S or s		
12	Free or attached earlobes	Free is dominant over attached.	E or e		

Laboratory Exercise 11 Genetics II: Understanding Inheritance **325**

	Characteristic	Dominant trait	Alleles	My Pheno-type	My Geno-type
13	Dimples	Dimples are dominant.	D or d		
14	Morton's toe—second toe longer than the big toe	Having Morton's toe is dominant.	M or m		
15	Achondroplasia—inability to produce growth factor receptor 3. Causes dwarfism (if shorter than 4'10")	Tall is dominant; Achondroplasia is recessive.	G or g		
16	Tongue rolling	Ability to roll the tongue is dominant. (A second dominant allele allows for lateral folding.)	U or u		
17	Colorblindness	Color blindness is sex-linked and recessive.	X^C or X^c		
18	Blood ABO type: A, B, AB, O	A and B are co-dominant.	I^A, I^B, i		
19	Blood Rh factor	Rh factor is dominant.	R or r		
20	Hair color—influenced by multiple genes including one for brown pigments and another for red pigments.	Brown is dominant over blonde; *not red* is dominant over red.	B or b N or n		

Source: Images: Row 2: Irina Bg/Shutterstock.com; 3: charless/Shutterstock.com; 4: Antonio Guillem/Shutterstock.com; 6: Paul Luyster; 7: Champion studio/Shutterstock.com; 8: Rommel Canlas/Shutterstock.com; 12a-b: aleks333/Shutterstock.com and L Julia/Shutterstock.com; 13: Cameron Whitman/Shutterstock.com; 14: Lisa Culton/Shutterstock.com; 16: AFPics/Shutterstock.com.

Observing more complicated genetic variations in humans

We can observe differences in hair color, skin color, and height among members of the human population. How can we represent the genes for these kinds of traits? Recall that these are traits that express *continuous variation*, or a full range of phenotypes.

Imagine a hypothetical gene we will call *SkinA* that contributes to skin tone (allele *A* is for dark skin, allele *a* is for light). Imagine that genes *SkinB* and *SkinC* are separate genes that also contribute to skin tone. A person with the genotype *AABBCC* would be very dark-skinned, while a person with a genotype *aabbcc* would be very light-skinned.

Consider a hypothetical couple, a man and a woman, who are each heterozygous for all three skin-color genes. What skin colors might be expressed in their children?

Step 1 — Determine traits and alleles:
   A, B, C = dark; a, b, c = light.

Step 2 — Determine parent genotypes and phenotypes:
   Both parents are heterozygous for all three genes: AaBbCc.

Step 3 — Determine gametes:
   (For three sets of genes, each parent can produce 8 possible gametes.)

   Father's possible gametes: _____ _____ _____ _____ _____ _____ _____ _____.

   Mother's possible gametes: _____ _____ _____ _____ _____ _____ _____ _____.

Use a pencil, or preferably, a colored pencil, to complete this activity. (Doing this with a pen will not work well.) Replace all capital letters with a dark colored square, and the lowercase letters with a white square. Use one color (any color) for all of the capital letters. Fill in the boxes below accordingly.

Father's possible gametes: ☐☐☐ ☐☐☐ ☐☐☐ ☐☐☐ ☐☐☐ ☐☐☐ ☐☐☐ ☐☐☐

Mother's possible gametes: ☐☐☐ ☐☐☐ ☐☐☐ ☐☐☐ ☐☐☐ ☐☐☐ ☐☐☐ ☐☐☐

A Punnett square containing colored squares instead of allele letters will help illustrate the possible range of skin color in their offspring.

Complete the Punnett square by shading in the correct number of boxes for each potential offspring to represent the dominant alleles they may inherit.

Use your finger or an eraser to smear/smudge the colored pencil in each of the 64 boxes of the Punnett square. This should provide a range of coloration from very dark to very light across the grid of the Punnett square. Just a few genes can produce a continuous variation across a wide phenotypic range.

How would the appearance of this Punnett Square change if one parent was homozygous dominant for all three genes, and the other was homozygous recessive for all three genes?

Making a Pedigree to Track Genetic Traits

The following is a "soap-opera" genetics puzzle. It is a bit entertaining and twisted, but it will hopefully help you implement the genetic principles covered in this lab. Work with a partner to unravel the story. Use a separate sheet of paper to work out the relationships. Create a pedigree using the standard symbols and notations, and answer the questions below. You may be asked to present your work to the class.

Assume eye color is Mendelian, and that B = brown, b_1 = blue, and b_2 = green alleles. Brown is completely dominant over blue and incompletely dominant over green. Blue and green are co-recessive alleles. Colorblindness is sex-linked.

- Zoe Gray is a brown-eyed girl with no common sense. Her father, Mr. Gray has blue eyes. Mrs. Gray has hazel eyes. There is no history of colorblindness in their family.
- During a rather wild and stupid weekend of drinking and debauchery, Zoe ended up sleeping with all three of the Lavender bothers, Alan, Bill, and Carl. As a result, she got pregnant.
- Nine months later she had a son she named Xavier who, as it turns out, has hazel eyes and is not colorblind. Zoe's parents, Mr. and Mrs. Gray, are understandably upset. They want the daddy, whoever he is, to take responsibility and provide child support.
- Alan Lavender is not colorblind and has blue eyes. Bill Lavender is colorblind and has green eyes. Carl Lavender is not colorblind and has brown eyes. They have a colorblind sister named Ellie with blue-green eyes. Their mother, Mrs. Lavender, is not colorblind and has blue-green eyes. Mr. Lavender, who ran off with a "dancer" a few years ago, had hazel eyes.
- One of the Lavender boys was conceived when Mrs. Lavender had an affair with Mr. Gray.

1. On an extra sheet of paper, draw a pedigree for colorblindness that includes everyone above. Label with genotypes.

2. Which of the Lavender boys is only half-brother to the other siblings? How do you know?

3. *Who's the daddy* of Zoe's baby? How do you know?

4. Was Mr. Lavender colorblind? How do you know?

Optional Challenge Pedigree

Visit https://mudcat.org/@displaysong.cfm?SongID=2991. On an extra sheet of paper create a pedigree for the crazy family described there! (You can also view a video version at https://www.youtube.com/watch?v=qu_Y1wQ923g)

Problem to Solve 11.3 Can I Determine if Observed Phenotypic Ratios Match Predictions?

This activity will compare *actual* number of offspring with different phenotypes to the *expected* percentages of phenotypes predicted by Punnett Squares.

This activity will utilize corn cobs. **DO NOT UNWRAP THE COBS**. Each corncob contains many kernels. Each kernel is a seed. A seed is actually an embryo or baby corn plant. So the next time you have corn for dinner, you are eating many babies. (But they are plant babies, so it is OK. Corn is also a fruit, not a vegetable. So are peas and beans. You will learn more about all of that in Lab Exercise 24!)

The inner part of the corn kernel is called the *endosperm*. It is mostly composed of either starch or sugar. Sweet corn kernels contain more sugar, and field corn kernels (typically fed to livestock or used for biofuels) contain more starch. But one cob can have both types of kernels. The difference is easily seen in dried corncobs. Starchy kernels appear smooth and full, where sugary kernels appear wrinkled and deflated. Starchy endosperm (S) is dominant over sugary endosperm (s).

The outer layer of the kernel is called the *aleurone*. It may or may not contain purple pigments. The allele for purple pigments (P) is dominant. The allele for colorless aleurone is recessive (p). If the aleurone does not produce pigments, the kernels will appear yellow to white. Figure 11.15 displays some of the many varieties of corn pigmentation. In this exercise, we will only deal with purple and yellow.

You will be given a cob with varying individual kernels on it. Each individual kernel will be smooth or wrinkled, and it will be purple or yellow. You will collect data by counting frequencies of different phenotypes and compare your actual data with the expected data. You will use a Chi Square test to determine if your results are *significantly different* from your prediction.

Figure 11.15 Corn phenotypes. Corn coloration may be influenced by several alleles.

Title: Genetics of corn

Goal: See if genes for corn kernel color and shape follow Mendelian inheritance patterns.

State null and alternative hypotheses:

H₀: _____ ($p > 0.05$)

H₁: _____ ($p < 0.05$)

Materials:
(Per pair of students)
Ears of corn as directed by instructor (Keep them sealed in plastic!)

Protocol:
Question to answer: Do the individual kernels on a corncob display actual phenotypic ratios that match predicted phenotypic ratios? The kernels were produced by self-pollination of a plant that was heterozygous for both color and shape (SsPp).

1. Determine the expected ratio (or percentage) of traits produced by self-pollination.

   Parent genotypes: _____ x _____

   Possible male gametes (pollen): _____ _____ _____ _____

   Possible female gametes (eggs): _____ _____ _____ _____

Laboratory Exercise 11 Genetics II: Understanding Inheritance

Title

Purpose

Continued from previous page.

Protocol

Punnett Square:
(Dihybrid cross)

Phenotypic ratios predicted by Punnett square:

Phenotype	Number predicted by Punnett square	Percentage
Smooth and purple	/16	
Smooth and yellow	/16	
Wrinkled and purple	/16	
Wrinkled and yellow	/16	

2. Determine *actual* phenotypic ratios. Work with a partner. One person should choose a row of kernels on the cob and state the phenotypes observed while working down that row to the end. Then they should go to the next row and repeat. The other partner should record tally marks in the appropriate boxes of the data table until *at least 100* total kernels have been recorded.

Continued from previous page.

3. Count the tally marks for each phenotype observed. Record the totals in the blanks provided.
4. Add the total of all phenotypes and record the grand total below the data table. If the grand total is not at least 100 kernels, count another row.
5. Divide the total for each phenotype by the grand total to get the percentages of each phenotype. Record the percentages in the Analysis Table.
6. Review Lab Skill 11.2 Using a Chi Square Test to Compare Actual Results with Expected Results.
7. Determine the degrees of freedom (df) for this experiment. There are four categories (smooth and purple, smooth and yellow, wrinkled and purple, wrinkled and yellow) so the df is number of categories minus 1, or 4 − 1 = 3.
8. Use the Chi Square test to compare your actual phenotypic ratios to your expected phenotypic ratios.
    a. Calculate the χ^2 value for each category.
    b. Total the χ^2 values.
    c. Use the Chi Square Values table in Appendix G to determine the p value.
9. State if you accept or reject the null hypothesis and the alternative hypothesis in the conclusions section.

Continued from previous page.

Protocol

Data:

Smooth purple	Smooth yellow
Total: _____	Total: _____
Wrinkled purple	**Wrinkled yellow**
Total: _____	Total: _____

Grand total of all phenotypes: _____

Analysis:

Phenotype	% Observed	% Expected	$\chi^2 = (o-e)^2/e$
Smooth and purple			
Smooth and yellow			
Wrinkled and purple			
Wrinkled and yellow			
		Total of all χ^2	

Approximate p value: _____

Continued from previous page.

Conclusions:

The null hypothesis is assumed valid unless the data produced by the experiment is significantly unlikely to have occurred if the null hypothesis were true. Based on the p value, will you reject the null hypothesis (and therefore accept the alternative hypothesis)? Clearly explain your rationale.

Did the data match one of the predicted parental crosses? If not, what might explain the difference?

Problem to Solve 11.4 Can I Win the Bloody Game?

The Bloody Game

This is an optional activity if time permits. The cards are a good study tool!

How to play:

This is a two-player game. One player must remove the following page from their manual and cut out the eight cards. You will need a score sheet (provided on the back of this page) and a six-sided die. Review the genetics of blood types before playing the game.

Choose who will be the first **Questioner**. The other player will be the **Responder**. When a six is rolled on the die, these roles will reverse.

The *Responder* must roll the die. The *Questioner* must follow the instructions below. When holding up a card, do not let your opponent see the other side or read through the card.

- If a one is rolled, the *Questioner* randomly chooses a card, holds it up so the **phenotype** side is visible to the *Responder* and asks: **What are all possible genotypes?** (In other words, what are all of the genotypes that can produce the phenotype shown?)
- If a two is rolled, the *Questioner* randomly chooses a card, hold it up so the **genotype** side is visible to the *Responder*, and asks: **What is their blood type?** (In other words, what phenotype is produced by the genotype(s) shown?)
- If a three is rolled, the *Questioner* randomly chooses a card, holds it up so the **phenotype** is visible to the *Responder*, and asks **Who can they donate blood to?** (In other words, what are all of the blood types to which this person could safely donate their blood?)
- If a four is rolled, the *Questioner* randomly choose a card, holds it up so the **phenotype** side is visible to the *Responder*, and asks **Who can they receive blood from?** (In other words, what are all of the blood types this person could safely receive if they needed a transfusion?)
- If a five is rolled, the *Responder* may choose a number 1 through 4, and the *Questioner* must randomly choose a card and ask the question for that number.
- If a six is rolled, switch roles! The *Questioner* must hand the cards and score sheet to the *Responder*. The responder must hand the die to the questioner. (If the last roll of the dice was a six, do not switch roles and just roll the dice again.)

After the *Responder* has answered, the *Questioner* must reveal the other side of the card and award points. If the *Responder* answered the question correctly, the *Responder* receives one point, indicated by checking off a blood droplet on the score sheet. If the *Responder* did not answer correctly or failed to give a complete answer, the *Questioner* receives one point.

Shuffle the card back into the deck and roll the die again. The first player to earn 10 points is the winner!

The Bloody Game Score Sheet

Die roll	Show	Action
1	Phenotype	Ask: What are all possible genotypes?
2	Genotype	Ask: What is their blood type?
3	Phenotype	Ask: Who can they donate blood to?
4	Phenotype	Ask: Who can they receive blood from?
5	-	Responder's choice of #1-4.
6	-	Switch roles (unless the last roll was a 6).

Each correct answer earns the *Responder* a point. Each incorrect/incomplete answer earns the *Questioner* a point. Indicate points by checking off the drops of blood. The first to earn 10 drops of blood is the winner!

Name of player 1 _____   **Name of player 2:** _____

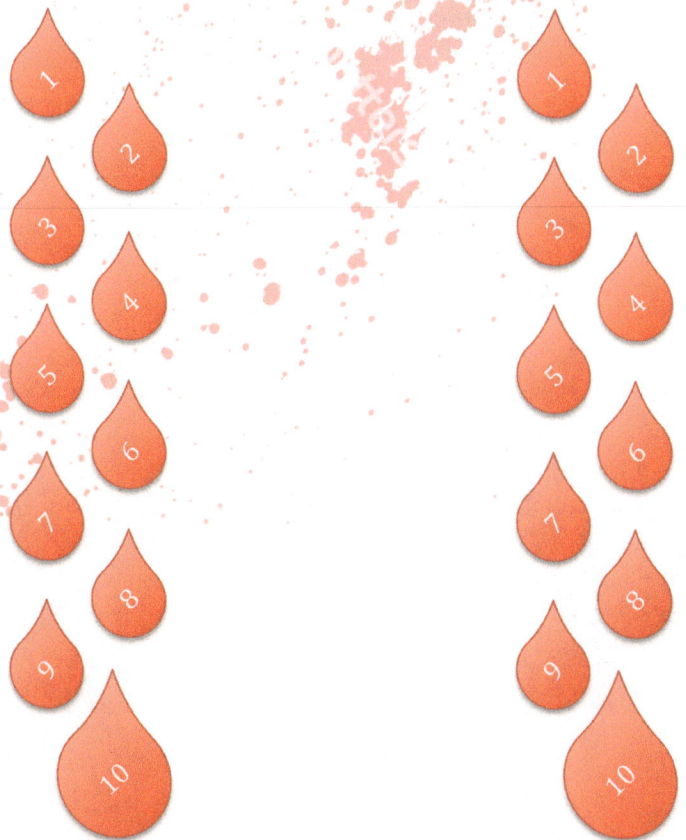

Genotypes
I^AI^Arr, I^Airr

Donates to: A+, A-, AB+, AB-
Receives from: A-, O-

Genotypes
I^AI^ARR, I^AI^ARr, I^AiRR, I^AiRr

Donates to: A+, AB+
Receives from: A+, A-, O+, O-

Genotypes
I^BI^Brr, I^Birr

Donates to: B+, B-, AB+, AB-
Receives from B-, O-

Genotypes
I^BI^BRR, I^BI^BRr, I^BiRR, I^BiRr

Donates to: B+, AB+
Receives from: B+, B-, O+, O-

Genotype
I^AI^Brr

Donates to: AB+, AB-
Receives from: A-, B-, O-

Genotypes
I^AI^BRR, I^AI^BRr

Donates to: AB+
Receives from: A+, A-, B+, B-, AB+, AB-, O+, O- (universal recipient)

Genotype
iirr

Donates to: A+, A-, B+, B-, AB+ AB-, O+, O- (universal donor)
Receives from: O-

Genotypes
iiRR, iiRr

Donates to: A+, B+, AB+, O+
Receives from: O+, O-

Phenotype
A+

Phenotype
A-

Phenotype
B+

Phenotype
B-

Phenotype
AB+

Phenotype
AB-

Phenotype
O+

Phenotype
O-

Name: _____ Date: _____

POST-LAB QUIZ 11: What have you learned?

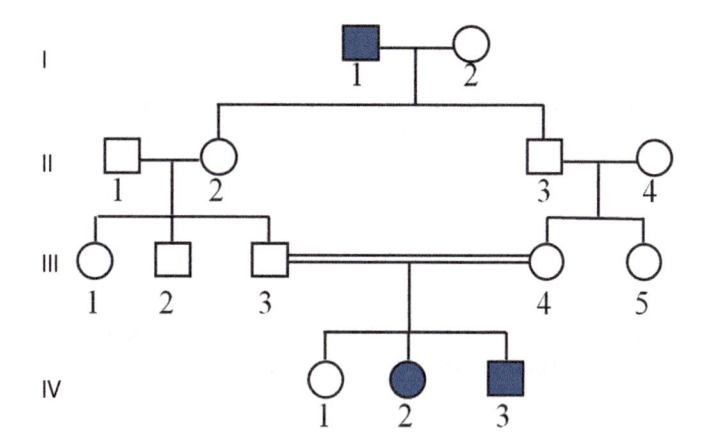

1. Use the pedigree to answer the following questions. Assume B and b represent the dominant and recessive alleles for a rare trait X. Individuals with the trait are indicated by filled boxes on the pedigree.
   a. Is trait X dominant or recessive? Why?
   b. Is the trait sex-linked? Why or why not?
   c. What is the likely genotype of individual II-2? How do you know?
   d. Which two individuals represent male carriers of trait X?
   e. How are III-3 and III-4 related?
   f. Why is inbreeding a bad idea (particularly in terms of genetics)?

2. A biologist's null hypothesis stated that a particular treatment would have no effect. His alternative hypothesis stated that it would have a significant effect. He conducted the experiment and analyzed the results. The p value was 0.1. Which hypothesis should he accept, and which should he reject? Why?

3. After a couple took their newborn to doctor for checkup, the husband noticed in the records that the baby's blood type was O-. He has A+ blood. His wife has B+. He claimed that his wife must have cheated on him and that the baby wasn't his. He said if the baby were his, it would have A+, B+, or AB+ blood. Did she cheat on him? How could you explain the child's genetics to this couple?

4. In the Human Mendelian Genetics activity, could you taste all three of the taste papers (PTC, thiourea, and sodium benzoate)? If so, you are a supertaster! Based on the class results from tasting these test papers, does every human experience the same flavors when we taste food? Provide a *genetic* explanation for different food preferences.

5. Why is it said that men inherit pattern baldness from their *mother's* father, not *their* father? What does that imply about pattern baldness? Why is pattern baldness far more common in men than women?

6. What is the probability that a woman will have a child (of either sex) with hemophilia if both her father and husband have it? (Show how you obtained your answer.)

7. A male black Labrador retriever and a female chocolate lab had puppies. The litter contained 3 black puppies, 2 brown puppies, and 1 yellow puppy (no black or brown pigment). What does this imply about lab coat color? What are the likely genotypes of the parent dogs? (Hint: review *epistasis*.)

8. Although traits like height, hair color, and skin color are genetic, why can't they be easily represented with Punnett squares?

9. Parents Joe and Sandy are expecting a child. Joe has type AB- blood. Sandy has O+ blood and her mother was O-. What are their genotypes? What are the genotypic and phenotypic probabilities for their child? Should Sandy be concerned about blood incompatibility during pregnancy? Why or why not?

10. Who can Joe (AB-) give blood to? If he needed a transfusion, which blood types could he receive?

Laboratory Exercise 12

Microbiology: Viruses, Bacteria, and Protists

Concepts to Understand

After completing this lab, the student should be able to:

1. Differentiate between viruses, bacteria, and protists.
2. Explain the characteristics and life cycle of viruses.
3. Describe the characteristics and classification of bacteria.
4. Describe the major features of various clades of protists.
5. Explain the role of microbes in ecology, food production, and medicine.

Skills to Learn

After completing this lab, the student should be able to:

1. Visually identify common microbes.
2. Use aseptic technique to culture microbes.
3. Use the gram stain to identify bacteria.

SEE ALSO:

Campbell Biology—Chapter 19, Section 19.1-19.2; Chapter 27, section 27.1, and 27.3; Chapter 28.

JoVE (video)—Decontamination for Laboratory Biosafety

YouTube—Gram Stain HD Animation

Others:

Important terms to know:

Agar	Endospore	Primary stain
Antibiotic	Envelope (viral)	Resistance (antibiotic)
Antiseptic	Filamentous	Rhizoid
Autoclave	Gram stain (pos./neg.)	Safranin
Binary fission	Helical	SARS-CoV-2
Capsid	Icosahedron (20)	Secondary stain
Chemoautotroph	Lobate	Selective media
Chemoheterotroph	Lysogenic	Sterilize
Counter stain	Lytic	Symbiosis (if not elsewhere)
Crystal Violet	Microbiota	Ubiquitous
COVID-19	Mordant	Umbonate
Culture	Pathogen	Undulate
Cyanobacteria	Peptidoglycan	Vector
Differential media	Photoautotroph	
Disinfectant	Plasmodium	

Pre-lab Preparation and Background Information

Microbiology is the branch of biology that deals with life at the microscopically invisible scale. Mankind was unaware of microbes until Anton Leeuwenhoek, in the late 1600s, invented the first simple microscope. Even then, it took another two hundred years before doctors, scientists, and biologists began to understand microbes, and especially the connection between microbes and diseases. Microbiology is a fairly new area of biology, and there is still a great deal of the microbial world yet unexplored.

Before the developments of the late nineteenth and early twentieth centuries, it was commonly held that mystical forces, unbalanced *humors* (body fluids), or *miasma* (bad air) caused diseases. Doctors did not wash their hands or medical instruments before or after surgery, and did not bother to clean wounds. Doctors even wore their unwashed, bloodstained, and soiled surgical smocks with pride, as they believed the amount of soiling displayed their level of experience.

The history of microbiology provides a nice example of how science evolves. In the late 1800s, several important researchers and doctors made discoveries that revolutionized biology and medicine. Louis Pasteur (French, 1822-1895), Robert Koch (German, 1843-1910), and Joseph Lister (English, 1827-1912) are typically credited as the major founders of "modern microbiology."

Pasteur was the first to clearly demonstrate the fallacy of *spontaneous generation*, the idea that life could arise from nonliving sources. He showed that if broth was sterilized and closed in a sealed container, it would not spoil (Figure 12.1). He suggested that killing microbes by heat, chemicals, or filtration could prevent rot and spoilage. Among his many discoveries, he developed some of the first methods of killing microbes we now commonly refer to as *pasteurization*. His methods revolutionized food production and storage.

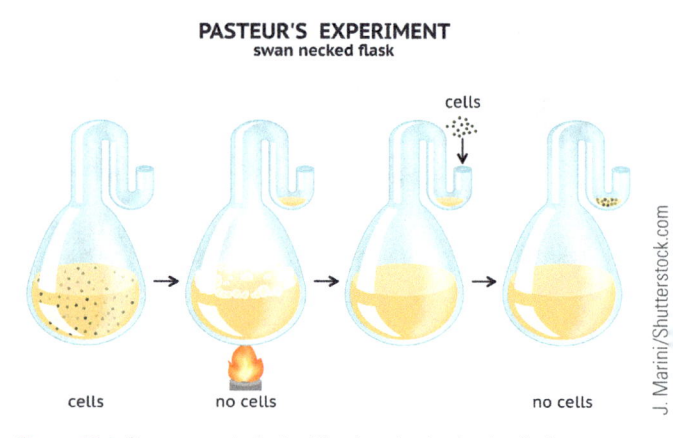

Figure 12.1 Pasteur revolutionized food production by developing pasteurization. He showed that heat, chemicals, and filters could rid food of microbes that caused milk and foods to spoil.

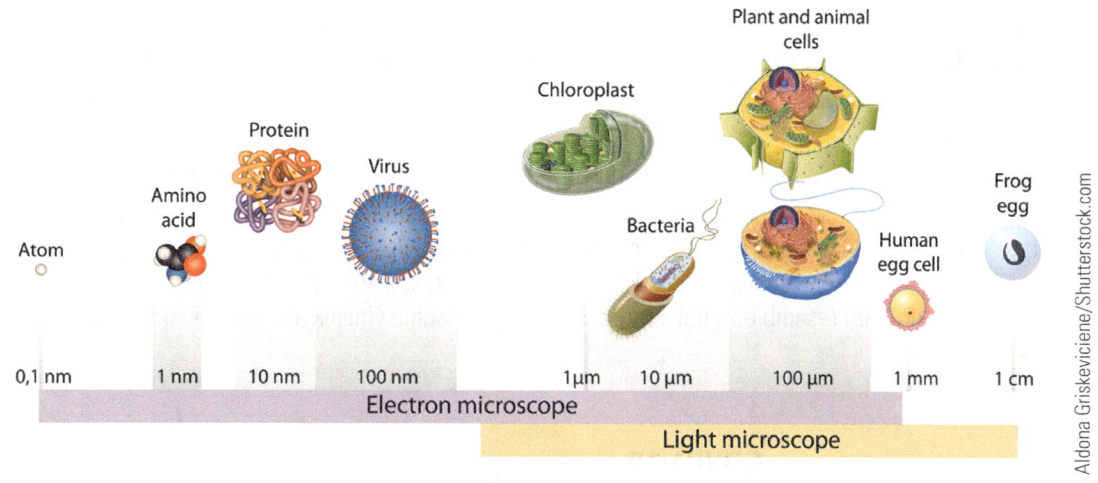

Figure 12.2 Viruses are little more than complex particles composed of proteins and nucleic acids. Bacteria are prokaryotic cells generally hundreds of times larger than viruses. Protists and other eukaryotic cells are tens to hundreds of times larger than bacteria. Note that as you move from left to right across the X-axis of the figure, the scale is logarithmic. Each unit is ten times larger than the previous one.

At about the same time, Koch showed that bacteria and other microbes are **pathogens**, or disease-causing agents. He demonstrated that bacteria caused specific diseases such as tuberculosis and cholera, not bad air or unbalanced humors. He isolated microbes from diseased animals, introduced them into other animals, and showed that the microbes were the disease-causing agents. He is typically credited as the founder of modern bacteriology. In England, Lister, a surgeon, was aware of Koch and Pasteur's work. He suggested that doctors would have much greater success keeping their patients alive if they washed their hands and used aseptic techniques. He was mocked, ridiculed, and dismissed by the doctors of the day, but eventually his theories were accepted and he is today credited with being the "father of modern surgery."

In the late 1800s, two researchers independently discovered that particles smaller than bacteria could cause disease. Dmitri Ivanovsky (Russian, 1864-1920) identified infectious particles that could pass through filters that blocked bacteria. A few years later, Martinus Beijerinck (Dutch, 1851-1931) verified these findings and named these infectious particles viruses. Ivanovsky and Beijerinck are considered two founders of modern virology.

Since these revolutionary discoveries in the late 1800s, the field of microbiology has grown and spread. Discoveries in microbiology have been applied to fields as diverse as treating infectious diseases, immunology, agriculture, food production, ecology, biofuel development, and genetics. And there are many areas of microbiology yet unexplored.

This lab will survey the three fundamental types of microbes: viruses, bacteria, and protists. They are generally invisible to the naked eye unless many collect in one place, such as a bacterial bloom, or they form colonies or simple multicellular structures, such as some photosynthetic protists. Other organisms such as fungi and microscopic animals are sometimes considered microbes due to their size, but they will not be covered in this lab exercise (Figure 12.2).

Viruses

Most people are aware that bacteria are **ubiquitous** (found almost everywhere). However, viruses are as ubiquitous as bacteria. In fact, some researchers suggest that *bacteriophages* (viruses that infect bacteria) outnumber all organisms alive on the planet combined, including bacteria. Viruses, however, are not truly alive. Recall that living organisms are composed of cells, contain DNA, metabolize, reproduce, respond to their environment, and maintain homeostasis. Viruses are incredibly small and simple, and they have few of these characteristics of life. Some viruses are smaller than a ribosome. Viruses must hijack a host cell, and the host cell does all of the work. Recall that a virus is little more than a protein container, or **capsid**, around a fragment of DNA or RNA. Some also have an external **envelope** of phospholipids. They are considered infectious particles, not cells. They are too small to see with a light microscope, so we will not go into great detail on them in this lab. In fact it wasn't until electron microscopes were developed in the 1930s that we could actually see viruses at all. But they are an important part of the microbial world. Humans tend to think of viruses only as disease agents. There are many diseases caused by viruses, but viruses tend to infect specific kinds of cells. Most viruses in nature infect bacteria and plants, not humans. But when a virus mutates or crosses from one species to another, as has recently happened with the *severe acute respiratory syndrome coronavirus 2* (**SARS-CoV-2**), the virus that causes **COVID-19**, the results can be catastrophic (Figure 12.3).

Most viruses exist in one of three basic shapes, based on the arrangement pattern of the proteins in the capsid. **Helical** viruses have a capsid composed of a tubular spiral of proteins that surround a DNA or RNA nucleic acid. A *polyhedral* (many sided) virus has a capsid in the form of a 20-sided **icosahedron**. Spherical viruses have an outer viral **envelope** composed of phospholipids "stolen" from the host cell as the virus exited the cell. This outer envelope allows the virus to more easily bind to future host cells. The SARS and influenza viruses are both enveloped. The bacteriophages have a more complex structure that resembles a helix topped by a capsid. Some viruses also contain a few enzymes (Figure 12.4).

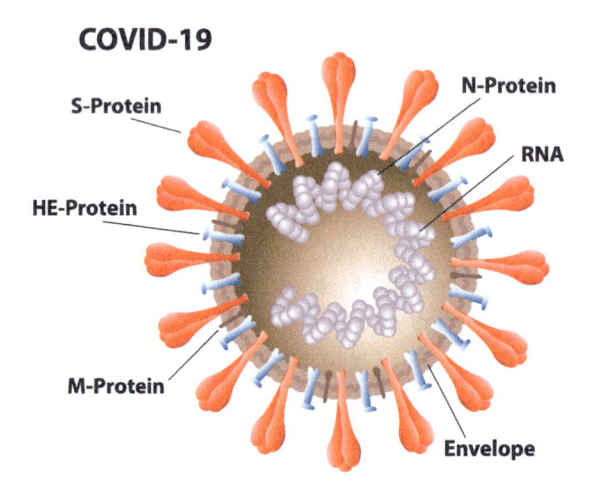

Figure 12.3 The SARS-CoV-2 virus that spread across the planet is thought to have originated from bats to humans in China. SARS viruses are not particularly unusual, but this strain has been particularly dangerous and contagious.

Figure 12.4 Virus shapes. Viruses are simply protein capsids, nucleic acids and in some cases an outer envelope of phospholipids.

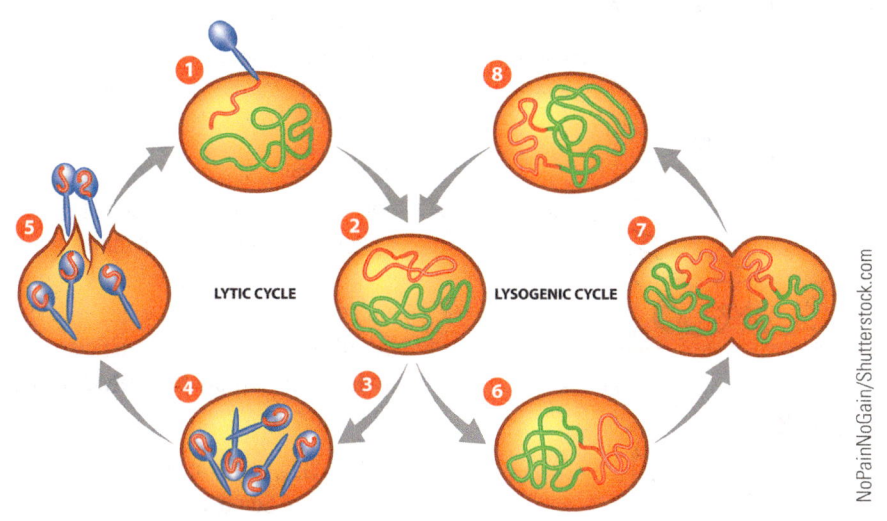

Figure 12.5 Lytic and lysogenic cycles of viruses. Viruses may spend time in the lysogenic cycle, hidden in the host cell, until a later outbreak. Lytic cycles occur when the host cell makes many copies of the virus until it bursts.

Viral activity is very limited. Viruses have no organelles. They can't act independently or reproduce outside a host cell. Most viruses can't last long outside a host. Heat, ultraviolet rays or sunlight, and chemical agents like bleach, acids, and bases will destroy most viruses. But if an intact virus comes in contact with a cell, the cell may take in the virus and the DNA or RNA of the virus will become part of that cell's "recipe book" for making proteins. The cell may either incorporate the viral nucleic acids into its own DNA, or it will use it to create copies of viral proteins and nucleic acids. The **lytic cycle** of a virus occurs when the cell becomes a virus "factory" and bursts, releasing the viruses to spread to other cells. The **lysogenic cycle** occurs when the host cell incorporates the nucleic acids until a later time (Figure 12.5). Cells with the incorporated viral DNA can reproduce and pass on the virus to daughter cells. Some lysogenic viruses can never be completely eliminated and produce periodic outbreaks. Viruses are typically classified by their nucleic acid content. Viruses may contain single- or double-stranded DNA or RNA.

Bacteria

Bacteria are found almost everywhere. Soil contains millions of kinds of bacteria, viruses, fungi, and other microbes. Even a handful of dirt may contain a complex, active, but invisible ecosystem. In fact, the unseen battle between bacteria and *bacteriophages* has resulted in bacterial defense strategies such as the *restriction enzymes* and *Cas9 enzymes* discussed in Lab Exercise 7. Bacteria are also in any open water, the air, and on most surfaces. They are in and on your body. In terms of numbers of cells, you have more bacterial cells in and on you than human cells by a factor of about 10, so by cell count, you are more bacteria than human. If you gain a little weight, blame it on your bacteria. Just kidding! Bacteria are much smaller than your eukaryotic cells. They do outnumber your cells by 10 to 1, but their contribution to your overall mass is minimal. Many of the bacteria that live in and on you are beneficial. In fact, current research in the human **microbiota** (microbes that live in the digestive tract) is revealing the importance of these microbes to human health. When these internal microbial populations are disrupted by oral medications, unusual eating habits, chemotherapy, or "cleanses," it may take months or years to reestablish them.

Unlike viruses, bacteria are true organisms. They are simple, but they exhibit all of the qualities of life. Recall from Lab Exercise 5 that bacteria make up the two prokaryotic domains of life, *Archaea*, and *Eubacteria* (or Bacteria). Recall that bacteria may generally described by their shape. Bacterial shapes include *coccus* (spherical), *bacillus* (rod), and *spirillum* (spiral). The prefix *strep* describes bacteria that form chains, the prefix *staph* describes bacteria that form clusters, and *diplo* and *tetra* describe pairs or groups of four cells, respectively. You may find it helpful to consult Lab Exercise 5, especially Figure 5.2, to review bacterial shapes.

Also recall the structures of a bacterial cell. Typical bacteria contain a nucleoid region of DNA that contains its circular chromosome. They reproduce by **binary fission**, a simple process whereby they copy their circular chromosome and split

BINARY FISSION

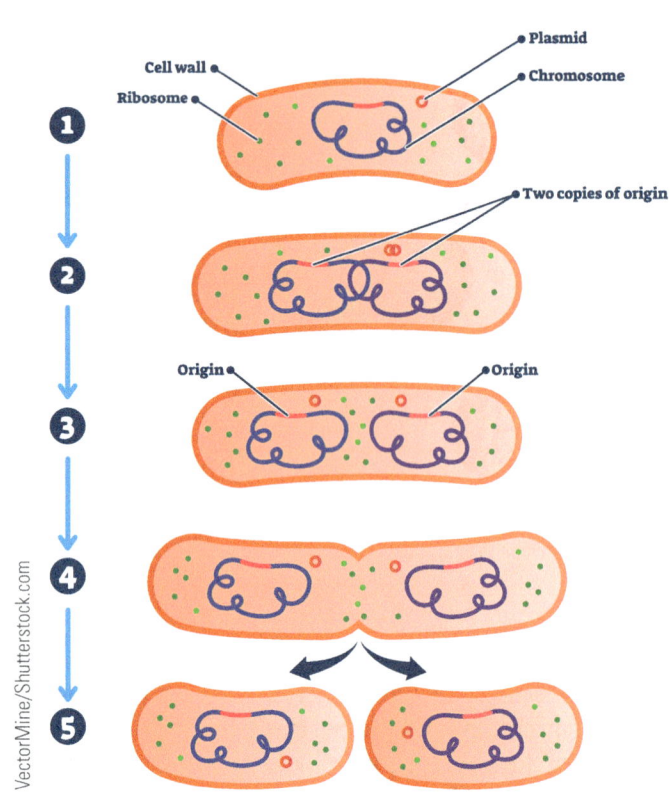

Figure 12.6 Bacteria can split by binary fission. The chromosome and plasmid are replicated and the cell splits into two identical clones.

into two new cells (Figure 12.6). Bacteria also contain ribosomes, plasmids, cell walls, and sometimes other structures such as pili, flagella, and a capsule. (See Lab Exercise 5, Figure 5.1). However, in order to really classify bacteria we need to go beyond the external shape.

Prokaryotes are classified by a number of characteristics. Cell shape alone is not enough to distinguish one species of bacteria from another. One major way bacteria are classified is their method of obtaining energy and metabolizing food molecules. Bacteria may be *obligate aerobes* (must have O_2), *obligate anaerobes* (are poisoned by O_2), or *facultative anaerobes* (can switch between aerobic and anaerobic environments).

Additionally, bacteria are classified by their source of energy and carbon. Some archaea bacteria are **chemoautotrophs** and are able to use chemicals as an energy source to utilize inorganic carbon. Some bacteria, like plants, are **photoautotrophs.** They use sunlight as an energy source to turn an inorganic carbon source like carbon dioxide into food molecules. These are called **cyanobacteria** (Figure 12.7). Two common examples are *Anabaena* (Figure 12.8) and Oscillatoria (Figure 12.9). Cyanobacteria may become a problem when they overgrow and choke out life in lakes and ponds. This is usually a result of overuse of fertilizers that are carried into bodies of water by rain and runoff (Figure 12.10).

CYANOBACTERIA

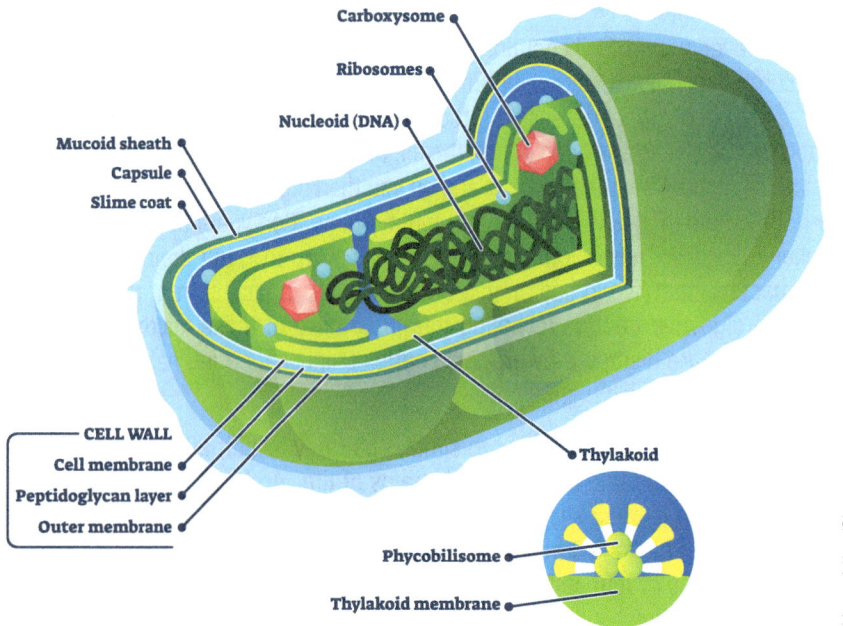

Figure 12.7 Cyanobacteria. The structure of a cyanobacteria in many ways resembles the chloroplasts of plants.

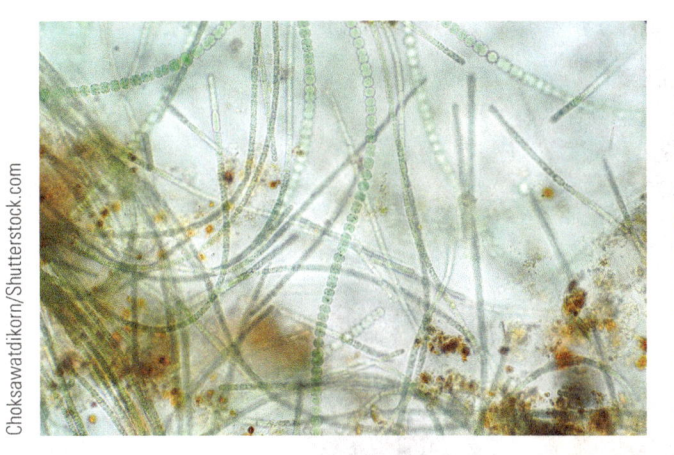

Figure 12.8 *Anabaena*, a common cyanobacteria. Note that cells do not exhibit a nucleus. These are colonial bacteria, not to be confused with plants or algae.

Figure 12.9 *Oscillatoria*, another common cyanobacteria. The long strands are composed of many side-by-side prokaryotic bacterial cells.

Figure 12.10 An overgrowth of cyanobacteria producing thick, green water. This can make it difficult for other aquatic organisms to survive.

Some older texts refer to cyanobacteria as *blue-green algae*. This is an obsolete (and misleading) term, as bacteria are not algae. The different types of algae are all eukaryotic protists, not bacteria. To prevent confusion, the term *blue-green algae* should be avoided.

Most non-photosynthetic bacteria are **chemoheterotrophs**. Like humans, they must find and ingest organic food molecules. But because they are simple, many bacteria are limited to which types of food molecules they can digest. Microbiologists capitalize on the metabolic requirements of bacteria to identify them. Unknown bacteria can be **cultured** (intentionally grown) on a nutrient-rich medium, and then spread on different kinds of media to further identify them (Figure 12.11).

Bacteria live virtually everywhere in the environment, but they also live in and on plants and animals. Many are harmless or even beneficial to a host. But some cause diseases. **Antibiotics** are medications that specifically treat bacterial infections. Antibiotics typically interrupt metabolic activities or damage bacterial cell structures, so they do not treat viral infections. But overuse of antibiotics has led to **resistant** strains of bacteria. When antibiotics are used indiscriminately, or prescribed inappropriately, they serve to speed bacterial adaptation. They act as a selecting agent for bacteria that can survive the antibiotic. Some antibiotics that used to be quite effective against bacterial infections now have a much weaker effect as bacterial strains have adapted to the antibiotics. Therefore, it is necessary to use the appropriate antibiotic to completely eradicate a bacterial infection rather than just wipe out the weakest and leave the strongest.

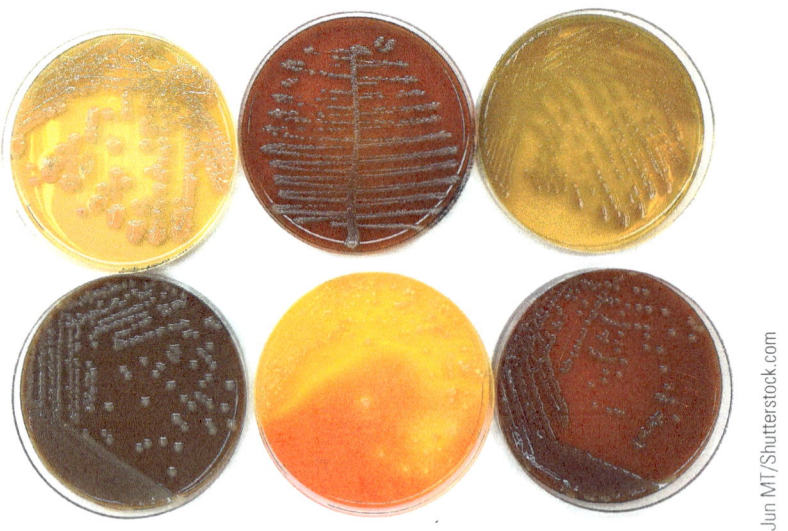

Figure 12.11 Petri dishes containing different types of nutrients for growing bacterial colonies.

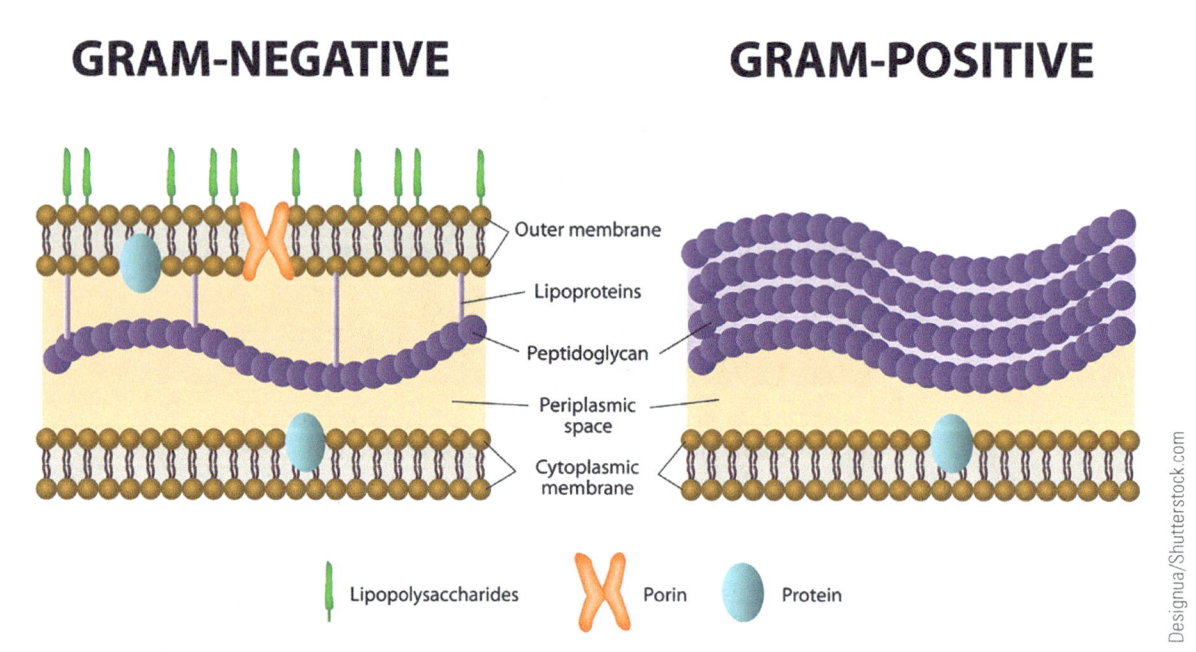

Figure 12.12 The cell wall structure of gram-positive and gram-negative bacteria. Gram-negative bacteria have a double membrane cell wall and less peptidoglycan. Gram-positive bacteria contain a thick peptidoglycan layer.

One of the most important ways to identify bacteria, particularly in a medical context, is a **Gram stain**. The Gram stain was developed by Hans Gram in the late 1800s to categorize bacteria based on their cell wall structure. Most bacteria contain some amount of **peptidoglycan** (a peptide-sugar compound) in their cell wall. The Gram stain will stain those bacteria with larger amounts of peptidoglycan a purple color. They are considered *gram-positive* bacteria. Other types of bacteria have a double-membrane structure and less peptidoglycan in their cell wall. When stained with the Gram stain they appear pink. These are considered *gram-negative* bacteria (Figure 12.12). This is important medically because certain antibiotics interact with peptidoglycan. Antibiotics that disrupt the peptidoglycan layer are far more effective on gram-positive bacteria than gram-negative. Under a microscope, Gram stained bacteria can be identified by their color. Figure 12.13 shows dark purple gram-positive bacteria on the left. The right image shows pink gram-negative bacteria.

Protists

All protists are eukaryotic, and most are unicellular organisms. They are much larger than bacteria and viruses. We typically like to categorize objects or organisms by their properties. But protists are frustratingly difficult to classify in this way. In fact, it is easier to classify protists by stating what they are *not*. The protist kingdom is incredibly diverse, and the classification of organisms in this group is in flux. Some are photosynthetic, or plant-like. Others are heterotrophic and mobile, much like tiny animals. Others have much in common with fungi. Some live in water, some in soil, and some are parasitic with complex life cycles. Protists are incredibly diverse, which is why their classification is problematic.

Protists are considered the simplest eukaryotic organisms. Protists are thought to have evolved from bacteria, including some of their organelles. Mitochondria and chloroplasts have much in common with bacteria. The **endosymbiotic theory** hypothesizes that an early eukaryotic cell engulfed an aerobic bacterium, and rather than digest it, utilized it as an organelle. Eventually the same occurred with a photosynthetic bacterium (Figure 12.14). These "adopted" bacterial components became the mitochondria and chloroplasts observed in modern protists. Modern plants, animals, and fungi all share these organelles, suggesting a pattern of evolution that explains the consistency of cellular structures across the diverse groups within Domain Eukarya.

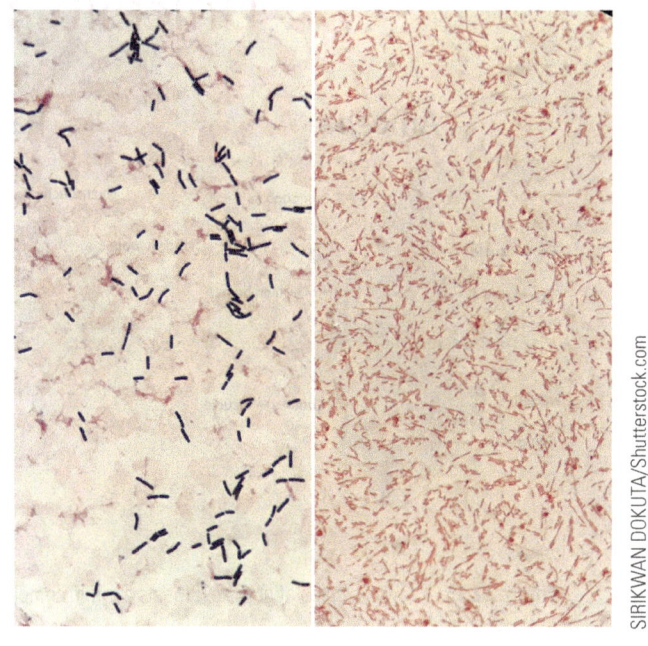

Figure 12.13 Gram-stained bacteria. Gram positive bacteria stain purple (left), and gram-negative bacteria stain pink (right).

Protists have complex organelles and structures that are absent in prokaryotes. Intracellular membrane-bound structures allow protists to compartmentalize cellular activities. Protists developed many unusual characteristics and strategies for survival, and they exist in virtually all aquatic and some terrestrial environments. All protists are eukaryotic. Most are unicellular, but some are colonial, and few, like brown algae (or kelp) are considered multicellular, though too simple to be grouped with the plants. Figure 12.15 illustrates a few common examples of aquatic protists found in lakes, streams, and ponds all over the world.

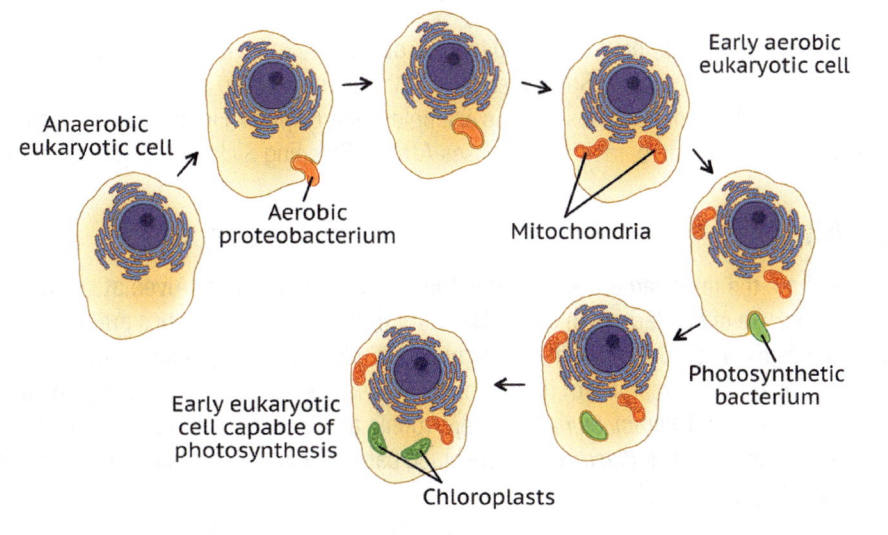

Figure 12.14 The endosymbiotic theory. This theory suggests that ancient eukaryotic cells engulfed bacteria, incorporated them, and utilized them as organelles.

Unicellular Organisms

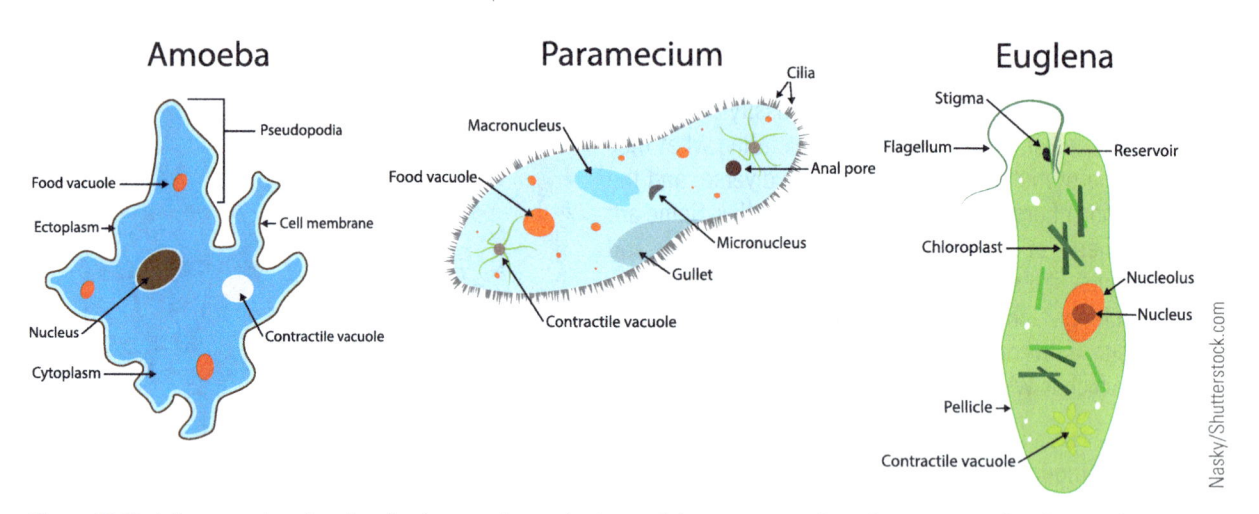

Figure 12.15 A few examples of protists. Protists contain complex intracellular structures and membranous organelles that are absent in prokaryotes.

As humans we tend to focus on terrestrial organisms, and those we can see with the naked eye. But microbes far outnumber multicellular organisms, and protists outnumber all of the other multicellular eukaryotes. Protists are as diverse as they are numerous. Some are photosynthetic, and generally referred to as *algae*. Protists serve as the basis of the marine food chain. The abundant marine photosynthetic protists are generally called *phytoplankton*, and marine heterotrophs are generally called *zooplankton*. However, these are loose terms and many protists do not easily fit either category. Some protists are both heterotrophic and autotrophic. Even though they are unicellular, they often exhibit amazing complexity. Some produce tiny shells, and others have light sensitive eyespots. Since protists are so diverse, it is difficult to state exactly what makes a protist a protist. Instead, it is easier to define a protist as a eukaryotic organism that is *not* a plant, animal, or fungus. As biologists have learned more about the genetic and molecular makeup of protists, the kingdom Protista is being abandoned. Different hypotheses have been suggested for classifying protists, but to date no definite system has been adopted. Instead, protists are grouped into four *supergroups*, each containing *clades* that best reflect their evolutionary origins.

Supergroup Excavata

This group is composed of protists that have a feeding groove, or "excavation," that goes down the side of their cell. They also have unique flagella and mitochondria. Some members of this diverse group are photosynthetic, some are heterotrophic, and some are parasitic. A common photosynthetic example is the *Euglena* (Figure 12.16a). A parasitic member of this group is *Trypanosoma*, the bloodborne parasite that causes African Sleeping Sickness (Figure 12.16b).

Supergroup SAR

The name SAR comes from the three large clades within this group: Stramenopila, Alveolata, and Rhizaria. These clades are grouped together because of similarities in their DNA, but their other characteristics are highly diverse, and the classification of these organisms is still under debate. This group includes the glassy, silica-shelled diatoms (Figure 12.17a). Ceratium and paramecium are also examples of this supergroup. *Ceratium* are a type of phytoplankton that appear to have flexible spikes (Figure 12.17b), and *Paramecium* are heterotrophic and covered with cilia (Figure 12.17c). An unlikely but important member of this supergroup is *Plasmodium*, the blood parasite that causes malaria. It is so small it can enter and infect human blood cells (Figure 12.17d).

Supergroup Archaeplastida

The third supergroup, Archaeplastida, is composed of red and green algae, and, since it is a supergroup and not a kingdom, it also includes land plants. The classification of this supergroup is somewhat odd in that it includes members of what used to be kingdom Protista and members of the kingdom Plantae. Examples of this group include *Spirogyra* (Figure 12.18a) and *Volvox* (Figure 12.18b). Both are photosynthetic and colonial. They and other similar types of green algae are thought to be the precursors to land plants. Archaeplastids serve as the base of many aquatic food chains.

Supergroup Unikonta

This group includes ameboid protists thought to be most like the ancestors of animals and fungi. Like the Archaeplastida supergroup, the Unikonta supergroup includes protists that resemble animals and fungi, as well as Kingdom Animalia and Kingdom Fungi. Unikonts move using **pseudopodia,** extensions of their cytoplasm (Figure 12.19).

VISUAL SURVEY OF THE PROTISTS
Supergroup Excavata

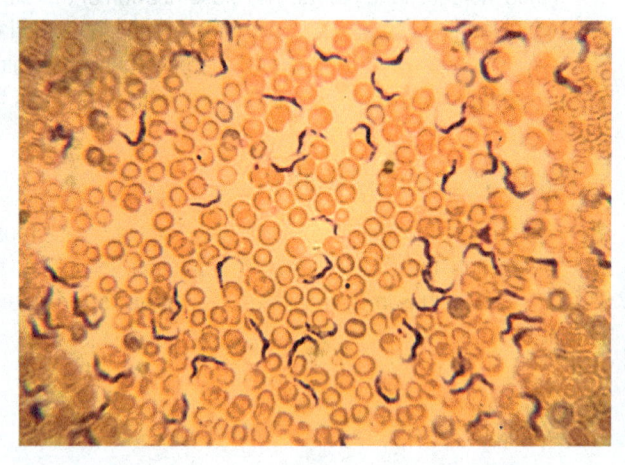

Figure 12.16 Members of supergroup Excavata. The photosynthetic *Euglena* (a) is a common aquatic protist. The blood parasite *Trypanosoma* causes African Sleeping Sickness, visible here as the purple worm-like cells among the red blood cells (b).

Supergroup SAR

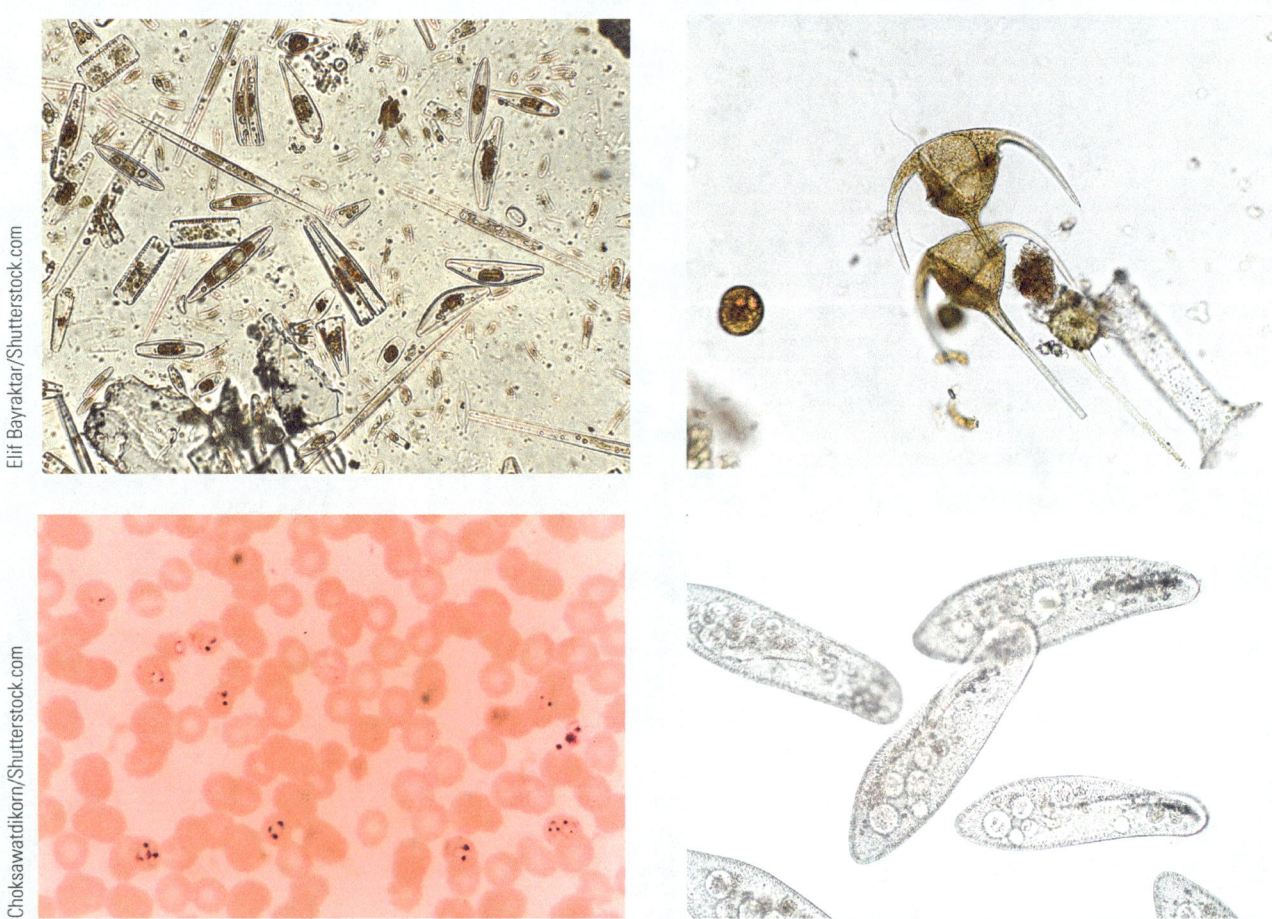

Figure 12.17 Examples of supergroup SAR. Diatoms come in many shapes and sizes, and have a two-part, glass-like silica shell (a), and *Ceratium* are common, spiky, photosynthetic protists (b). *Paramecium* are fast-moving, heterotrophic, ciliated, protists (c), and *Plasmodium*, the dark purple specks, are tiny blood parasites that cause malaria (d).

Laboratory Exercise 12 Microbiology: Viruses, Bacteria, and Protists 353

VISUAL SURVEY OF THE PROTISTS
Supergroup Archaeplastida

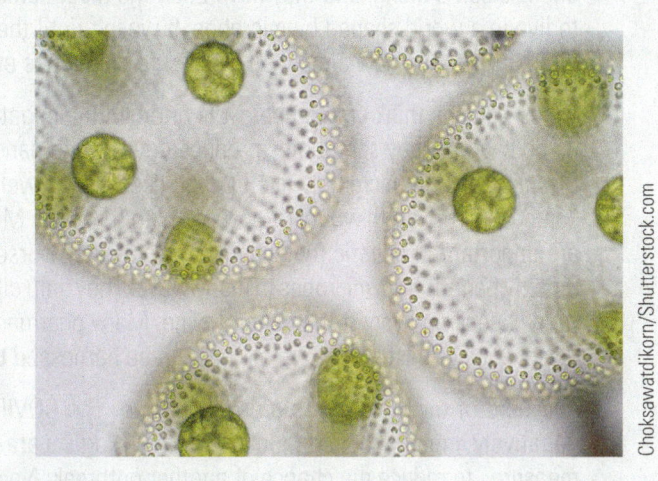

Figure 12.18 Examples of supergroup Archaeplastida. Spirogyra (a) is easily recognized by the spiral pattern of chloroplasts in the long colonies of cells joined end to end. Volvox (b) forms spherical colonies with daughter colonies forming within the sphere.

Supergroup Unikonta

Figure 12.19 Example of supergroup Unikonta. This *Amoeba proteus* resembles a splatter of goo. The paramecium (green) may become the amoeba's meal.

The Importance of Microbes

Viruses, bacteria, and protists are easy to dismiss or forget about since they are virtually invisible to the naked eye. But microbes play an incredibly important role in biology. This planet belonged to microbes well before multicellular plants and animals came along, and they have been the most numerous organisms on the planet since life began. Humans evolved to live in a world shaped by microbes. It wasn't until the evolution of cyanobacteria that this planet even had an oxygen atmosphere. And now almost all life on the planet has evolved to live in an oxygen-rich environment.

When humans think of microbes, it is often with a negative perspective. Microbes do cause disease, spoil food, and give us body odor. They sometimes produce devastating pandemics that kill or harm thousands, or even millions, of humans. But microbes are incredibly useful and necessary as well. Many microbes decompose and recycle nutrients. Without decomposers, the earth would be covered with the dead. Microbes keep much of the natural ecosystem in balance. Microbes are important in aquatic food chains. For better or worse, microbial diseases tend to keep populations in check. Humans directly benefit from microbes in many ways. They help digest our food, and they are used in many areas of food production, from cheese and yogurt to beer and wine. Many pharmaceuticals were discovered from microbial compounds. And recent discoveries of bacterial enzymes that can be harnessed to manipulate DNA have revolutionized genetics.

The field of microbiology is an exciting one. The COVID-19 pandemic is a contemporary lesson on how microbes can negatively impact the world. But the microbiologists are on the front line, developing treatments, cures, and preventative measures to reduce the chance of another outbreak. Now, more than ever, the world needs skilled microbiologists.

Laboratory Exercise 12 Microbiology: Viruses, Bacteria, and Protists   355

Lab Skill 12.1 Preventing Microbiological Contamination

Microbes are too small to see, and everywhere. So what can we do to reduce their spread, especially when we grow them on purpose in a microbiology lab? This is an important question, since many microbes cultured in a lab are potentially dangerous to humans. The solution is to follow **aseptic techniques**, or protocols designed to reduce contamination. Contamination can occur in multiple ways. Microbes may contaminate a cell culture in the lab or food in your refrigerator. But microbes can also contaminate laboratory surfaces if a container of microbes is spilled or broken. Contamination may also include microbes spreading to our own bodies. If we fail to take proper precautions, microbes in the lab can make us deathly ill.

Practical Rules for Reducing Contamination
- Keep surfaces and glassware clean, uncluttered, and organized.
- Clean surfaces with appropriate antimicrobial chemicals.
- Wear PPE, especially gloves and eyewear. When removing protective equipment, be careful to avoid cross-contaminating other surfaces.
- Label petri dishes, test tubes, and any cultured media well to ensure anyone can recognize what it is.
- Dispose of bio-hazardous waste, especially microbial cultures, in bio-hazard bags that will be appropriately handled.
- Autoclave lab equipment and biological waste to destroy microbes. An **autoclave** is a piece of laboratory equipment that uses alternating heat and pressure to kill microbes without destroying the glassware or instruments placed in it (Figure 12.20). It may be small enough to fit on a countertop, as large as a refrigerator (Figure 12.21). In industrial settings, autoclaves may be the size of a room.
- Read safety warnings on and follow safety precautions for all bio-hazardous materials.
- Be aware of damaged equipment including seals and other protective barriers intended to block the spread of microbes. Report and replace damaged equipment immediately.
- Do not pour liquids containing bio-hazardous organisms down the drain.
- Never eat or drink in the lab.
- Avoid touching your face or head while wearing contaminated gloves.
- Use common sense!

Figure 12.20 Autoclaves sterilize instruments and glassware.

Chemical Agents Used to Prevent Contamination

We often utilize a number of chemicals to reduce or eliminate microbes, but some are more effective than others. It is important to know what the labels on antimicrobial agents actually mean.

Disinfectant—a chemical that eliminates dangerous or disease-causing microbes on *non-living* surfaces. However, many bacteria can produce endospores (dormant forms of the organism) that are highly resistant to disinfectants. Disinfectants do not kill all microbes.

Antiseptic—a chemical that eliminates microbes on *living surfaces*, such as an antiseptic ointment used on a wound or during surgery. Antiseptics are typically not as strong

Figure 12.21 Larger autoclaves can process larger pieces of lab equipment and more material at once. Since they are pressurized when operating, they must be sealed with a cranking handle that makes them look like a bank safe.

or as effective as disinfectants since they could otherwise harm the living tissue to which they are applied. Disinfectants should not be applied to living tissue, since they can be more concentrated and harmful to living cells.

Antimicrobial—a general term for compounds that kill or inhibit microbes.

Antibiotic—a compound often derived from a fungus that is absorbed into the body to kill bacteria. Antibiotics work to *kill* bacteria (*bactericidal*) or to slow their growth (*bacteriostatic*). Antibiotics are generally designed to interfere with metabolic activities of bacteria, or damage bacterial cell walls. For this reason, they are typically ineffective against viruses or non-prokaryotic organisms. Some antibiotics are very specific to certain types of bacteria.

Sterilization—eradicate all microbes completely. This is typically done with heat, strong chemicals such as bleach, intense ultraviolet light, and different forms of radiation. Microbes are hard to kill, and what damages a microbe can also damage other kinds of cells. Sterilizing techniques should be used with caution.

In microbiology research and in medicine, there is a fine line between preventing contamination and promoting bacterial resistance. It takes common sense and thoughtful laboratory practices to reduce risk of infection and contamination to humans while at the same time preventing the emergence of new resistant "superbugs."

Lab Skill 12.2 Culturing Microbes in a Petri Dish

Microbes, especially bacteria, are often grown in shallow petri dishes containing a thin layer of some type of nutrient media (Figure 12.22). The bottom of the petri dish is typically filled with **agar** (a version of the same *agarose* gel used for electrophoresis) that contains nutrients and sometimes other compounds. Sometimes microbes may be grown in test tubes containing agar or liquid broth.

By choosing the types of nutrients included in the agar, microbiologists can grow specific microbes that utilize those nutrients and inhibit the growth of others. Sometimes the agar is infused with compounds toxic to certain types of microbes but not others. The agar may also contain *indicators* that can show chemical changes caused by the metabolic activity of certain microbes. Some agar even contains blood cells that can indicate the presence *hemolytic* (blood cell bursting) bacteria.

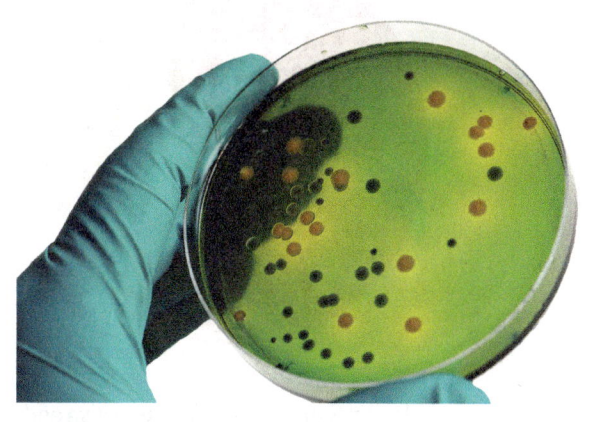

Figure 12.22 A petri dish containing bacterial colonies growing on TCBS agar. This particular green media supports the growth of *Vibrio* bacteria, a strain of bacteria that can infect animal intestines. The color and appearance of the colonies on the media help further identify the strains.

There are many types of growth media. A nutrient-rich medium contains a variety of nutrients that most microbes will consume. This type is ideal for cultivating a wide variety of microbes or microbes of unknown types. A **selective medium** contains compounds that inhibit some microbes, or contains a specific type of nutrient that encourages the growth of microbes that utilize that nutrient. A **differential medium** contains compounds that produce visual clues to help identify the types of bacteria present. And some complex types of media can be both differential and selective. Microbiologists routinely use many types of media to identify and culture specific microbes. It is beyond the scope of this course to describe them all, but Figure 12.23 provides a sample of the colorful variety of media used in the microbiology laboratory.

Identifying characteristics of colonies

The type of media on which a bacteria grows is an important first step in determining the species of a bacteria. But individual colonies have identifiable characteristics as well. In Figure 12.24, microbes collected from an air filter were cultured on nutrient agar. The colonies, which appear as dots or circles, are the offspring of an individual microbe that has been allowed to grow on the plate during an incubation period. Typically a single microbe can produce a colony like those shown here within a few days. If allowed to continue to grow, the microbes will eventually spread out over the entire plate, depending on the available nutrients, temperature of incubation, and interspecies competition.

Note that the individual colonies vary in color, texture, and shape. Some colonies have a smooth **margin**. (In biology, *margin* is used to describe the edge of something, like you might describe the margin of a paper.) Some colonies have irregular, **lobed** (with rounded projections), or **undulate** (wavy) margins. Although it is difficult to see from directly above a colony, some colonies are raised, some are flat, and some are **umbonate**, with raised centers. Some **crenelated** colonies appear to have curly layers. Colonies that look "fuzzy" are **filamentous**, and may be fungal rather than bacterial. Some **rhizoidal** colonies form branching extensions. All of these characteristics also help microbiologists determine microbial types. See Figure 12.25 for illustrations of the common morphological terms.

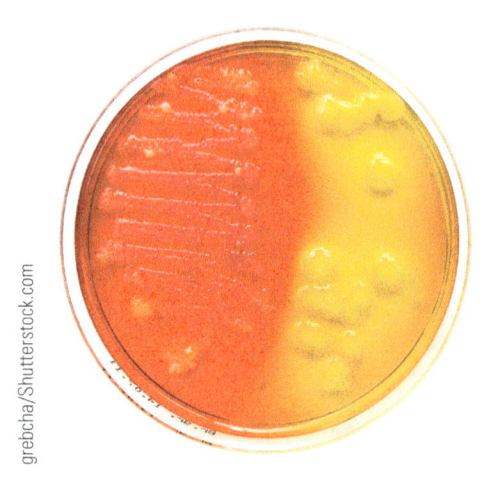

a.) Manitol Salt Agar (MSA) is selective and differential for S. aureus.

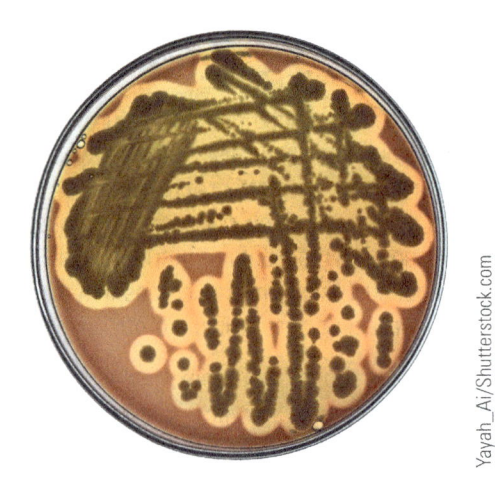

b.) Blood Agar (BA) is differential for hemolytic (blood cell-disrupting) bacteria.

c.) MacConkey Agar (MAC) is selective for gram-negative bacteria and differential for lactose fermenters.

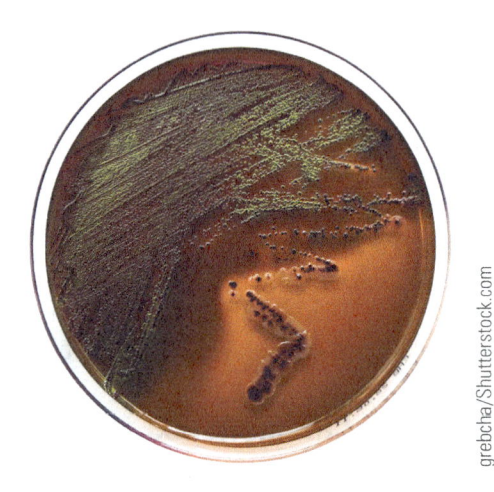

d.) Eosin Methylene Blue (EMB) is selective for gram-negative bacteria and differential for E. coli and lactose fermenters.

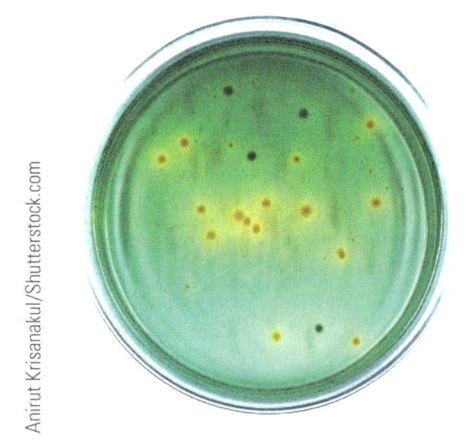

e.) Thiosulfate Citrate Bile Salts-Sucrose Agar (TCBS) is selective for Vibrio cholerae

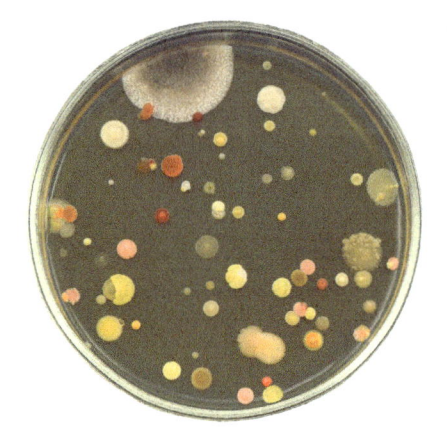

f.) Trypticase Soy Agar (TSA) is non-selective and nutrient rich, promoting the growth of most microbes.

Figure 12.23 Types of media used to culture and identify bacteria and other microbes.

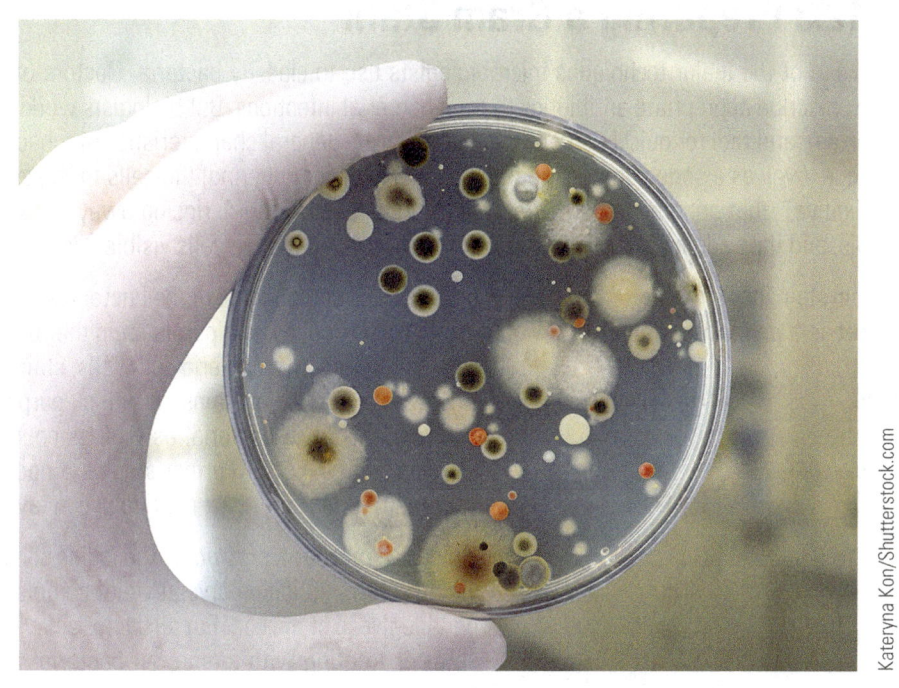

Laboratory Exercise 12 Microbiology: Viruses, Bacteria, and Protists **359**

Figure 12.24 Microbial colonies on a nutrient-rich medium. Note the different properties of various colonies.

Smooth   Undulate (wavy)   Lobed   Irregular

Filamentous   Rhizoidal (root-like)   Crenelated (curly layers)

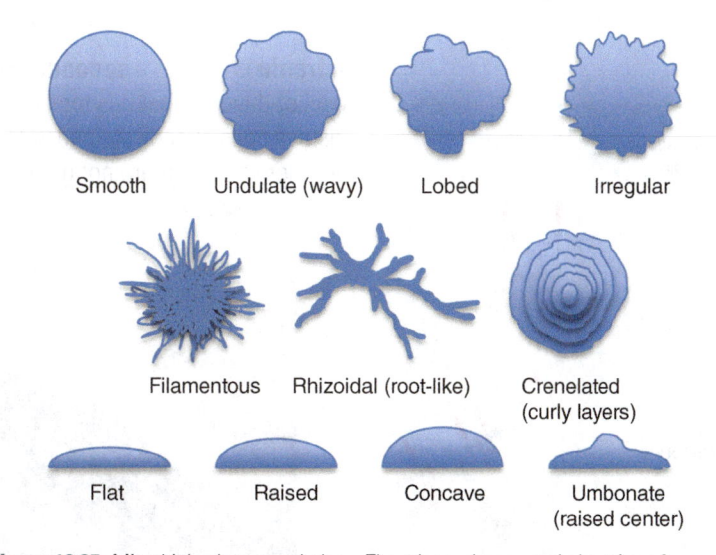

Flat   Raised   Concave   Umbonate (raised center)

Figure 12.25 Microbial colony morphology. The edges, shapes, and elevation of the colonies can help a microbiologist identify the organism.

Lab Skill 12.3 Preparing a Gram Stain

The Gram stain is one of the major techniques microbiologists use to classify bacteria. Doctors often rely on the Gram stain to help them prescribe appropriate antibiotics to treat bacterial infections. But biologists working in many fields may find the Gram stain a useful tool for quickly determining at least the general characteristics of a bacterial population. The Gram stain technique involves a series of steps that include "fixing" (or adhering) the cells to the slide, adding a primary stain, adding a "fixative" (that adheres the stain to the gram-positive bacteria), rinsing away the stain to decolorize the gram-negative cells, and then adding a secondary stain to make gram-negative cells visible (Figure 12.26).

Recall that the Gram stain identifies strains of bacteria based on the components of the bacterial cell wall. The Gram stain uses **crystal violet** as a **primary stain**. Crystal violet, as its name implies, is a purple stain that binds to the peptidoglycan in gram-positive bacteria cell walls, staining them dark purple. However, crystal violet stains everything purple, both gram-positive and gram-negative bacteria, as well as countertops, fingers, and clothing (Figure 12.27). Further steps are needed to decolorize the bacteria that lack the thick peptidoglycan layer of the cell wall.

Iodine is a **mordant**, or *fixative*. The iodine combines with the crystal violet to form an iodine-crystal violet complex that adheres to the peptidoglycan of the cell. But all bacteria have some peptidoglycan, so after treating with iodine, all bacterial cells will continue to appear purple.

Alcohol (or acetone) is used as a *decolorizer*. The alcohol rinses away the crystal violet and iodine from all gram-negative bacteria, leaving the gram-negative bacteria colorless. However, gram-negative bacteria would be very difficult to see under the microscope if left this way, so a final step is needed.

Safranin is used as a **secondary stain**. It is a pink stain that binds to all of the cells, but it is too light to change the color of the gram-positive cells already stained with crystal violet. Thus any bacteria (or other cells) that are not gram-positive will appear pink (Figures 12.28 and 12.29).

Figure 12.26 The basic steps of the Gram stain procedure.

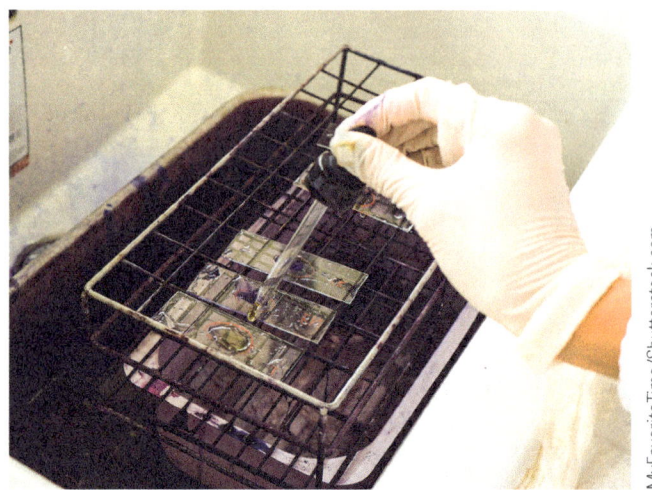

Figure 12.27 The Gram stain. Identifying bacteria as gram-positive or gram-negative helps narrow down the type of bacteria present, and when treating a bacterial infection, helps determine the appropriate antibiotic to prescribe.

Laboratory Exercise 12 Microbiology: Viruses, Bacteria, and Protists **361**

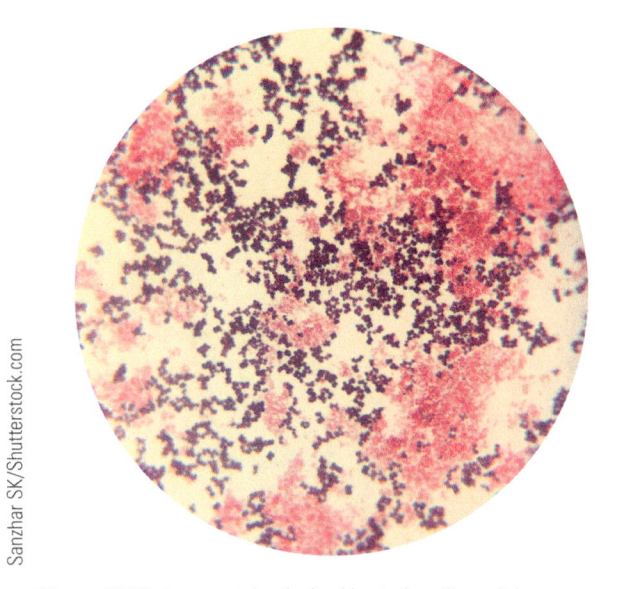

Figure 12.28 A gram stain of mixed bacteria will result in gram positive bacteria appearing purple, and gram negative bacteria (or any other cells present) appearing pink.

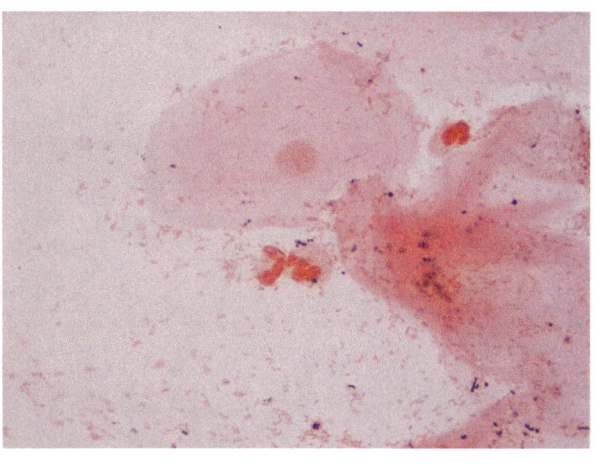

Figure 12.29 Gram stain. This slide shows tiny gram-positive and gram-negative bacteria, as well as some eukaryotic cells also stained pink by safranin. Note the difference in size between bacterial cells and eukaryotic cells.

Name: _____   Date: _____

PRE-LAB QUIZ 12: Are you ready to proceed?

1. What is "pasteurized" milk? What does pasteurization mean?

2. What exactly are viruses made of? Are they prokaryotic or eukaryotic organisms? Why?

3. What is "antibiotic resistance"? What can we do about it?

4. How does the endosymbiosis theory explain the evolution of mitochondria and chloroplasts found in so many higher organisms?

5. What is the major structural difference between gram-negative and gram-positive bacteria? Why would a doctor want to know if a strain of bacteria causing an infection was one or the other?

6. When doing a gram stain, iodine serves as a *mordant*. What does that mean?

7. Why are protists no longer in their own kingdom?

8. What are the four supergroups of eukaryotic organisms? Which contains plant-like protists and plants? Which contains animal and fungus-like protists, animals, and fungi?

9. Explain the difference between an antiseptic, antibiotic, and disinfectant. If you are rinsing your mouth with mouthwash to "kill the germs that cause bad breath," which are you using?

10. A fellow student was looking at a petri dish containing bacterial cultures and stated that a bacterial colony had an *undulate margin* and *umbonate elevation*. What does this mean?

Problem to Solve 12.1 Can I Visually Recognize Types of Microbes?

Viruses are an important part of the microbial world, but they are too small to be seen with a light microscope. They will not be observed in this lab activity.

You have already seen a prepared slide of mixed bacteria in Lab Exercise 5. You may wish to view that slide again. In this activity, we will look at photosynthetic prokaryotes, the cyanobacteria. Since they are colonial, you will not need the oil immersion lens to see them.

Then you will look at a few examples of simple eukaryotes. Representatives of each supergroup are provided. Do not worry about trying to visually recognize common traits among organisms in the same supergroup. In many cases, they are grouped not by their appearance, but by their genetic and molecular similarities. Instead, just become familiar with the names of the supergroups, their general characteristics, and these specific examples of common protists.

When sketching, record the name of the organism and the supergroup to which it belongs. Sketch and label cellular structures, including flagella, cilia, chloroplasts, and pseudopods. It is helpful to look at diagrams and photographs in your textbook as you work through these slides.

When you make the wet mount slides, try to get enough matter on the slide that you can see particles with your naked eye, but do not get so much matter that the coverslip is unable to rest flat on the slide. It helps to press lightly on the coverslip to spread out the particles and reduce the depth of field for any moving organisms. The Protoslo or glycerin will help slow down fast-moving protists. If you look closely in both the aquatic and soil samples, you should see bacteria, protists, and even small animals such as insect larvae, tiny worms, or even a *tardigrade* (water bear).

Title: Microbes Under the Microscope

Purpose
Goal: Be able to Recognize Types of Microbes

Materials:
Prepared slides of
- Anabaena
- Oscillatoria
- Euglena
- Trypanosoma
- Diatoms
- Ceratium
- Paramecium
- Plasmodium
- Spirogyra
- Volvox
- Amoeba

Pond water and soil samples
Blank slides and coverslips
Weigh boat
2 Transfer pipettes
Protoslo® or glycerin

Protocol:

- Observe the two cyanobacteria slides. Do not use oil immersion for these slides.
  a. Sketch what you observe on the following pages.

- Observe the nine protist slides.
  a. Sketch each specimen and label it.
  b. Include details such as nuclei, noticeable organelles, shells, flagella, cilia, and pseudopods.
  c. Indicate the magnification viewed when sketched.

- Make a wet mount of pond water. Use the transfer pipette to collect a sample from the bottom of the container or from an area with plant debris.
  a. Place a drop of Protoslo or glycerin on the slide to slow down the microorganisms.
  b. Add a coverslip. Do not add stain.
  c. Observe under the microscope. Look for moving microbes.

- Collect s small soil sample from an area of rich soil and place it in the weigh boat.
  a. Add a few drops of water.
  b. Transfer a few drops to a slide.
  c. Add a drop of protoslo or glycerin.
  d. Add a coverslip. Do not add stain.
  e. Observe under the microscope. Look for moving microbes.

Continued from previous page.

Organism: _____
Description: _____
_____
Magnification of sketch _____

Organism: _____
Description: _____
_____
Magnification of sketch _____

Organism: _____
Description: _____
_____
Magnification of sketch _____

Organism: _____
Description: _____
_____
Magnification of sketch _____

Laboratory Exercise 12 Microbiology: Viruses, Bacteria, and Protists

Title

Purpose

Continued from previous page.

Protocol

Organism: _____
Description: _____
_____
Magnification of sketch _____

Organism: _____
Description: _____
_____
Magnification of sketch _____

Organism: _____
Description: _____
_____
Magnification of sketch _____

Organism: _____
Description: _____
_____
Magnification of sketch _____

Researcher signature | Date | Supervisor/Instructor signature | Date

Continued from previous page.

Organism: _____
Description: _____
_____
Magnification of sketch _____

Organism: _____
Description: _____
_____
Magnification of sketch _____

Organism: _____
Description: _____
_____
Magnification of sketch _____

Organism: _____
Description: _____
_____
Magnification of sketch _____

Laboratory Exercise 12   Microbiology: Viruses, Bacteria, and Protists

Continued from previous page.

Organism: _____
Description: _____
_____
Magnification of sketch _____

Organism: _____
Description: _____
_____
Magnification of sketch _____

Organism: _____
Description: _____
_____
Magnification of sketch _____

Organism: _____
Description: _____
_____
Magnification of sketch _____

Problem to Solve 12.2 Can I Observe the Ubiquity and Diversity of the Microbes Around Me?

It is helpful to do this procedure the week before covering this laboratory exercise.

Since bacteria and other microbes are invisible, we tend to forget that they are around us all the time. We may assume that a quick wash of the hands or wipe of a paper towel eradicates them. But this is not true. Microbes are everywhere. Even after washing your hands repeatedly, bacteria are present in the external layers of your skin. Wiped surfaces can still teem with bacteria. Fortunately, microbes that live in these environments generally survive best there, and they will not do as well trying to live in you. It is only when bacteria opportunistically invade open wounds or other parts of our bodies that they become a problem.

In this activity your goal is to pick two locations in or near the laboratory room that you think contain microbes. You may use locations in the building, just outside the building, or surfaces of your personal affects, like your backpack, purse, cellphone, keys, or wallet.

As ubiquitous as microbes are, not all surfaces are great sources of microbial diversity. Very dry surfaces, surfaces exposed to intense sunlight, and surfaces that are cleaned regularly are likely poor sources of microbes.

Once you have determined the locations you will use to collect your samples, record the locations in the *Goal* section of the following lab notebook page. Then hypothesize which of the two locations will have the greater microbe diversity. Record this in the *Hypothesis* blank, and proceed with the protocol.

Laboratory Exercise 12 Microbiology: Viruses, Bacteria, and Protists

Title: Culturing Microbes

Goal: Choose two locations that should have plenty of microbes. Hypothesize between the two which will have the greatest diversity of microbes.

Location A: _____   Location B: _____

Hypothesis: _____

Materials:
One petri dish containing TSA (trypticase soy agar)
Package of two sterile swabs
Permanent marker
Tape

Protocol:
1. Obtain a petri dish containing nutrient agar. Use aseptic technique when handling it. Use gloves. Keep it closed until ready to spread culture.

2. Label the bottom of the petri dish (the half containing the agar.)

    a. Draw a line down the center.

    b. Label one side A and the other B as shown.

    c. Record your initials, date, and locations around the perimeter of the bottom so your writing will not obscure colonies when viewed from the bottom.

Continued from previous page.

Protocol

3. Tear open the package of swabs at the bottom end. Remove one swab and place a few drops of tap water on the end to moisten it. Do not touch the end of the swab to anything until ready to collect your sample.

4. Rub the swab across the surface of your first chosen location. Twist swab as you rub to fully to coat the end of the swab with any bacteria present.

5. Set the petri dish on a flat surface. With one hand, open the lid just enough to insert the swab. Wipe swab across one half of the agar surface in a zigzag pattern, being careful not to press down too hard on the agar. Then turn the petri dish 90° and repeat to ensure complete coverage, twisting the swab slightly as you rub.

6. Dispose of the swab in the trash. Repeat the process with the second swab at the second location on the other half of the petri dish.

7. After collecting both cultures, close and tape the petri dish shut. Since you do not know what will grow, keep it sealed. Incubate upside down at 37 °C.

Continued from previous page.

Data:

Do not open the petri dish. Bacteria have had a chance to grow in large numbers, and some may be pathogenic.

Observe the cultures of bacteria on your petri dish through the top or bottom of the dish. Try to distinguish the number of different types of colonies present on the dish. You should look for characteristics described in Lab Skill 2.2. Examine color, shape of colony, if edge of colony is smooth or irregular, if colony appears raised, and any other interesting characteristics. As you observe a colony, circle it (by writing on the bottom of the petri dish) and give it a number. Record observations for each unique type of colony in the data table.

Side A	Source of culture:		
Colony number	Color	Type of margin (edge)	Other visible features (See Figure 12.25.)

Continued from previous page.

Side B	Source of culture:		
Colony number	Color	Type of margin (edge)	Other visible features (See Figure 12.25.)

Conclusions:

Was your hypothesis supported?

Which location had the greatest diversity?

List at least three factors that may have contributed to a greater diversity of microbes in this location compared to the other.

Which characteristics were most useful for distinguishing one type of microbe from another?

Problem to Solve 12.3 Can I Identify the Types of Bacteria Living in My Mouth?

You have already collected cells from your cheeks for previous laboratory activities. In this case, you will be looking at the other residents of your mouth—your bacteria. Bacteria live on and between your teeth. Regularly practicing oral hygiene can reduce the likelihood of getting cavities, but the bacteria are always present. The bacteria population may vary, as they come and go with the different foods and beverages we consume and when we practice oral hygiene. But it is likely that some gram-positive and gram-negative bacteria are present in your mouth right now.

This lab activity will allow you to practice the gram-stain technique. If you do it carefully, some bacteria should appear purple, and others pink. It should not surprise you to also find some familiar-looking eukaryotic cheek cells as well. If you recently ate something, you may even find food particles and other interesting debris on your slide.

Title: Gram stain procedure

Goal: Collect a sample of bacteria from teeth and gums, differentiate between gram-positive and gram-negative bacteria, and determine bacterial shapes.

Materials:
(Per student)
- Clean blank slide
- Wax pencil
- 2 Toothpicks
- Microscope
- Immersion oil

(Per class—multiple stations)
- Staining tray and rack
- Hot plate set at 80 °C
- Crystal violet in dropper or wash bottle
- Safranin in dropper or wash bottle
- Iodine in dropper or wash bottle
- Alcohol or acetone in wash bottle
- Distilled water in wash bottle

Protocol:
1. Use the wax pencil to draw a quarter-sized circle in the center of the microscope slide.
2. Place two drops of distilled water in the circle on the slide.
3. Use a toothpick to vigorously scrape around the teeth and gum line.
4. Dab end of toothpick in the water on the slide, and spread the water and bacteria around to fill the circular area.
5. Throw toothpick in the trash.
6. Repeat the procedure with the second toothpick, using different teeth.

Title

Purpose

Continued from previous page.

Protocol

7. "Heat fix" the bacteria to the slide. Set the slide on the hotplate until the water has evaporated. Check to ensure slide is not too hot to touch—it should be warm but not hot. It will take a few minutes for the water to completely evaporate. This "fixes" or adheres the bacterial cells to the slide. Ensure the side is dry before proceeding or the bacteria will wash off.

8. Place the slide on the rack in the area provided. The stains for this lab will permanently stain clothing. Wear gloves, safety glasses, and, if desired, an apron or lab coat.

9. Place a few drops of crystal violet on the slide until the wax circle is completely filled. Allow the stain to sit on the cells for at least one minute.

10. Leave the slide on the rack, but raise one end of the slide. Use a wash bottle of distilled water to gently rinse the slide. Do not spray water directly on the circular area. Instead, spray above the circular area and let the water wash over the circle until most of the stain has washed off. There should still be a purple haze on the slide.

11. Leave the slide on the rack, and add iodine to the slide until the circular area is filled. Let the iodine sit on the slide for at least one minute.

12. Repeat the rinse procedure described in step 10.

13. Use a dropper or wash bottle to slowly rinse the slide with alcohol. Apply only a few drops at a time. Too much alcohol will decolorize the gram-positive bacteria. The slide should still have a light purple haze.

14. Leaving the slide on the rack, add safranin to the slide until the circular area is filled. Let the safranin sit on the slide for at least one minute.

15. Gently rinse the slide with distilled water as described in step 10.

16. Use a Kimwipe or paper towel to gently blot excess water from the slide. Do not wipe the slide. Allow the slide to dry before observing under the microscope.

17. Examine the slide under the microscope. Start with the scanning objective and work up to high power.

Continued from previous page.

18. Once bacteria are visible on high power, add a drop of immersion oil directly to the slide. (Note that you are not using a coverslip, so if you make a mistake or have not found your bacteria already, there is no way to wipe the oil from the slide and return to a lower-power objective. Only add the immersion oil after you are certain you see bacteria.)

19. Sketch your results.

Data and Observations:

Organism: _____
Description: _____
_____
Magnification of sketch _____

Conclusions:
Were gram positive and gram-negative bacteria present?

What bacterial shapes are visible?

Are any other cells or substances visible on the slide? If so, what?

Problem to Solve 12.4 Can I Keep my Classmates from Infecting Me?

This is a fun, optional activity to do if time permits.

Although microbes are beneficial and important to the biological world, microbes are also responsible for many diseases. Table 12.1 summarizes some of the major viral, bacterial, and protistan diseases that have affected the human population. Some like cholera and leprosy can be effectively treated today, and some, like smallpox, are considered eradicated. But all have had a devastating effect on human populations at one time or another.

Table 12.1 Human diseases caused by viruses, bacteria, and protists. They are listed in alphabetical order. This is in no way a comprehensive list, but they are diseases that have had a significant impact on human populations in the past or are still a health problem to this day.

Major diseases caused by viruses	Major diseases caused by bacteria	Major diseases caused by protists
Chickenpox	Anthrax	African Sleeping Sickness
Dengue	Bacterial meningitis	Amoebiasis
Ebola	Botulism	Babesiosis
Hantavirus	Cholera	Giardiasis
Hepatitis	Diphtheria	Leishmaniasis
HIV	Gonorrhea	Malaria
Influenza	Leprosy	Toxoplasmosis
Measles	Pertussis	
Mumps	Shigellosis	
Poliomyelitis	Streptococcal Pneumonia	
Rabies	Syphilis	
Rotavirus	Tetanus	
SARS-CoV-2	Tuberculosis	
Smallpox	Typhoid Fever	

Microbes are transmitted by a number of routes, including aerosol droplets (coughing, sneezing), direct contact, oral contact (usually from contaminated food or water), fomites (inanimate objects that carry microbes, such as clothing, shoes, trash cans), vectors (insects that carry microbes from host to host, such as mosquitos and ticks), and zoonotic transmission (from animal to human, such as variants of the flu derived from swine, or birds, or COVID-19, which originated in bats).

In this activity you will be given a role to play, and you will simulate the transmission of an unknown disease by trading samples of water with other students (much like the transfer of body fluids). See if you can identify any trends that suggest which populations of humans have a higher or lower risk of infection.

Title: How Microbes Spread and Cause Disease

Goal: Model the spread of a disease through a population, and determine if certain behaviors increase or decrease the rate of transmission.

Materials:
(Per class)
Enough test tubes for each student in the class
Distilled water
10% NaCl solution
Phenolphthalein indicator solution (dropper bottle)
Slips of paper, each containing a role listed on the data table—one per student
Transfer pipette—one per student

Instructor Setup:
Label all test tubes with a number.
Fill all but one test tube half full of water.
Fill the remaining test tube half full with a 10%-20% NaOH solution.
Note the tube number that contains the NaOH solution.

Protocol:
1. Each student should receive a slip of paper containing their ROLE/GOAL.

2. Each student should also receive a half-filled, numbered test tube and a transfer pipette.

3. When the activity starts, use the dropper to transfer liquid from your test tube to another student's tube and allow the other student to do the same as directed by your slip of paper. Transferred amounts should be around one ml in volume.

4. Some transfers are mandatory, some are by agreement, and some are one-way transfers only. Follow the instructions on the slip of paper.

5. When ready, start the activity and a 5-minute timer.

Protocol

Continued from previous page

6. Attempt to accomplish the goal stated on your slip of paper. You may share your ROLE/GOAL with other students, but you do not have to. You can also lie about it.

7. When 5 minutes have passed, stop the activity. The instructor will then act as the doctor, providing an infection test to each student by adding a few drops of *phenolphthalein* to their test tube.

8. A bright pink color will be a positive indication of infection.

9. Record infections on the data table.

10. The instructor should reveal who had the original "infected" test tube.

11. If possible, track the path of transmission.

12. Repeat the activity as time allows, with new ROLES/GOALS, new tubes, and a new secret tube to start the spread of disease.

13. Look for patterns in the data and draw conclusions.

Role and Goal:	After Trial 1, indicate if infected	After Trial 2, indicate if infected.	After Trial 3, indicate if infected	After Trial 4, indicate if infected	Total infections
SICK! In need of a blood transfusion. Receive fluid from 2 willing donors, but do not reciprocate. You may also do a regular transfer with one other willing partner.					
PLAYA #1. Transfer fluids with as many partners as possible (at least 4).					
PLAYA #2. Transfer fluids with as many partners as possible (at least 4).					
PLAYA #3. Transfer fluids with as many partners as possible (at least 4).					
PLAYA #4. Transfer fluids with as many partners as possible (at least 4).					
DRUGS! #1. Any other drug user you encounter MUST transfer fluids with you. SHARING NEEDLES! Also transfer with 2 other willing partners.					
DRUGS! #2. Any other drug user you encounter MUST transfer fluids with you. SHARING NEEDLES! Also transfer with 2 other willing partners.					
CHEATER! #1. Transfer fluids with one person at least twice, at least one minute apart. Also transfer with one other willing partner at any time.					
CHEATER! #2. Transfer fluids with one person at least twice, at least one minute apart. Also transfer with one other willing partner at any time.					
CHEATER! #3. Transfer fluids with one person at least twice, at least one minute apart. Also transfer with one other willing partner at any time.					

CHEATER! #4. Transfer fluids with one person at least twice, at least one minute apart. Also transfer with one other willing partner at any time.					
COMMITTED #1. Transfer with any <u>one</u> person only, 1-3 separate times at least a minute apart. They can be anyone who agrees. They do not need to be committed.					
COMMITTED #2. Transfer with any <u>one</u> person only, 1-3 separate times at least a minute apart. They can be anyone who agrees. They do not need to be committed.					
COMMITTED #3. Transfer with any <u>one</u> person only, 1-3 separate times at least a minute apart. They can be anyone who agrees. They do not need to be committed.					
COMMITTED #4. Transfer with any <u>one</u> person only, 1-3 separate times at least a minute apart. They can be anyone who agrees. They do not need to be committed.					
BZZZZ #1. You may pick at random up to 4 people. They MUST trade fluids with you. **MOSQUITO BITE!**					
BZZZZ #2. You may pick at random up to 4 people. They MUST trade fluids with you. **MOSQUITO BITE!**					
SICK #1. Stand next to someone for 8-10 seconds. If they don't notice you, they MUST let you transfer fluid to them. **COUGH!** (They don't reciprocate.) You may also transfer with anyone else who agrees.					
SICK #2. Stand next to someone for 8-10 seconds. If they don't notice you, they MUST let you transfer fluid to them. **COUGH!** (They don't reciprocate.) You may also transfer with anyone else who agrees.					
BAD NURSE. (Gloves? What gloves?) Transfer with anyone who is **SICK** or a **DRUG** user. They MUST transfer. You may also transfer with anyone else who agrees.					
EWWW! You are a nasty bathroom. Stand somewhere near the middle of the room. Anyone who comes within 3 feet of you MUST transfer with you. *Fomite?* What's that?					

Name: _____ Date: _____

POST-LAB QUIZ 12: What have you learned?

1. What were at least three major differences between the slides of cyanobacteria and the slides of protists?

2. Which of the supergroups contains protists with lobed pseudopodia? What are these types of organisms generally called?

3. List at least five ways microbes are beneficial to humans.

4. Some microbial diseases are transmitted by vectors. Name two protists covered in this lab exercise that are parasitic pathogens, and name the vectors that transmit them.

5. Clearly explain the difference between *differential* and *selective* media.

6. If a doctor is treating a patient with what seems to be a bacterial infection, which should they do first, and why? (A.) Do a Gram stain. (B.) Grow the bacteria on a differential medium. (C.) Grow the bacteria on a nutrient-rich medium. (D.) Prescribe broad-spectrum antibiotics to treat the infection.

7. Jake collected bacteria from two surfaces and cultured them in a petri dish on nutrient-rich agar like you did in this lab exercise. Then he selected one interesting-looking colony and conducted a gram stain test on those bacteria. When he looked at them under the microscope, all of the cells appeared purple. Which *two* of the following may explain this. Explain why you think so. (A.) He used too much safranin when conducting the Gram stain. (B.) He failed to heat-fix the bacteria. (C.) He used too little iodine. (D.) He used too much crystal violet. (E.) He used too little alcohol. (F.) He did the Gram stain correctly, but the colony was made up of all one type of bacteria.

8. If you purchased a medical facemask that claimed to have a pore size of 10 μm, would this be an effective barrier to a virus like SARS-CoV-2? Explain your reasoning.

9. Fifty years ago, penicillin and many other antibiotics were far more effective for treating bacterial infections than they are now. Why? Explain clearly.

10. What can be done to reduce the problem described in question 9 above?

Appendix A

Metric and SI Units

Prefixes for metric units commonly used in biology

Prefix	Meaning	Exponential Notation
mega	one million	10^6
kilo	one thousand	10^3
milli	one-thousandth	10^{-3}
micro	one-millionth	10^{-6}
nano	one-billionth	10^{-9}
pico	one-trillionth	10^{-12}

Commonly used metric units for volume, mass, size/distance, and number of atoms/molecules

Volume	Symbol	Equivalent
liter	l	
milliliter	ml	10^{-3} l
microliter	µl	10^{-6} l
Mass	**Symbol**	**Equivalent**
kilogram	kg	10^3 g
gram	g	
milligram	mg	10^{-3} g
microgram	µg	10^{-6} g
Size	**Symbol**	**Equivalent**
meter	m	
centimeter	cm	10^{-2} m
millimeter	mm	10^{-3} m
micrometer (micron)	µm	10^{-6} m
nanometer	nm	10^{-9} m

Molecules	Symbol	Equivalent
mole	mol	
millimol	mmol	10^{-3} mol
micromole	μmol	10^{-6} mol
nanomol	nmol	10^{-9} mol
picomole	pmol	10^{-12} mol

International System of Units typically used in scientific research

Base Units	Abbreviation	Quantity measured
liter/litre	l	volume/capacity
meter/metre	m	size/length
second	s	time
kilogram	kg	mass
mole	mol (M)	number of molecules
Kelvin	K	thermodynamic temperature
Celsius	C	temperature

Conversion formulae

Common Conversions from imperial units to metric units	
1 inch = 2.54 cm	1 cm = 0.39 inches
1 fluid ounce = 28.413 ml	1 ml = 0.035 fluid ounces
1 gallon = 4.546 liters	1 liter = 0.264 gallons
1 ounce = 28.35 grams	1 gram = 0.035 ounces
Converting temperature scales	
Celsius to Fahrenheit	°F = (°C * 1.8) + 32
Fahrenheit to Celsius	°C = (°F − 32)/1.8
Celsius to Kelvin	°K = °C + 273.15
Kelvin to Celsius	°C = °K − 273.15

Appendix B
Scientific Notation and Significant Figures

Scientific Notation is a way of writing and working with very large or very small numbers. All real numbers can be written as $m \times 10^n$ where m represents a real number times 10 raised to the nth power.

A positive exponent next to a number means that many of a number multiplied together. Since scientific notation involves powers of 10, we will only look at those here.

$$10^1 = 10$$
$$10^2 = 10 \times 10 \text{ or } 100$$
$$10^3 = 10 \times 10 \times 10 = 1{,}000$$

You can move the implied decimal point after the "10" to the right as many times as the exponent, adding zeroes as you go, to convert a power of 10 into a standard format.

$$\text{So } 10^{12} = 1{,}000{,}000{,}000{,}000.$$

Exponents can also be negative. A negative exponent indicates that the number is a reciprocal or can be inverted as a fraction to eliminate the negative sign.

$$10^{-1} = 1/10^1 = 1/10 = 0.1$$
$$1 \times 10^{-2} = 1/10^2 = 1/100 = 0.01$$
$$1 \times 10^{-3} = 1/10^3 = 1/1{,}000 = 0.001$$

You can convert a "10" with a negative exponent to decimal format by moving the decimal place to the left as many times as the exponent.

For example, 10^{-12} as a decimal is 1 preceded by 10 zeroes and the decimal point. (You have to move the decimal point before the zero and 1 and then to the left 10 more times).

$$10^{-12} = 0.00000000001$$

Therefore, any very large number can be written in scientific notation simply by moving the decimal point to after the first non-zero digit, counting how far left or right it moved, and adding a power of 10 to reflect that movement. If you moved the decimal point to the left because the number was very large, the exponent on the 10 will be positive. If you move the decimal point to the right because the number was very small, the exponent on the 10 will be negative.

Standard form of number	Scientific notation
1,230,000,000,000	1.23×10^{12}
0.00000000000123	1.23×10^{-12}

Scientific Notation on a Calculator

Calculators often use "e" to represent the 10× term. So if you multiply large numbers and get an answer that looks like "1.23e12," this actually means 1.23×10^{12}.

Significant figures are the digits of a number that contribute to the *precision* of the number.

- All digits other than zero are always significant. Zeroes are significant in certain situations.
- In *non-decimal* numbers, zeroes between non-zero digits are significant, but zeroes that come after all of the non-zero digits are not.

    Examples:
        4, **4**00, and **4** × 10^{12} all have one significant figure (highlighted).
        345, **34,5**00,000, and **304**0 all have three significant figures.

- In *decimal* numbers, all zeroes that come after any non-zero digits are significant.

    Examples:
        0.**4**, .000000**4**, and **4** × 10^{-12} all have one significant figure.
        0.**34**, 0.0000**34**, and **3.4** × 10^{-12} all have 2 significant figures.
        40.0, 0.000**400**, and **4.00** × 10^{12} all have three significant figures.

The measuring instrument determines the precision of a measurement. For example, you can't measure with micrometer precision using a ruler marked in centimeters and millimeters. But when we use a calculator to manipulate numbers, we sometimes get answers that have many numbers after the decimal. The calculation suggests a level of precision that may not be there. The precision of a calculated answer is limited by the precision of the numbers used.

- When adding or subtracting decimals, the answer *cannot* have more significant figures to the right of the decimal point than the least precise number in the calculation.

    Examples:
        23.4 + 5 = 28.4 ⟶ 28 (Since the 5 does not have a decimal point, the answer is rounded to a whole number.)
        23.4 + 5.0 = 28.4 (Since the zero after the 5 is significant, the answer is not rounded.)
        23.4 − 0.559 = 22.841 ⟶ 22.8 (Since 23.4 only has one significant number after the decimal point, the answer must be rounded to the same level of precision.)

- When multiplying or dividing numbers, the answer must have no more significant figures than the *least* number of significant figures in the numbers used in the calculation.

    Examples:
        (23.4 × 5.6789)/78.00 = 1.70367 ⟶ 1.70 (Must be rounded since 23.4 has only three significant figures.)
        $2.345 \times 10^{-4} \times 2.1$ = 0.00049245 or 4.9245×10^{-4} ⟶ 4.9×10^{-4} (Must be rounded since 2.1 has only two significant figures.)
        108.0/6.3 = 17.14285714285714 ⟶ 17 (Must be rounded since 6.3 has only two significant figures.)

Appendix C

Atoms, Isotopes, and Ions

Atomic Structure

Atoms are composed of three fundamental parts: **protons, neutrons,** and **electrons**. Protons are located in the nucleus, have a positive charge, and have a mass of **one atomic mass unit (amu).** Neutrons are also found in the nucleus, are neutral (having no charge), and have a mass of one amu. Electrons spin around the nucleus in an area referred to as the **electron cloud**. The mass of an electron is only 1/2000 of a proton or neutron. Electrons are so small that their mass contributes almost nothing to the atom's mass, so they are typically given a mass of zero amu. The differences between the subatomic particles are summarized in Table C.1.

Table C.1 A comparison of subatomic particles.

Subatomic Particle	Mass	Charge	Location
Proton	1 amu	Positive +	In the nucleus
Neutron	1 amu	Neutral 0	In the nucleus
Electron	0 amu	Negative -	Orbitals around the nucleus

Atomic mass

Figure C.1 provides a rough schematic of an atom. Based on the information provided, this atom has three protons, four neutrons, and three electrons. Since electrons don't count toward mass, this atom has a mass of 7 amu.

QUANTUM MECHANICAL MODEL

Bohr Model
Electron Orbits

Quantum Mechanical Model
Electron Clouds (Orbitals)

Electron
Negatively charged particles
Atomic mass 0

Neutron
Particles that contain no charge
Atomic mass 1

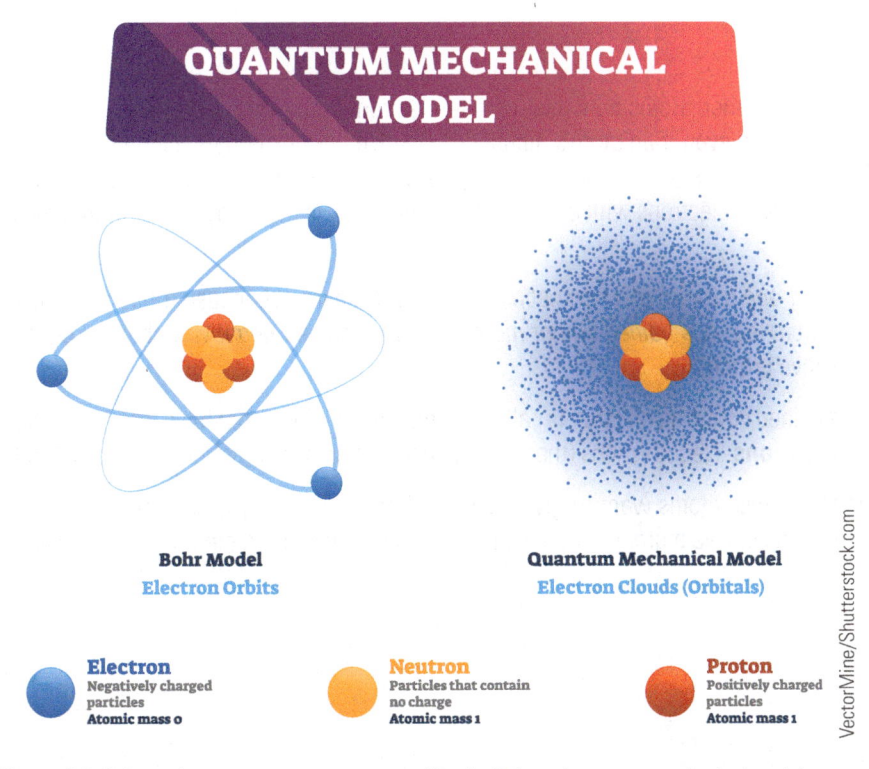

Proton
Positively charged particles
Atomic mass 1

Figure C.1 Subatomic components as represented by the Bohr and quantum mechanical models.

Electrons

If an atom is uninvolved in any chemical reactions, the number of electrons will equal the number of protons. This causes the atom to be neutrally charged, as the positive charges of the protons are balanced by the negative charges of the electrons.

The atom shown on the left in Figure C.1 is neutrally charged since it has three positive protons and three negative electrons.

Electrons are arranged in **energy levels** often described as circular **orbitals** around the nucleus similar to planets circling the sun. It is an over-simplification to think of electrons as traveling in circles, but it works well enough here. Like the orbits of planets around the sun, each orbital or energy level is larger the further from the nucleus it is. More electrons can occupy larger energy levels.

Electrons fill energy levels in a predicable way. The first energy level can hold only 2 electrons. The next can hold only 8. The third can hold 18. At higher energy levels, electrons do not fill the energy levels in order, and the fourth energy level begins to fill after eight electrons occupy the third.

Those electrons in the outermost energy level are **valence electrons**. When atoms come near each other, valence electrons of one atom can interact with valence electrons of other atoms. Most of chemistry deals with the interactions of valence electrons in different atoms.

Atoms are identified by type according to their number of protons. An **element** is a substance made of one kind of atom in which all of the atoms have the same number of protons. Unless the atom is undergoing a nuclear reaction (a reaction that affects the nucleus of the atom), the number of protons will never change. Biochemistry does not deal with *nuclear* reactions!

Isotopes

However, naturally occurring atoms of the same element can differ in their number of neutrons. This rarely affects the properties of the element, only the mass of that particular atom. These different forms of an element based on the differences in neutron number are called **isotopes**. Isotopes having very high numbers of neutrons can be unstable and even radioactive. See Figure C.2 for examples of hydrogen and carbon isotopes.

Ions

As atoms get near each other, atoms may gain or lose electrons to become more stable. The *stability* of an atom is not the same as its *charge*. So even if an atom's number of positive protons equals the number of negative electrons, it may be unstable. To become more stable, it may want to *give* electrons to another type of atom, *take* electrons from another type of atom, or even *share* electrons with another atom. Chemistry is the study of the interactions between atoms as they seek to be more stable.

An atom that has *lost one or more electron* becomes *positively* charged because it has more positive protons than negative electrons. An atom that has *gained one or more electron* becomes *negatively* charged because it has more negative electrons than positive protons.

Any atom that has gained or lost an electron is called an **ion**. Positive ions (that have lost one or more electrons) are specifically referred to as **cations**. Negative ions (that have gained one or more electrons) are specifically referred to as **anions**.

Once you understand which atoms want to give, take, or share electrons, you can predict how elements will react when combined. If atoms can achieve a greater stability by exchanging or sharing electrons, they will, and this produces chemical bonds.

Appendix C   Atoms, Isotopes, and Ions   393

ISOTOPES

3 Isotopes of Hydrogen

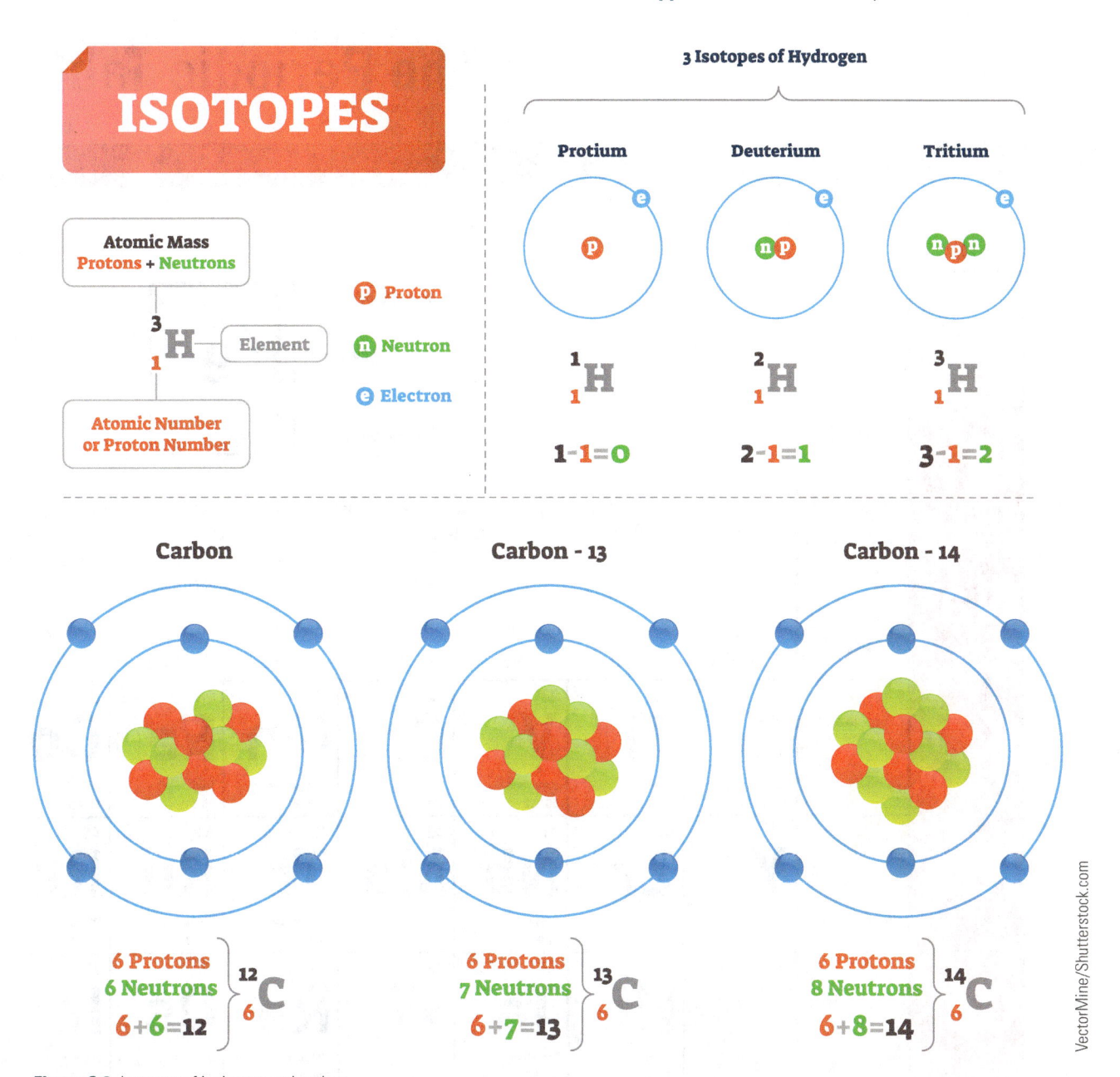

Protium — Deuterium — Tritium

Atomic Mass
Protons + Neutrons

$^{3}_{1}H$ — Element

p Proton
n Neutron
e Electron

Atomic Number
or Proton Number

$^{1}_{1}H$ $^{2}_{1}H$ $^{3}_{1}H$

1−1=0   2−1=1   3−1=2

Carbon
6 Protons
6 Neutrons
6+6=12
$^{12}_{6}C$

Carbon − 13
6 Protons
7 Neutrons
6+7=13
$^{13}_{6}C$

Carbon − 14
6 Protons
8 Neutrons
6+8=14
$^{14}_{6}C$

Figure C.2 Isotopes of hydrogen and carbon.

Summary

Protons are positively charged, found in the nucleus of an atom, and can't be given or taken in a chemical reaction (unless it is a nuclear reaction). The number of protons determines the type of atom or element.

Neutrons are neutral, found in the nucleus of the atom, contribute to the mass of the atom, and their number can naturally vary in different isotopes of an element.

Electrons are negatively charged, found in orbitals around the nucleus, and are arranged in energy levels. The outermost electrons are involved in chemical reactions. When atoms gain or lose electrons, they become charged ions.

Appendix D

The Periodic Table o

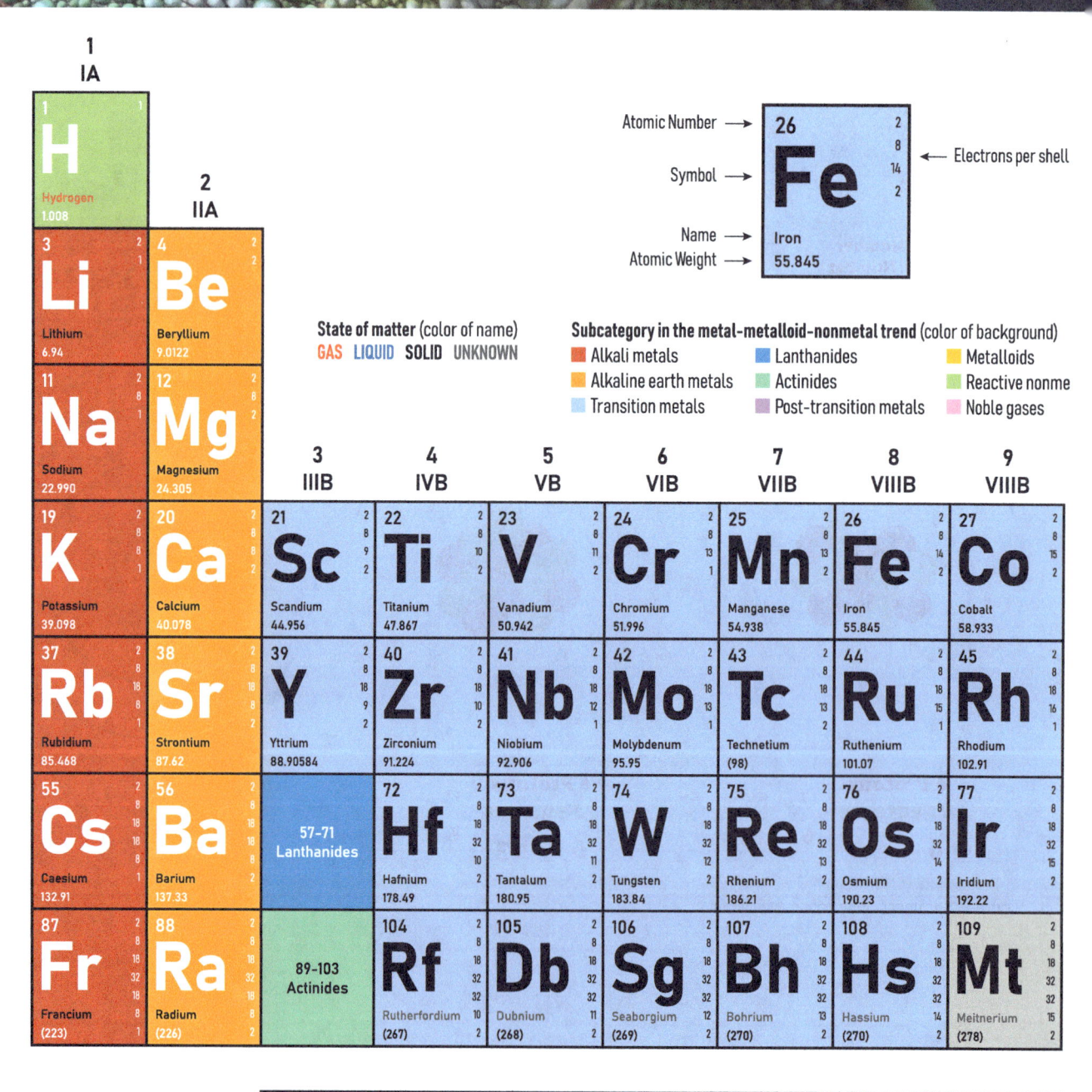

1 **IA**								
1 **H** Hydrogen 1.008	**2** **IIA**							
3 **Li** Lithium 6.94	4 **Be** Beryllium 9.0122							
11 **Na** Sodium 22.990	12 **Mg** Magnesium 24.305	**3** **IIIB**	**4** **IVB**	**5** **VB**	**6** **VIB**	**7** **VIIB**	**8** **VIIIB**	**9** **VIIIB**
19 **K** Potassium 39.098	20 **Ca** Calcium 40.078	21 **Sc** Scandium 44.956	22 **Ti** Titanium 47.867	23 **V** Vanadium 50.942	24 **Cr** Chromium 51.996	25 **Mn** Manganese 54.938	26 **Fe** Iron 55.845	27 **Co** Cobalt 58.933
37 **Rb** Rubidium 85.468	38 **Sr** Strontium 87.62	39 **Y** Yttrium 88.90584	40 **Zr** Zirconium 91.224	41 **Nb** Niobium 92.906	42 **Mo** Molybdenum 95.95	43 **Tc** Technetium (98)	44 **Ru** Ruthenium 101.07	45 **Rh** Rhodium 102.91
55 **Cs** Caesium 132.91	56 **Ba** Barium 137.33	57–71 Lanthanides	72 **Hf** Hafnium 178.49	73 **Ta** Tantalum 180.95	74 **W** Tungsten 183.84	75 **Re** Rhenium 186.21	76 **Os** Osmium 190.23	77 **Ir** Iridium 192.22
87 **Fr** Francium (223)	88 **Ra** Radium (226)	89–103 Actinides	104 **Rf** Rutherfordium (267)	105 **Db** Dubnium (268)	106 **Sg** Seaborgium (269)	107 **Bh** Bohrium (270)	108 **Hs** Hassium (270)	109 **Mt** Meitnerium (278)

Atomic Number → 26
Symbol → **Fe**
Name → Iron
Atomic Weight → 55.845
← Electrons per shell

State of matter (color of name)
GAS LIQUID SOLID UNKNOWN

Subcategory in the metal-metalloid-nonmetal trend (color of background)
- Alkali metals
- Alkaline earth metals
- Transition metals
- Lanthanides
- Actinides
- Post-transition metals
- Metalloids
- Reactive nonme
- Noble gases

57 **La** Lanthanum 138.91	58 **Ce** Cerium 140.12	59 **Pr** Praseodymium 140.91	60 **Nd** Neodymium 144.24	61 **Pm** Promethium (145)	62 **Sm** Samarium 150.36	63 **Eu** Europium 151.96
89 **Ac** Actinium (227)	90 **Th** Thorium 232.04	91 **Pa** Protactinium 231.04	92 **U** Uranium 238.03	93 **Np** Neptunium (237)	94 **Pu** Plutonium (244)	95 **Am** Americium (243)

ements and How to Use It

■ Unknown chemical properties

10 VIIIB	11 IB	12 IIB	13 IIIA	14 IVA	15 VA	16 VIA	17 VIIA	18 VIIIA
								2 **He** Helium 4.0026
			5 **B** Boron 10.81	6 **C** Carbon 12.011	7 **N** Nitrogen 14.007	8 **O** Oxygen 15.999	9 **F** Fluorine 18.998	10 **Ne** Neon 20.180
			13 **Al** Aluminium 26.982	14 **Si** Silicon 28.085	15 **P** Phosphorus 30.974	16 **S** Sulfur 32.06	17 **Cl** Chlorine 35.45	18 **Ar** Argon 39.948
28 **Ni** Nickel 58.693	29 **Cu** Copper 63.546	30 **Zn** Zinc 65.38	31 **Ga** Gallium 69.723	32 **Ge** Germanium 72.630	33 **As** Arsenic 74.922	34 **Se** Selenium 78.971	35 **Br** Bromine 79.904	36 **Kr** Krypton 83.798
46 **Pd** Palladium 106.42	47 **Ag** Silver 107.87	48 **Cd** Cadmium 112.41	49 **In** Indium 114.82	50 **Sn** Tin 118.71	51 **Sb** Antimony 121.76	52 **Te** Tellurium 127.60	53 **I** Iodine 126.90	54 **Xe** Xenon 131.29
78 **Pt** Platinum 195.08	79 **Au** Gold 196.97	80 **Hg** Mercury 200.59	81 **Tl** Thallium 204.38	82 **Pb** Lead 207.2	83 **Bi** Bismuth 208.98	84 **Po** Polonium (209)	85 **At** Astatine (210)	86 **Rn** Radon (222)
110 **Ds** Darmstadtium (281)	111 **Rg** Roentgenium (282)	112 **Cn** Copernicium (285)	113 **Nh** Nihonium (286)	114 **Fl** Flerovium (289)	115 **Mc** Moscovium (290)	116 **Lv** Livermorium (293)	117 **Ts** Tennessine (294)	118 **Og** Oganesson (294)

64 **Gd** Gadolinium 157.25	65 **Tb** Terbium 158.93	66 **Dy** Dysprosium 162.50	67 **Ho** Holmium 164.93	68 **Er** Erbium 167.26	69 **Tm** Thulium 168.93	70 **Yb** Ytterbium 173.05	71 **Lu** Lutetium 174.97
96 **Cm** Curium (247)	97 **Bk** Berkelium (247)	98 **Cf** Californium (251)	99 **Es** Einsteinium (252)	100 **Fm** Fermium (257)	101 **Md** Mendelevium (258)	102 **No** Nobelium (259)	103 **Lr** Lawrencium (266)

Humdan/Shutterstock.com

A Tour of the Periodic Table

The **periodic table of elements** is a chart of the known elements that organizes the atoms in seven **periods** (rows) and eighteen **groups** (columns) according to some of their fundamental properties.

The red "zigzag" line on the right side of the table separates the **metals** (all elements to the left of the line) from the **non-metals** (all elements to the right). Most elements are metals. Metals tend to be electron donors in chemical reactions. Many important elements in living organisms are metals, including sodium, potassium, iron, and magnesium. Non-metals tend to be electron recipients in chemical reactions. The elements in the column on the far right of the periodic table are called **noble gases**, and they are *inert* (unreactive) elements. All other elements are unstable, or reactive, and will bond with other elements to become more stable. (Other, more specific categories of elements are indicated by their color on the table.)

Atomic Number, Atomic Mass, and Mass Number

Each box on the periodic table contains several important items of information. The periodic table here shows the 118 currently known elements. (However, only 92 are naturally occurring.) Each box of this table gives the **atomic number**, **symbol**, **element name**, and the **atomic mass** for each element. Notice that the symbols do not always match the name. Some symbols refer back to an earlier Latin name for that element. For example, *Fe* refers to iron because the name for iron in Latin was *ferrum*.

Every element is unique based on the number of protons in the atoms of that element. Hydrogen atoms always have one proton, helium atoms always have two, and so on. The **atomic number** indicates the number of protons in all atoms of that element. The atomic number is usually above the symbol. In normal (non-nuclear) chemical reactions, the number of protons in an atom cannot be altered.

Since protons and neutrons each have a mass of one **atomic mass unit (amu)**, the mass of an atom should be just the total of the number of protons and neutrons in it. Iron, for example, has 26 protons, and 30 neutrons. It should have a mass of 56 amu. However, the **atomic mass** is 55.845 amu. The atomic mass can be rounded off to the nearest whole number (in this case 56), which is called the **mass number**. The mass number does give the mass of the most common form, or isotope, of the element, and also the total number of protons and neutrons in it. (Note that electrons are so small that they do not significantly contribute to an element's mass. The presence of a fractional value in the atomic mass is *not* due to the mass of the electrons.)

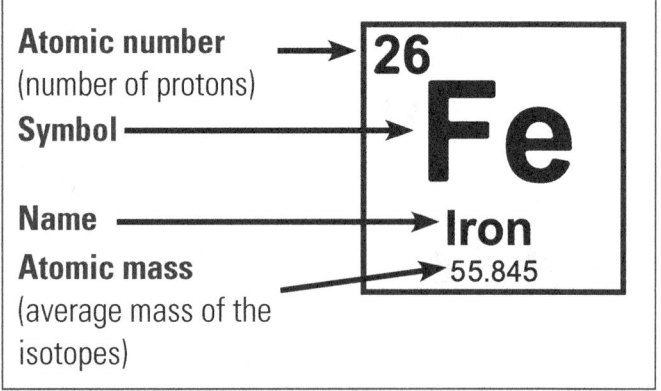

Figure D.1 Typical Information for an element provided on the periodic table.

Iron, for example, has four isotopes: iron-54, iron-56, iron 57, and iron-58. The most common isotope of iron is iron-56, but if you *average* their masses, weighted by the percent abundance of each isotope in nature, the *average* atomic mass of all isotopes is 55.845 amu. This is why atomic masses listed on the periodic table are often not whole numbers.

Periods and Groups

Each **period** (row) of the periodic table generally represents the energy level (or sub-level) being filled with electrons as you move from one element to the next in a period. This periodic table lists the number of electrons in each energy level (shell) for each atom. As you go across a period, you can see how additional electrons fill the energy levels. Elements in the first **group** (column) on the periodic table have one electron in their **valence** (outer) energy level. Those in the last group, group 18, are the **noble gases**. These atoms have full outer energy levels. The groups in between have partially filled energy levels.

The first energy level holds only two electrons. Hydrogen has only one electron. Hydrogen is in period 1 and group 1, indicating that it has one valence electron, its only electron. However, helium is also in period 1. It has two electrons, thus filling its only energy level. Helium is placed in the last group—group 18—to represent this. It is a noble gas. Lithium has three electrons, but since the first two electrons are in the first energy level it has only one valence electron in its *outer* energy level, placing Lithium in group 1 under hydrogen. Neon, with 10 electrons, has 2 electrons in the first energy level, and in the second, filling it completely. Neon, like helium, has a full outer energy level and is therefore also a noble gas (group 18).

Atoms with the same number of valence electrons, arranged in the same group of the periodic table, tend to have very similar properties. Therefore, they tend to form the same kinds of chemical bonds. By locating the column an element is in, one can quickly determine if that atom has a full outer shell. Those in group 1 are unstable and would become more stable if they could lose their one valence electron. Those in group 2 have two valence electrons that, if removed, would make them more stable. Those in group 3 have three valence electrons to lose. Those in group 17 are missing one valence electron and would be more stable if they could gain an electron. Those in group 16 need two electrons, and so forth.

Think of elements on the left side of the table as electron givers, and those on the right as electron takers (except for the last group—the noble gases). Most elements involved in biochemistry can be found in the top four periods of the periodic table, where chemical behavior tends to be predictable. As you go down the table, the behavior of the elements tends to become more complex, but these elements are not typically part of biochemical reactions.

The most important aspect of using the periodic table is identifying what any atom "wants." Atoms are most stable with a full outer energy level. **Chemistry** is simply the study of how atoms gain, lose, or share valence electrons with other atoms to create more stable compounds.

Appendix E

How to Write a Scientific Paper

Scientific papers are not like papers you write for English or history classes. Scientific papers are intended to convey factual scientific information, so they should not be confused with opinion articles, persuasive arguments, fictional stories, or other forms of literature.

All scientific papers should be written in a *very specific style as described below*. This format (much like the procedural rules in a courtroom) ensures that the information will be presented in a logical way, the reader will understand the evidence, and they will reach the appropriate conclusion. This style often seems formal and dry, but the purpose is to convey facts, not entertain.

All background information stated as fact should be *cited* (but not *quoted*) to show where the information came from. All sources should be peer-reviewed journal articles, governmental websites, or experts in the field. *Typical Google searches will not give you this type of information.* Use library databases or Google Scholar to find peer reviewed articles. If using websites, only use sites whose URL ends with.edu or.gov, such as the National Institute of Health website at NIH.gov. Consult your instructor for help on accessing articles from the TCC database if necessary.

Title

The title should explicitly state what the paper is about. Sometimes the title may be long. It is better to be lengthy and specific than short and vague. For example, "Hemoglobin oxygen affinity and regulating factors of the blood oxygen transport in canine and feline blood" is <u>WAY</u> better than something as vague as "Cat blood." Your title should have the first word capitalized, but everything else in lowercase.

Overall Format

Your paper must be typed, double spaced, with 1 inch margins, in 12 point font (preferably Times or *sans serif* fonts like Arial or Calibri—a font you would see in a scholarly work or textbook.) It should be in APA format.

It must have a title page with the title, your name, the date it is turned in, and your class and instructor. Each following page should have a heading that states the first few words of the title and your last name, for example "Hemoglobin oxygen affinity—Luyster." The bottom of each page should be numbered.

A few other general formatting tips:

Scientific names are always italicized, with genus capitalized and species lowercase—e.g. *Homo sapiens*.

Do not refer to yourself as "I" or your group as "we." Everything in your paper should be written in *third person*. You can refer to yourself as "the researcher" or "the author."

Your paper should be written in past tense. Do not write the paper, especially the Materials and Methods section, as a series of instructions. It is a historical record or report of what you or your team already did and what you discovered.

This is a scientific paper, not a persuasive article, editorial, or creative writing assignment. Do not try to make it entertaining, funny, persuasive, or a story—keep to the facts. Avoid slang, jokes, emojis, or inappropriate abbreviations (LOL).

A scientific paper is typically presented in 6 sections. Each section should be preceded with the following headings in bold font: **Abstract, Introduction, Materials and Methods, Results, Discussion and Conclusion, References**.

Abstract

The Abstract, though first in the paper, *should be written after the rest of the paper has been written*. It is a brief summary of the paper so that the reader can get a good idea of what the paper is about without reading the whole thing. You can't write a summary of a paper you have not yet written, so it must be written AFTER your paper has been written, and then inserted at the beginning. It should be a *brief paragraph* that states a) the goal of the study, b) the hypothesis, c) a short overview of the experiment, d) and what was concluded from the results.

Most common mistakes: making the abstract too long, not including all four things it should, not writing it in third person, or simply being too vague to make any sense to the reader.

Introduction

The Introduction should provide the reader with the background information needed to understand the problem and give the rationale for the hypothesis. It should be several paragraphs (sometimes many pages) long, with each paragraph describing a key concept the reader needs to know. These concepts should build an argument for why the researcher is doing the study and justify the hypothesis. Usually, the hypothesis and prediction are stated at the end of the introduction.

Direct quotes should not be used. Paraphrase relevant background information in your words. However, *all factual statements should have citations.* Any time a factual statement is made that is not general knowledge (i.e. your non-scientist family members probably don't know it), you *must* indicate where you obtained this information. (See references section for how to do this.)

Most common mistakes: not citing sources, using direct quotes, plagiarizing, writing in too familiar a style, not stating a hypothesis or prediction, and not providing enough information to back up the hypothesis and prediction.

Materials and Methods

The Materials and Methods section simply describes what materials were needed and what steps were done to conduct the experiment. If you wrote out your protocols and kept a good notebook, this is a very easy section to write. *This section must be in **past tense***. It should *NOT* be given as steps to follow. It should be a report of what you did. But it should be written in **third person**. You must refer to yourself as "*the researcher*" or "*the author.*" Do not use "I" or "we" in your paper.

Do not talk about problems you encountered or your results in this section. The materials and methods section should *ONLY* be a description or list of materials needed and the specific steps that were taken to conduct the experiment. Cover any problems or mistakes in the discussion section.

Most common mistakes: writing instructions to follow instead of an historical account, using first person, not including enough detail for a reader to be able to replicate the experiment, not listing all important materials or equipment used, and inappropriately including results or discussion information.

Results

The Results section should clearly present the data gathered from the experiment. The results should be presented in data tables and/or graphs, but they also need to be summarized in words. Figures (graphs) and tables should be identified with "Figure 1…" or "Table 1…" followed with a title and a description, just like they are in textbooks. All numerical data must have units. This section should *NOT* include a discussion of the *meaning* of the results. Interpreting the data occurs in the next section.

Most common mistakes: not providing a summary paragraph for data, not giving units, not giving data tables or graphs appropriate titles or descriptions, not labeling graphs and tables, fudging data to try to match the hypothesis, and jumping into the interpretation of the data too soon.

Discussion and Conclusion

The Discussion and Conclusion section is where you, the author, tie everything together. This is the most important part of a scientific paper. This section should start with a **paraphrase of the hypothesis and prediction** and a **brief summary of the data collected**. Then you should explain the **meaning of the data** as it relates to your prediction. Then you should offer a **clear conclusion that shows how your data did or did not support your hypothesis**. This is your big chance to convince the reader of your findings and **what they mean**.

This section is much like the final argument a lawyer might give a jury before they decide a verdict. This is where you...

- remind the reader of the goals (problem, hypothesis, and prediction),
- remind them of the main steps taken in the investigation (experimental methods),
- present a short summary of the evidence (experimental data) and what it means, and
- draw your conclusions based on that evidence.

If an experiment was performed and the data did NOT support the hypothesis, it could still be a valuable experiment! The purpose of any experiment is to see if a hypothesis is correct. When the results don't support the hypothesis, that discovery can help others know what *doesn't* work. Additionally, that experiment may help you or other researchers develop an even better hypothesis! ***Having an unsupported hypothesis does not mean you have done bad science or written a bad paper!***

The ***main*** purpose of the discussion/conclusion is to answer the question—was the hypothesis right? However, this is also where you should also discuss experimental errors, ideas for follow-up experiments, possible uses of your research findings for other research, or ethical implications that may need to be considered.

Most common mistakes: not clearly restating hypothesis and how the data either supported or didn't support it, suggesting that data from one experiment "proves" something, introducing new background information that should have been in the introduction, focusing only on experimental errors, providing no implications of what the results of the research mean, and making statements that contradict other sections of the paper (e.g. changing the hypothesis).

References

The References section is simply a list of the sources used to write the paper. All of the factual statements made in the introduction should come from reliable sources, and these sources must all be listed here. In scientific journals, they are usually listed in APA format and in alphabetical order. Articles should be cited appropriately. Journal articles should be cited as articles, not as web pages. When web pages from appropriate institutions are used, citations (the URL) must lead to the specific page from which the information was retrieved, not a general home page for the institution.

When making a factual statement, the appropriate way to cite it is to paraphrase the statement in your own words followed by open parenthesis, the author's last name, comma, year of publication, close parenthesis, period. For example: Scientific papers follow a specific format (Luyster, 2020). Then the article written by Luyster should be listed alphabetically in the References section. It should be cited like this:

Luyster, P. (2020). Teaching Undergraduate Biology Majors How to Write Scientific Papers. *Journal of Biology Information* *1*(1), 123–145.

As a science major, it is likely that most or all of your future papers will be written in APA, so it is well worth getting hold of an APA guide. The Purdue OWL website is an excellent resource for figuring out how to list references in APA format.

https://owl.purdue.edu/owl/research_and_citation/apa_style/apa_style_introduction.html

Most, if not all, information gained from other sources (and the appropriate citations) should be in the introduction section of the paper. The following sections should focus on what you as the researcher have done or discovered. There should be little need to cite sources in the methods, results, or discussion and conclusion sections.

A common mistake students make is that they try to write a paper first, and then they hunt for sources to cite afterwards. Not only is this a terrible way to write a legitimate scientific paper, it takes a lot more time. Instead, find the background information you need *before* doing any writing, record the citation for each source as you go, and refer to that information and cite it appropriately as you write the paper. In many cases, you can download the article and the citation in APA format directly from the library web page. Then you can use the articles (and cite them) as needed when you start to write the paper.

Plagiarism is a major mistake, and can result in a failing grade, disciplinary action, or even removal from an educational program. **Cutting and pasting text or copying material from any source *is plagiarism*.** If you are not sure if you are plagiarizing, ask your instructor for help. Many instructors require students to submit their work through websites like TurnItIn.com that will check for plagiarism against published work and other student papers. If you plagiarize, you will probably get caught.

Most common mistakes to avoid: not citing sources correctly in the text of the paper, using MLA format instead of APA, using webpage URLs when the article should be cited directly, using poor quality sources, using unnecessary direct quotes, not citing all factual statements, and plagiarizing.

Appendix F

An Application of Stoichiometry and Dimensional Analysis

Stoichiometry is the mathematical calculation of products in a reaction using **dimensional analysis**, or unit conversions. Most chemistry courses require students to learn how to use stoichiometry and dimensional analysis to work chemistry problems, but sometimes it is difficult to see the practical application. However, stoichiometry can be used to explain the equation used to calculate the alcohol by volume (ABV) content of hard cider. This appendix will not try to teach stoichiometry or dimensional analysis. You will learn these techniques in your chemistry courses. Here, we will just see how they can be used in biology.

In Lab Exercise 9, we saw that the percent of ethanol (or ethyl alcohol) in a batch of fermented cider can be estimated by measuring the specific gravity before and after fermentation, finding the difference, and multiplying that number by 131.25.

$$ABV = (\text{initial specific gravity} - \text{final specific gravity}) * 131.25$$

But where does the "131.25" come from? If we understand a little chemistry, and can use stoichiometry, we can figure it out. (Note that the equation above provides a rough approximation for calculating percent alcohol. The calculations here will produce a similar multiplication factor, but it will not be exactly the same. The conversion of mass of ethanol to volume of ethanol is a little more involved than what is used here.)

First, review the following equations:

Cellular respiration: $C_6H_{12}O_6 + 6\ O_2 \longrightarrow 6\ CO_2 + 6\ H_2O + \text{energy for ~32 ATP}$

Alcohol fermentation: $C_6H_{12}O_6 \longrightarrow 2\ CH_3CH_2OH\ (\text{ethanol}) + 2\ CO_2 + \text{energy for 2 ATP}$

Cellular respiration is aerobic. It requires oxygen. Alcohol fermentation is anaerobic. It only occurs in the absence of oxygen. Yeasts are facultative anaerobes. If oxygen is present, they will produce ATP via cellular respiration. When oxygen is not available, yeasts can use the less efficient process of fermentation to produce ATP.

There are some assumptions we will need to make as we work this problem. They are rarely all true in the actual practice of brewing hard cider.

Assumptions:

- No oxygen is available during fermentation; the sugar is converted to ethanol rather than CO_2 and H_2O.
- The sugar is pure glucose.
- *All* of the sugar is converted to ethanol.
- *All* of the ethanol remains in the cider; none of it evaporates.
- *All* of the carbon dioxide bubbles out of the solution.
- No other ingredients settle out of the solution to further reduce the density.

Necessary Conversion Factors:

- In fermentation, 1 mole of glucose yields 2 moles of ethanol and 2 moles of carbon dioxide.
- The ratio of moles of carbon dioxide produced to moles of ethanol produced is 1:1.
- 1 mole of carbon dioxide = 44.01 g of carbon dioxide.
- 1 mole of ethanol = 46.07 g of ethanol.
- The density of ethanol is 0.789 g/ml. (This does vary a bit depending on other factors.)

Appendix F An Application of Stoichiometry and Dimensional Analysis

Based on our assumptions, the change in density is primarily due to the conversion of glucose into carbon dioxide, which bubbles out of the solution during fermentation. The rest of the glucose is converted into alcohol, which remains in the solution.

If the *change in specific gravity equates to grams of carbon dioxide lost*, then we can convert change in specific gravity to percent alcohol as follows:

Conceptual Explanation:

Start with the assumption that the change in specific gravity equals the grams of carbon dioxide produced per milliliter of solution. Then convert grams of carbon dioxide to moles, then moles of carbon dioxide to moles of ethanol, then to grams of ethanol, then to volume of ethanol. Multiply by 100 to express the final answer as a percentage.

Calculation:

$$\frac{\text{Grams } CO_2}{\text{ml } H_2O} \cdot \frac{1 \text{ mole } CO_2}{44.01 \text{ g } CO_2} \cdot \frac{1 \text{ mole ethanol}}{1 \text{ mole } CO_2} \cdot \frac{46.07 \text{ g ethanol}}{1 \text{ mole ethanol}} \cdot \frac{1 \text{ ml ethanol}}{0.789 \text{ g ethanol}} \cdot 100 = 132.6$$

Appendix G

Values of the Chi Square Distribution

The Chi Square test can be used to compare actual data with expected data.

The Chi Square formula:

$$\chi^2 = \sum \frac{(o-e)^2}{e}$$

Σ = the sum of
o = observed values
e = expected values

Calculate the χ^2 value. Determine degrees of freedom (df). On the appropriate df row, find the approximate χ^2 value on the row. Go up to the top row to determine the p value.

p df	0.995	0.975	0.2	0.1	0.05	0.025	0.02	0.01	0.005	0.002	0.001
1	0.000	0.001	1.642	2.706	3.841	5.024	5.412	6.635	7.879	9.550	10.828
2	0.010	0.051	3.219	4.605	5.991	7.378	7.824	9.210	10.597	12.429	13.816
3	0.072	0.216	4.642	6.251	7.815	9.348	9.837	11.345	12.838	14.796	16.266
4	0.207	0.484	5.989	7.779	9.488	11.143	11.668	13.277	14.860	16.924	18.467
5	0.412	0.831	7.289	9.236	11.070	12.833	13.388	15.086	16.750	18.907	20.515
6	0.676	1.237	8.558	10.645	12.592	14.449	15.033	16.812	18.548	20.791	22.458
7	0.989	1.690	9.803	12.017	14.067	16.013	16.622	18.475	20.278	22.601	24.322
8	1.344	2.180	11.030	13.362	15.507	17.535	18.168	20.090	21.955	24.352	26.124
9	1.735	2.700	12.242	14.684	16.919	19.023	19.679	21.666	23.589	26.056	27.877
10	2.156	3.247	13.442	15.987	18.307	20.483	21.161	23.209	25.188	27.722	29.588
11	2.603	3.816	14.631	17.275	19.675	21.920	22.618	24.725	26.757	29.354	31.264
12	3.074	4.404	15.812	18.549	21.026	23.337	24.054	26.217	28.300	30.957	32.909
13	3.565	5.009	16.985	19.812	22.362	24.736	25.472	27.688	29.819	32.535	34.528
14	4.075	5.629	18.151	21.064	23.685	26.119	26.873	29.141	31.319	34.091	36.123
15	4.601	6.262	19.311	22.307	24.996	27.488	28.259	30.578	32.801	35.628	37.697
16	5.142	6.908	20.465	23.542	26.296	28.845	29.633	32.000	34.267	37.146	39.252
17	5.697	7.564	21.615	24.769	27.587	30.191	30.995	33.409	35.718	38.648	40.790
18	6.265	8.231	22.760	25.989	28.869	31.526	32.346	34.805	37.156	40.136	42.312
19	6.844	8.907	23.900	27.204	30.144	32.852	33.687	36.191	38.582	41.610	43.820
20	7.434	9.591	25.038	28.412	31.410	34.170	35.020	37.566	39.997	43.072	45.315

Generally, a p value greater than 0.05 indicates there is no significance difference between actual and expected values. The null hypothesis should not be rejected. A p value less than 0.05 indicates there is a statistical difference between actual and expected values. The null hypothesis should be rejected.